Springer Series in Optical Sciences   Volume 39

Edited by Arthur L. Schawlow

Springer Series in Optical Sciences

Editorial Board: J.M. Enoch D.L. MacAdam A.L. Schawlow K. Shimoda T. Tamir

Optical and Laser
Remote Sensing

Editors:
D. K. Killinger and A. Mooradian

With 294 Figures

Springer-Verlag Berlin Heidelberg GmbH 1983

Dr. Dennis K. Killinger
Dr. Aram Mooradian

Massachusetts Institute of Technology, Lincoln Laboratory, Lexington, MA 02173, USA

ISBN 978-3-662-15736-7   ISBN 978-3-540-39552-2 (eBook)
DOI 10.1007/978-3-540-39552-2

Library of Congress Cataloging in Publication Data. Main entry under title: Optical and laser remote sensing. (Springer series in optical sciences ; v. 39). Includes index. 1. Atmosphere–Remote sensing. 2. Atmosphere–Laser observations. 3. Atmospheric chemistry–Remote sensing. 4. Laser spectroscopy. 5. Optical radar. I. Killinger, D. K. (Dennis K.), 1945- . II. Mooradian, Aram, 1937- . III. Series. QC871.06 1983   551.5'028'7   83-420

2153/3130-543210

Preface

The field of optical and laser remote sensing has grown rapidly in recent years. This dynamic growth has been stimulated not only by technological advances in lasers, detectors, and optical system design, but also by the potential application of remote sensing systems to a wide variety of atmospheric measurements. Optical and laser remote sensing can allow single-ended measurement capability not offered by conventional point-detection techniques. While many past measurements have been associated with laboratory research, practical systems have recently been developed which are capable of remotely detecting, measuring, and tracking a wide range of molecular and atomic species in the atmosphere with concentrations of parts per billion and at ranges over 100 km.

This book is a compilation of papers which represent an overview of the present state of development of optical and laser remote sensing technology. The subjects covered include both passive and active remote sensing techniques in the UV, visible, and IR spectral regions, related laser and detector technology, and atmospheric propagation and system analysis considerations. While the papers do not constitute an exhaustive treatment of the excellent research being conducted in this field, they are representative of the wide diversity of present efforts. It is hoped that the reader will gain a general understanding of the current research in optical and laser remote sensing as well as an overview of current systems development.

Lexington, Massachusetts   *D.K. Killinger · A. Mooradian*
December, 1982

Contents

Part 3   UV–Visible DIAL Techniques

Part 4   Atmospheric Propagation and System Analysis

Part 8   Lidar Technology

Part 1

IR Differential-Absorption LIDAR (DIAL) Techniques

1.1 Airborne Remote Sensing Measurements With a Pulsed CO$_2$ Dial System

Jack L. Bufton, Toshikazu Itabe*, and David A. Grolemund**

Goddard Space Flight Center, Greenbelt, MA 20771, USA

A lidar instrument based on compact, pulsed carbon dioxide (CO$_2$) lasers has been developed for airborne remote sensing of atmospheric trace species at infrared wavelengths. It was designed for differential absorption lidar (DIAL) measurements using backscatter of laser pulse energy from the ocean and terrain surface in order to infer trace specie column content. The instrument is now operational on the NASA/Wallops Flight Center P3 aircraft. A flight test program started in July 1981 has produced results on instrument performance, backscatter data statistics, and target signatures. We will present these results and discuss their implications for remote sensing with this type of instrument.

An optical and electronic schematic of the lidar instrument is given in Figure 1. The laser transmitter subsystem is composed of two, transversely excited atmospheric (TEA) CO$_2$ lasers. Nominal laser output is 100 mjoule in a 100 nsec pulse in the 9–11 μm wavelength region. Each laser has a 0.75 m folded cavity with a pair of copper electrodes and a flashboard in each cavity arm. Ultraviolet pre-ionization and a spark gap-triggered discharge provide operation up to 10 pps. Laser gases circulate in the sealed cavity transverse to the optical axis. We prefer flowing rather than sealed–off operation in order to maximize laser output energy. The lasers and high voltage power supply are enclosed in a nitrogen–filled, pressurized case to prevent high voltage

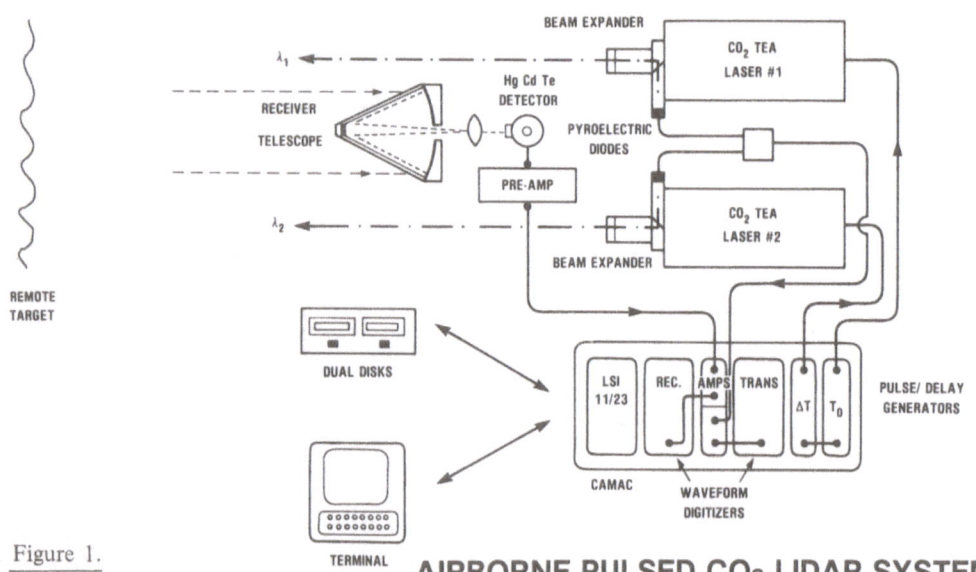

Figure 1.

AIRBORNE PULSED CO$_2$ LIDAR SYSTEM

---

*T. Itabe is an NAS/NRC Research Associate.
**D. Grolemund is employed by the Bendix Field Engineering Corp.

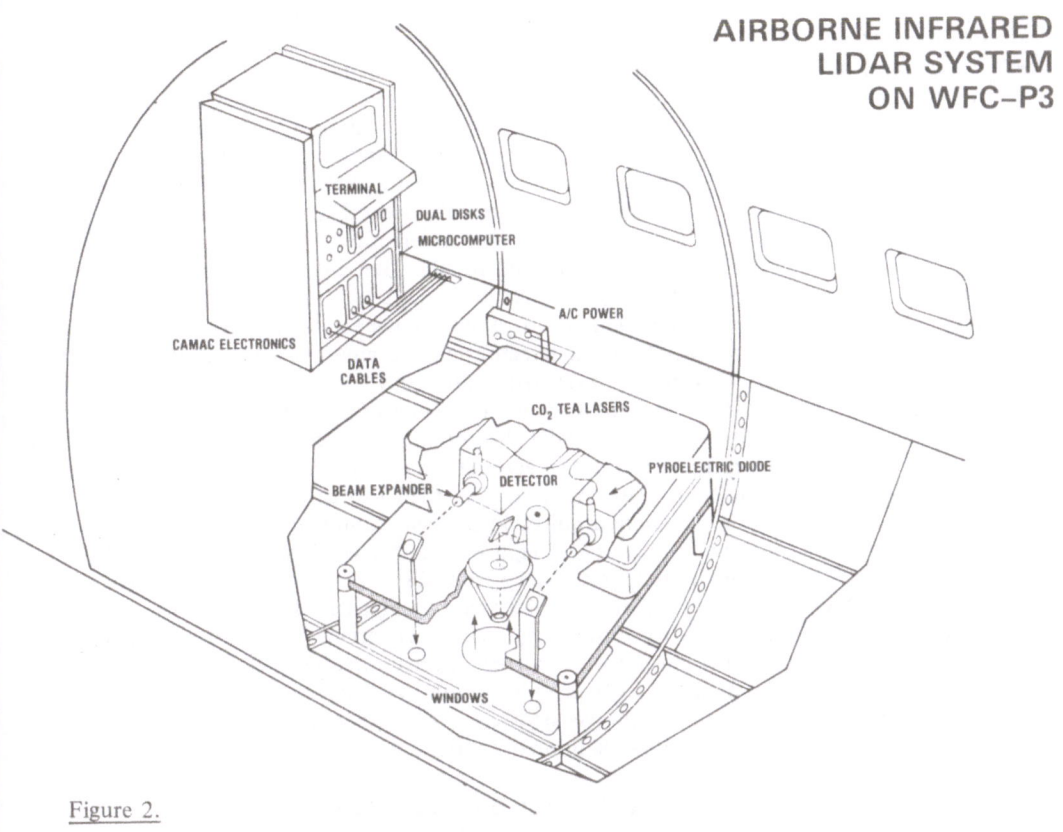

TERMINAL

DUAL DISKS

MICROCOMPUTER

A/C POWER

CAMAC ELECTRONICS

DATA
CABLES

CO$_2$ TEA LASERS

PYROELECTRIC DIODE

BEAM EXPANDER

DETECTOR

WINDOWS

Figure 2.

corona and arcing above 1 km altitude. In operation each laser is grating–tuned to one of the two wavelengths required for the differential absorption measurement. One laser is also triggered 25 μsec to 50 μsec after the other to provide a near–simultaneous dual–wavelength measurement with only one detector and one set of amplifier/digitizer electronics. Output laser energy is monitored with pyroelectric photodiodes and then beam divergence is reduced to about 1 mrad in 2x beam expander telescopes. Both lasers are mounted horizontally in the aircraft and their output beams are directed at nadir with turning mirrors. A diagram of the flight configuration is included as Figure 2. Note that the lidar instrument is located beneath the cabin floor and the electronics are rack–mounted above. This provides a compact, flexible installation that is compatible with simultaneous use of the same aircraft by several different experiments. The turning mirrors are adjusted to align each laser beam to the center of the receiver field–of–view. A 0.18 m diameter telescope serves as the collector telescope and a 25 mm focal length field lens is located before the 1 mm square HgCdTe photoconductive diode detector. Field–of–view is about 3.7 mrad. Detectivity is about 5×10^{10} cm Hz$^{1/2}$/watt and the bandwidth of detector and amplifier electronics is 1 MHz. The flight instrument occupies approximately 1 cubic meter and together with its rack of electronics weighs 350 kg.

Data collection electronics evolved in the flight tests to the use of two 10 MHz digitizers for continuous sampling of the 1 MHz bandwidth transmit and receive pulse waveforms. A microprocessor (LSI 11/23) packaged in CAMAC is used for control of waveform digitizers, disk storage of data, and statistical analysis of results. Both receive pulses are corrected for the digitizer pedestal and are normalized by the transmit pulse energy. Digitizer samples have 8–bit resolution

3

and are separated by 100 nsec. They are summed over $5\,\mu sec$ gate intervals around each pulse in order to compute pulse energy. Statistics including mean and standard deviation of the transmit and receive pulse energies, correlation between backscatter at the two wavelengths, and differential absorption ratio for each pulse–pair are computed during each data run.

Eight flight missions were conducted between July 1981 and May 1982 for a total instrument operation time of about 11 hours. During this period the instrument was operated at 1 km to 3 km altitude and differential absorption data were collected on several different wavelength pairs near $9.5\,\mu m$ over ocean and terrain surfaces. Direct–detection signal-to-noise ratio (S/N) for the pulse energy measurement was at least 60 db for the 2 km altitude where most data were acquired. Data quality was not limited by S/N but by backscatter variability, particularly with terrain targets where a large dynamic range was required. Since amplifier gain was set for the peak backscatter signal level, quantization and detector noise became an important consideration for low backscatter signals. We typically measured a 20 db and a 9 db dynamic range respectively for terrain and ocean backscatter data, and estimate the worst case energy measurement error as 5%. Prior to each flight, ground–test differential absorption data were acquired with a 1 km horizontal path between the aircraft and a tree–line across the runway. These data were the most stable and quantization and detector noise errors were less than 1% of the measured pulse energies. Beginning with flight #3 data were acquired in the near–simultaneous dual–wavelength mode. Sequential measurements of dual–wavelength backscatter were made on previous flights.

We found that the maximum backscatter came from the ocean surface at nadir. This result is opposite to what one would expect from consideration of Fresnel reflection coefficients for water and typical handbook values for terrain (vegetation) reflection. We attribute this result to the specular nature (glint) of the natural water surfaces near normal incidence angle. The glint is caused by small–scale wave structure in the laser footprint. The smallest waves called capillaries act like arrays of mirrors that confine the backscattered laser radiation to a cone of 20 to 40 degrees. The distribution of wave slopes is primarily a function of wind speed. Terrain reflec-

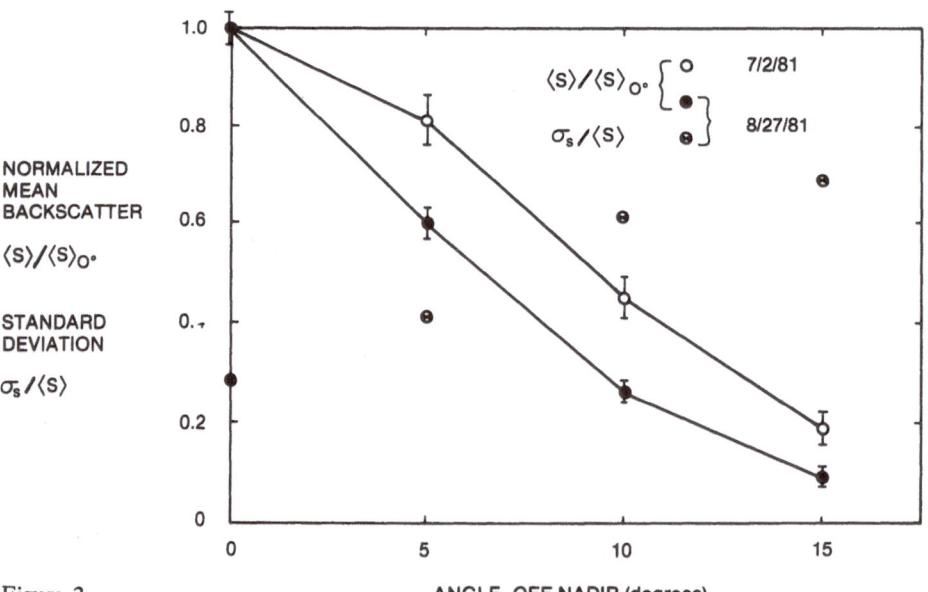

OCEAN SURFACE BACKSCATTER STATISTICS
9.5μm

Figure 3.

RATIO OF OCEAN TO TERRAIN BACKSCATTER
9.5 μm

	LASER LINE	7/2/81	8/27/81	10/14/81	3/11/82
			OCEAN/TERRAIN RATIOS		
MEAN BACKSCATTER:	P(14)		4.65 (± 0.06)		3.00 (± 0.01)
	P(16)			3.22 (± .09)	
	P(22)	2.87 (± 0.09)			
	P(24)		3.48 (± 0.05)	2.89 (± .06)	2.71 (± 0.01)
NORMALIZED STANDARD DEVIATION:	P(14)		0.28		0.39
	P(16)			0.24	
	P(22)	0.42			
	P(24)		0.36	0.43	0.36
DIFFERENTIAL ALBEDO:	P(14)/P(24)		1.34 (± 0.08)		1.103 (± 0.001)
	P(16)/P(24)			1.29 (± 0.03)	

Figure 4.

tivities are associated with more uniform angular backscatter patterns and are often approximated as Lambertain. As a result water surfaces exhibit a substantial directional gain over terrain features near normal incidence. We measured the angular backscatter pattern of the ocean for angles up to 15 degrees off-nadir as shown in Figure 3 for data from flight #1 and 2. In this figure each point indicates the mean backscatter pulse energy for several hundred pulse measurements, normalized by the measurement at nadir. The error bars indicate the expected standard deviation of the mean from the measured standard deviation for each data set and the assumption of independent Gaussian–distributed pulse energy measurements. Normalized standard deviation (σ) for the flight #2 data set is also plotted and shows almost a linear increase with off-nadir angle. Note that these values are an appreciable fraction of the mean backscatter.

Ratios of ocean to terrain backscatter are given in Figure 4 for mean backscatter, normalized standard deviation, and differential albedo. These ratios were formed only for data sets in which the aircraft altitude, electronic gain, and laser alignment remained constant for the water and terrain target passes. Data were collected at a variety of P branch CO_2 laser lines from P(14) to P(24) near 9.5 μm. Mean reflection from the ocean at nadir for all wavelengths was a factor of 2 to 5 times greater than for terrain surfaces. The terrain targets were primarily vegetation, either in the form of tree tops, fields, or crops. There were occasional road surfaces, house tops, and villages, but almost no areas of bare soil. Ratios of σ values indicate that terrain backscatter was a factor of 2 to 3 times more variable than ocean backscatter. This is attributed to rapid changes in terrain features during the data runs. A standard data run consisted of 100 pulse–pair measurements at a rate of about 2-3 pulses/sec. For a typical aircraft velocity of 100 m/sec, a data run would span about 4km of terrain and significant variations in albedo would be expected.

Differential albedo data for the P(14)/P(24) and P(16)/P(24) wavelength pairs were acquired respectively during flights #2 and 4. The ratios (ocean/terrain) of these quantities were quite similar showing 30% change in differential absorption for the two target types. Data for flight #6 for P(14)/P(24) show only a 10% difference when the ocean ratio is divided by the terrain ratio. We attribute this reduction to a lack of vegetation in the terrain targets in this March data set. Similar data have been reported in the literature. Shumate et al. (ref. 1) report differential albedo effects near 9.5 μm based on analysis of terrain material reflectivity and their field experience with an airborne CO_2 lidar system. Boscher and Lehmann (ref. 2) report a 35% increase in

5

the P(14)/P(24) ratio for sand relative to water targets in laboratory measurements. Additional support for changes in differential albedo for water relative to terrain is reported by Petheram (ref. 3). His analysis shows as much as a 10% increase in the P(14)/P(24) ratio for aerosol backscatter as the relative humidity changes from 30% to 96%. The implication of these data sets, analysis, and the data reported here is that differential absorption lidar measurements are a strong function of target type near $9.5\,\mu m$. Significant errors in trace specie concentration estimates will result. This is especially true in airborne measurements where target albedo is a rapidly changing function of time. The analysis of Petheram shows that this effect must also be considered for aerosol backscatter lidar systems in the $9-10\,\mu m$ region. We have taken a limited amount of aerosol backscatter data on the P(14) and P(24) wavelengths with a larger, more powerful ground-based CO_2 lidar similar to the airborne system. These results show significant differential absorption ratio variations among aerosols, aerosol layers, and cloud structures.

DUAL-WAVELENGTH CORRELATION COEFFICIENT
VS. BACKSCATTER STANDARD DEVIATION

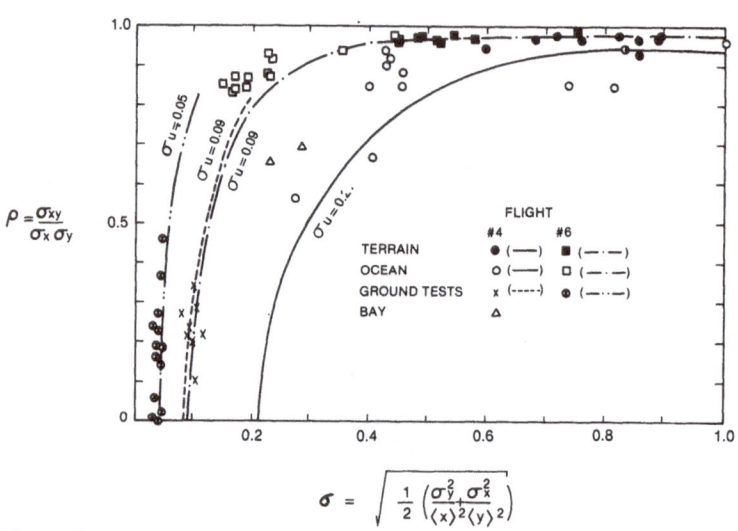

$$\rho = \frac{\sigma_{xy}}{\sigma_x\,\sigma_y}$$

$$\sigma = \sqrt{\frac{1}{2}\left(\frac{\sigma_y^2}{\langle x\rangle^2} + \frac{\sigma_x^2}{\langle y\rangle^2}\right)}$$

NORMALIZED STANDARD DEVIATION

Figure 5.

A correlation analysis was performed on dual-wavelength backscatter associated with flights #4-8 in order to analyze system performance. Dual-wavelength cross-correlation coefficients ρ are plotted in Figure 5 as a function of normalized standard deviation σ for the P(16) and P(24) data of flight #4 and the P(14) and P(24) data of flight #6. It is apparent that flight data, particularly terrain backscatter, are very well correlated ($\rho \approx 1$), and that they are associated with strong modulation ($\sigma = 0.5$ to 1.0). Lower σ values of ocean backscatter have somewhat lower correlation. Ground-test data are nearly uncorrelated and have the lowest ρ values. We feel that the ground-test data are determined primarily by an uncorrelated noise component, while the high variability of the ocean and terrain backscatter in the flight data tend to mask this component and drive ρ toward unity. These dual-wavelength data agree well with the dual-laser (single-wavelength) results reported by Killinger and Menyuk (ref. 4) for backscatter from diffuse and specular targets over horizontal paths with similar CO_2 TEA lasers. Their analysis considered an additive, uncorrelated noise variance to be present in all data. By definition this quantity is $\sigma_U^2 \equiv \sigma^2 - C_{XY}$ where $\sigma^2 = 1/2(\sigma_X^2/\langle X\rangle^2 + \sigma_Y^2/\langle Y\rangle^2)$ for data from laser X and Y, C_{XY} is cross-covariance for the two lasers, and $\rho = C_{XY}/\sigma_X\,\sigma_Y$ is the cross-correlation coefficient.

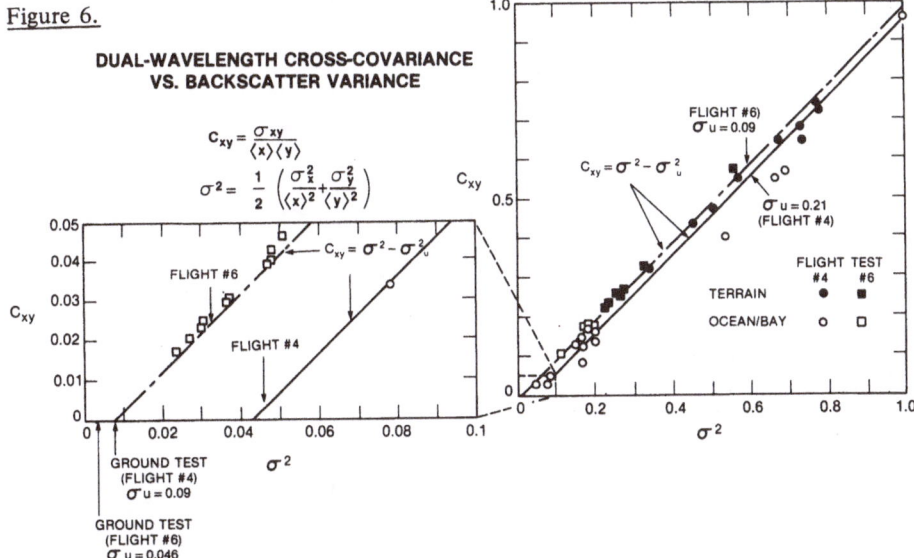

Figure 6.

**DUAL-WAVELENGTH CROSS-COVARIANCE
VS. BACKSCATTER VARIANCE**

$$C_{xy} = \frac{\sigma_{xy}}{\langle x \rangle \langle y \rangle}$$

$$\sigma^2 = \frac{1}{2} \left(\frac{\sigma_x^2}{\langle x \rangle^2} + \frac{\sigma_y^2}{\langle y \rangle^2} \right)$$

All variances and standard deviations are normalized by the mean backscatter. We use the term cross–covariance because of the delay between dual–wavelength lasers pulses. Pulse delay was set at $50\,\mu\sec$ in flight #4 data and then reduced to $25\,\mu\sec$ for flights #5–8.

We have fit the equations above to our data and computed least–square values of σ_U^2 for the ground–test and flight data. These results are plotted in Figure 6 in terms of C_{XY} vs. σ^2 and the region near the origin has been expanded to show how σ_U^2 causes an offset along the σ^2 axis. The flight #4 σ_U results of 0.09 and 0.21 are remarkably close to the Killinger and Menyuk values of 0.104 and 0.193 respectively for diffuse and specular targets. Our ground–test data were acquired with a diffuse target and our flight data are specular in type. The factor of two reduction in σ_U in flight #6 data may be due to improvements we made in the receiver optics after flight #4 by placing the detector focusing lens in the focal plane of the 0.18 m collector telescope. This provided a field lens with field–of–view increase to 3.7 mrad and a reduction in sensitivity to laser alignment or pointing angle fluctuations. A further reduction in σ_U by a factor of two was noted in all ground–test data sets from their respective flight data sets. This may be due to better temporal correlation present in that part of the ground–test σ caused by atmospheric turbulence. Typical ground–test σ values for either wavelength were in the range 0.03 to 0.12. Estimates of atmospheric turbulence–induced scintillation at $9.5\,\mu m$ for a 1 km horizontal path in daytime and a 0.18 m diameter collector range from 0.01 to 0.03. Some of the observed σ could be the result of turbulence. Low wind speeds of a few m/sec transverse to the ground–test optical path, predict a temporal correlation time of at least a millisecond. The $25\,\mu\sec$ to $50\,\mu\sec$ separation between dual–wavelength pulses should have effectively "frozen" the scintillation pattern and produced near unity scintillation correlation for the two pulses. Hence scintillation would contribute to C_{XY} and not to σ_U. In the airborne data however, an effective transverse wind speed of 100 m/sec provided by the aircraft velocity would reduce temporal correlation to the $100\,\mu\sec$ level which was comparable to the initial value of $50\,\mu\sec$ used to separate the dual–wavelength pulses. As a result scintillation effects may be included in the airborne σ_U values.

Another source of fluctuations in laser backscatter is target–induced speckle. This is a separate process from target–induced albedo fluctuations which are caused by movement of the aircraft in the few tenths–of–a–second between laser pulse pairs. In the ground–test data albedo changes could be caused by pulse–to–pulse pointing angle fluctuations or mode pattern changes.

7

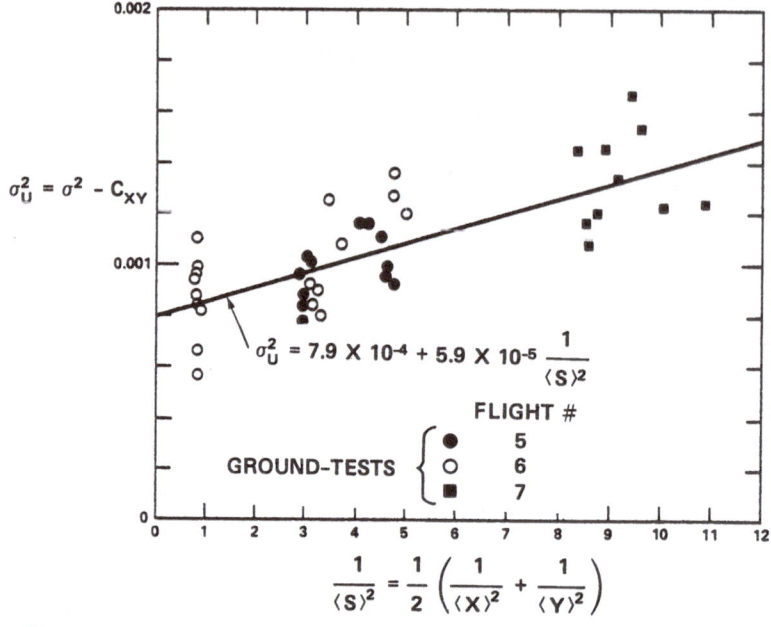

DEPENDENCE OF UN-CORRELATED NOISE VARIANCE ON MEAN BACKSCATTER SIGNAL

$\sigma_U^2 = \sigma^2 - C_{XY}$

$\sigma_U^2 = 7.9 \times 10^{-4} + 5.9 \times 10^{-5} \dfrac{1}{\langle S \rangle^2}$

FLIGHT #

GROUND-TESTS
- ● 5
- ○ 6
- ■ 7

$$\frac{1}{\langle S \rangle^2} = \frac{1}{2}\left(\frac{1}{\langle X \rangle^2} + \frac{1}{\langle Y \rangle^2}\right)$$

Figure 7. (RELATIVE UNITS)

Speckle results from the coherent combination at the receiver of backscatter from various portions of the laser "footprint" on the rough target surface. Since each portion of the "footprint" is a slightly different distance from the receiver there are significant optical phase differences and an interference or speckle pattern is produced in the receiver plane. The largest radial separations in the laser "footprint" give the smallest speckle structure in the receiver plane. Estimates for our CO_2 lidar system and the ground–test and flight propagation paths show that spatial averaging in our receiver telescope aperture, as well as temporal averaging in the laser pulse width of short term (nsec) TEA laser temporal fluctuations, should reduce speckle to very low σ values (0.01 or less) within each laser pulse. Any residual speckle would however contribute to σ_U^2 since speckle effects, as opposed to scintillation effects, rapidly decorrelate with laser wavelength separations.

We have further investigated the source of σ_U^2 in our data by plotting ground–test data for flight #5, 6, and 7 in Figure 7 as $\sigma^2 - C_{XY}$ vs. $1/\langle S \rangle^2$ where $1/\langle S \rangle^2 = 1/2(1/\langle X \rangle^2 + 1/\langle Y \rangle^2)$ is the inverse of the square of mean backscatter $\langle S \rangle$. We fit a least–square line to the data with the result $\sigma_U^2 = 7.9 \times 10^{-4} + 5.9 \times 10^{-5}/\langle S \rangle^2$. This indicates that the lowest value of σ_U^2 is 7.94 $\times$ 10^{-4} and that higher σ_U^2 is correlated with lower mean backscatter. Both laser pointing fluctuations and speckle produce σ_U^2 independent of backscatter signal strength. Detector noise and quantization noise begin to dominate at low signal levels and σ_U^2 increases at the expense of C_{XY}. The signal–to–noise performance of our DIAL system is proportional to σ_U^{-1} and maximum performance is obtained at large signal levels (i.e. operation near the origin in Figure 7). Our lowest detector S/N was associated with the ground–test data of flight #7, and these data have the highest σ_U^2 values.

The major implication for CO_2 TEA laser DIAL systems that comes out of this correlation analysis is that an uncorrelated noise level of about $\sigma_U = 0.025$ is the lowest achievable value in our differential absorption data to date. This noise exists irrespective of detector S/N level or the removal of all correlated modulation. We agree with Killinger and Menyuk (ref. 4) that σ_U

8

is most likely due to the pulse–to–pulse variations in TEA laser pointing and mode structure. It may however be due in part to residual speckle noise. In the flight data scintillation decorrelation at the $25\,\mu sec$ delay time would contribute to σ_U. The uncorrelated noise will require time-averaging of successive laser pulses in order to achieve DIAL measurements accurate to the 1% level. This is a necessity for even modest accuracy trace specie measurements because the trace specie differential absorption is usually only a fraction of the mean backscattered energy. In most cases averaging over more than one pulse–pair is possible and can be traded–off against desired spatial resolution and laser repetition rate. Additional efforts to identify the source of uncorrelated noise in TEA laser DIAL measurements are desirable; but at the present time we consider the differential albedo effect with terrain, water, and aerosol targets to be the limiting factor in application of CO_2 TEA laser systems to atmospheric trace specie measurement.

It should be possible in future applications to retain the advantages of the efficient CO_2 TEA laser source and reduce the effects of differential albedo variations. Nonlinear infrared crystals promise the ability, with frequency–doubling and sum–frequency generation, to shift trace specie measurement from the $9-11\,\mu m$ region to $4-6\,\mu m$. Byer (ref. 5) has reported crystal properties and frequency shifting methods and is now growing candidate crystal materials. Menyuk et al. (ref. 6) have already demonstrated remote–sensing with frequency–doubled CO_2 TEA lasers. A major source of differential albedo variations near $10\,\mu m$ is the significant absorption of this wavelength radiation by water and terrestrial materials. Near $5\,\mu m$ absorption effects are weaker and less variable with wavelength. This is demonstrated in the study of aerosol extinction by Shettle and Fenn (ref. 7). They report scattering as approximately 10 times more important than absorption near $5\,\mu m$ while these processes are comparable in effect near $10\,\mu m$. Their results are presented as a function of aerosol model and relative humidity. We feel that CO_2 DIAL measurement accuracy could improve in the $4-6\,\mu m$ region if adequate absorption lines exist for the desired trace specie.

REFERENCES

1. M. S. Shumate, R. T. Menzies, W. B. Grant, and D. S. McDougal, "Laser Absorption Spectrometer: Remote Measurement of Tropospheric Ozone," Applied Optics, 20, pp. 545–553, 15 Feb. 1981.
2. J. Boscher and F. Lehmann, "Experimentelle Untersuchungen der physikalischen Grundlagen zur Fernmessung von Boden und Vegetationsfeuchte durch aktive Infrarot – Reflexionsspektroskopie mit Hilfe der CO_2 – Lasertechnik," Report # BMFT-FB-W 80-037, Battelle Institute, Frankfurt, Germany, Dec. 1980.
3. J. C. Petheram, "Differential Backscatter from the Atmospheric Aerosol: The Implication for IR Differential Absorption Lidar," Applied Optics, 20, pp. 3941–3946, 15 Nov. 1981.
4. D. K. Killinger and N. Menyuk, "Remote Probing of the Atmosphere Using a CO_2 DIAL System," IEEE J. of Quantum Electronics, QE-17, pp. 1917–1929, Sept. 1981.
5. R. L. Byer and R. L. Herbst, "Parametric Oscillation and Mixing," Chpt. 3 in Nonlinear Infrared Generation, Topics in Applied Physics, Vol. 16, Springer-Verlag, New York, 1977.
6. N. Menyuk, D. K. Killinger, and W. E. DeFeo, "Remote Sensing of NO Using a Differential Absorption Lidar," Applied Optics, 19, pp. 3282–3286, Oct. 1980.
7. E. P. Shettle and R. W. Fenn, "Models of Aerosols of the Lower Atmosphere and the Effects of Humidity Variations on Their Optical Properties," Report # AFGL-TR-79-0214, Air Force Geophysics Lab., Hanscom AFB, Mass., 20 Sept. 1979.

1.2 Differential-Absorption Measurements With Fixed-Frequency IR and UV Lasers

K.W. Rothe, H. Walther[+], and J. Werner
Sektion Physik der Universität München, D-8046 Garching, Fed. Rep. of Germany

INTRODUCTION

The growing necessity of pollution monitoring and control has been recognized in recent years. The well-established chemical methods in widespread use suffer from a major drawback in that they only allow point monitoring measurements, which in many cases is not sufficient. Laser techniques do not have this disadvantage since they can determine the mean concentration of a species not only over a certain path length but also over a certain area or volume. In addition, pulsed high-power lasers can be used together with the differential-absorption technique to obtain a complete range-resolved map of the distribution of a species over fairly large distances or areas. Furthermore, it is of interest to compare observed distributions of gaseous components with theoretical models. This allows evaluation of the diffusion parameters and the total mass emission of a source, these obviously being key quantities for determining the environmental impact.

In the following a selection of our differential-absorption measurements with fixed-frequency infrared and ultraviolet lasers is presented, including range-resolved measurements of ethylene distributions over the area of a refinery with a pulsed CO_2 laser, long-path absorption measurements of gas concentration and temperature in the exhaust of a garbage and gas-burning power plant with a cw HF/DF laser, and investigation of the stratospheric ozone layer with a pulsed XeCl excimer laser.

FIELD MEASUREMENTS OF ETHYLENE PROPAGATIONS

In many cases a detailed mapping of different constituents over large areas is required. With such a range-resolved picture of the concentration distribution it is possible to detect and localize emission sources over large urban and industrial areas. In addition, the observed expansion of a species can be compared with theoretical diffusion models. The propagation of a species over long ranges can thus be predicted.

For these purposes a mobile LIDAR system has been developed. It is equipped with a pulsed CO_2 laser, which has a pulse energy of 1.3 J and a repetition rate of up to 70 Hz. The individual laser lines are selected by a grating, which is controlled by a stepper motor. Fast switching between the different laser lines is thus possible. Before being emitted into the atmosphere the beam is expanded by a factor

---

+) Also Max-Planck-Institut für Quantenoptik, Garching, West Germany

of 8. This reduces the divergence of the laser beam to about 0.4 mrad, a value significantly smaller than the field of view of the receiving optics. The latter consists of a spherical mirror with a diameter of 60 cm and a focal length of 240 cm which focusses the backscattered light onto a fast infrared detector. The detector signal is amplified, stored in a transient recorder as function of time and then transferred to a computer for further data evaluation. The computer also performs wavelength control of the laser as well as the mechanical steering of the whole LIDAR system, which can be rotated hydraulically around the horizontal and the vertical axes.

One of the key factors for specifying air quality is the pollution concentration. In many cases, however, the total mass emission rate from the area of a factory must be known since this quantity is mostly the basis of legal regulations. Evaluation of the LIDAR data by means of propagation methods gives significant information on the mass emission rate. This will be discussed in the following.

The diffusion of a plume from its source is mainly caused by the wind and - perpendicular to it - by atmospheric turbulent diffusion. Measurements should therefore be performed in a vertical plane perpendicular to the direction of the wind at the lee-side of the emission source. The concentration profile from such a vertical scan is shown in Fig. 1 by lines of equal concentration (left), and by a three-dimensional representation (right). This figure gives the ethylene concentration in a plane at the periphery of a refinery. For further evaluation both the horizontal (section A) and the vertical (section B) profiles through the center of mass of this distribution were determined and compared with a simple theoretical profile /1/. The results are shown in Fig. 2. The dashed lines give the measured concentration profile, and the solid lines represent a least squares fit of the theoretical curves to these data points.

Apart from describing the diffusion of a plume, LIDAR measurements can also give information on the total mass emission rate. As all the pollution is transported through the plane of observation, the total mass emission rate is simply given by the product of the wind velocity and the average concentration over this plane. With a wind velocity of 3 m/s these measurements give an ethylene output from the refinery of 250 kg per day.

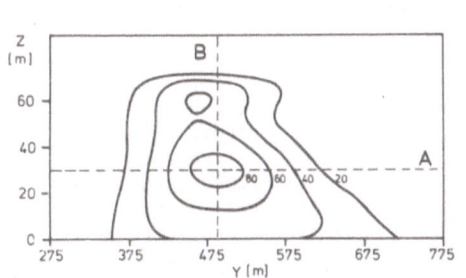 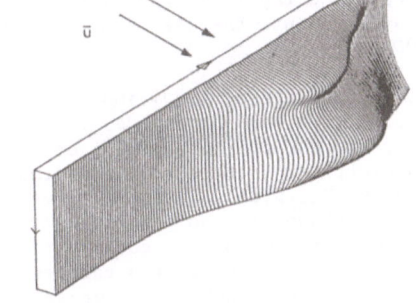

Fig. 1: Measurement of the ethylene distribution in a vertical plane at the periphery of a refinery (left) and a three-dimensional representation (right). The numbers on the left give the ethylene concentration in ppb.

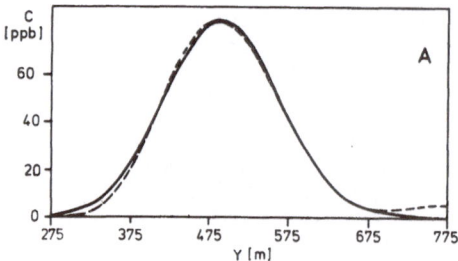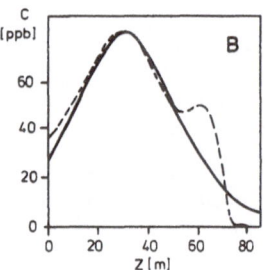

Fig. 2: Horizontal (left) and vertical (right) concentration profile
as measured (dotted line) and fitted (solid line) to a simple
propagation model /1/.

MEASUREMENTS OF CONCENTRATION AND TEMPERATURE
IN THE EXHAUST OF A POWER PLANT

It is often believed that environmental protection and industrial
interest are contradictory. Laser measurements of industrial emission
show, however, that the data are also useful for finding leakages and
for controlling and optimizing industrial processes. This helps to save
energy and raw materials and to decrease the environmental pollution
as well.

A system suitable for this specific application has to fulfil at least
two basic conditions: Firstly, it must respond to rapidly changing conditions fast enough to give reliable results and provide a signal for
process control. Secondly, it must be simple and sturdy enough to with-
stand rough conditions in industrial environments.

We constructed a simple HF/DF laser system which was operated for three
months in the aggressive atmosphere of the smoke stack of a gas and gar-
bage-burning power plant. Owing to the high stream velocity of the exhaust
gases and the rapidly changing dust content, the measurements on the different laser lines have to be performed in a rather short time interval.
A fast line selection mechanism is therefore required. At first the
laser was operated with broad-band mirrors, and a fast-scanning monochromator was used to select the individual laser lines. Later on this
could be further improved and simplified by removing the monochromator
and replacing the laser end mirror by a fast scanning grating. With
both arrangements the spectrum could be scanned with frequencies of up
to 200 Hz.

Owing to the extreme conditions inside the smoke stack (70 torr water
vapor, 20 torr carbon dioxide, temperature 400 K) the influence of
temperature and self-broadening on the absorption coefficients must
be carefully considered in the evaluation. We used the most recent
version of McClatchey's tape /2/, which was updated with more accurate
data from Flaud /3/ and Watkins /4/. As the number of laser lines by
far exceeds the number of components, a least squares fit procedure has
to be applied, which is also capable of determining the temperature
inside the smoke stack very precisely. As an example, Fig. 3 shows the
normalized χ^2 as a function of temperature. It can be seen from the
insert in the right part of Fig. 3 that the temperature can be determined with a relative accuracy of better than 0.1 K. This is sufficient for most practical applications.

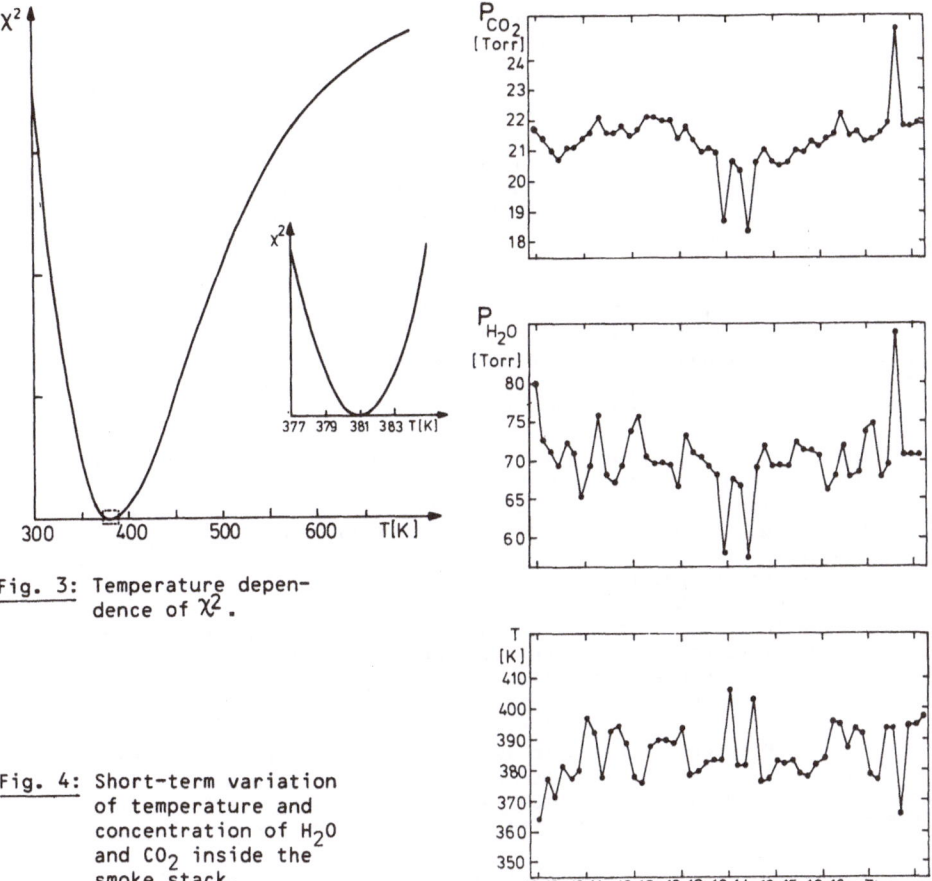

Fig. 3: Temperature dependence of χ^2.

Fig. 4: Short-term variation
of temperature and
concentration of H_2O
and CO_2 inside the
smoke stack.

Typical examples of the short-term variation of temperature and concentration of water vapor and carbon dioxide inside the smoke stack
are shown in Fig. 4. There is a strong correlation between water vapor
pressure and temperature, as well as a weaker correlation between each
of them and the carbon dioxide concentration. This is a typical result
of the combustion processes in garbage and gas-burning power plants.

These experiments demonstrate the feasibility of fixed-frequency infrared lasers for fast and sensitive measurements of concentration and
temperature even in rough environments.

RANGE-RESOLVED INVESTIGATION OF THE STRATOSPHERIC OZONE LAYER

In spite of its low concentration in the atmosphere ozone is very important for most stratospheric processes. The stratospheric temperature
profile, for example, is mainly due to the absorption of solar UV radiation by ozone; most of the dangerous UV radiation therefore does not
reach the ground. A possible reduction of the ozone layer may therefore

be noxious to life on earth. Photolysis of freon is considered to be the main reason for such a reduction.

Unfortunately, there are strong natural variations in the ozone concentration, comprising short annual as well as longer variations caused by the cycle of the sun-spots. This is the reason why the long-term trend of the global ozone concentration is still uncertain despite the fact that there are many methods of measuring ozone. They are either ground-based or balloon, rocket, or satellite-borne and comprise optical as well as chemical methods.

The main disadvantage common to all these methods is the fact that they are not capable of providing height-resolved, continuous, and precise data of the ozone concentration: chemical methods are balloon or rocket-borne and consequently incapable of providing continuous data; the rather precise Dobson method only gives the integral concentration; the Umkehr (inversion) technique is not precise enough.

LIDAR measurements in the UV spectral region might be a solution to this problem — provided they are accurate enough. First measurements with frequency-doubled dye lasers were performed by Gibson et al. /5/, and later on by Megie et al. /6/, achieving an UV energy of 30 mJ with a repetition rate of 0.25 Hz. But even this was not satisfactory, since a time interval of several hours was required to get an accuracy of 30 %. This also holds for the measurements of Uchino et al. /7/; they used a XeCl excimer laser at 308 nm without having a second frequency – necessary as the reference line for the differential-absorption technique. Instead of this they used the data of a balloon sonde (launched the next day) as a reference.

Our ozone LIDAR concept takes into account the requirements for a high-power laser as well as for a proper reference line. The apparatus is shown in Fig. 5. We use a XeCl laser with a pulse energy of 130 mJ and a repetition rate of up to 100 Hz. Its radiation is focussed into a high-

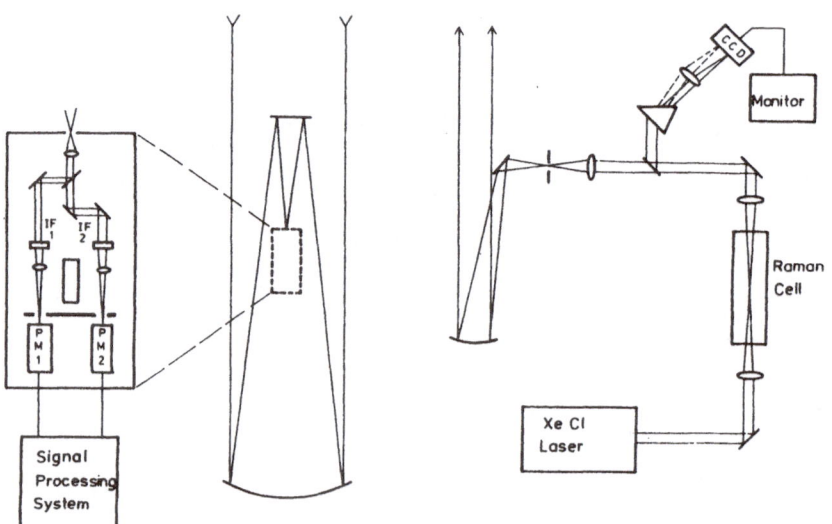

Fig. 5: Scheme of our ozone LIDAR.

pressure methane cell to generate the reference line by stimulated Raman scattering. Typical conversion rates are 15 % at a pressure of 35 atm and a focal length of 125 cm. This is more than sufficient since the Raman-shifted reference line at 338 nm is absorbed significantly less by ozone than the unshifted XeCl laser line at 308 nm.

Before being expanded and emitted into the atmosphere, a small fraction of the light is extracted in order to monitor the laser intensity and the conversion rate with an array of photodiodes (CCD).

The backscattered light is focussed by a spherical mirror (diameter 60 cm, focal length 240 cm) onto the entrance stop of the detection unit. The two wavelengths are separated by a dichroic beam splitter and two interference filters. The photomultipliers must be protected against the very strong intensities backscattered from the first few kilometers; a chopper wheel therefore rotates in front of the multi-pliers, covering them during and shortly after emission of the laser pulse. The photons backscattered from the stratosphere are counted, stored in a fast memory every 66.7 ns and afterwards transferred to a computer for further data evaluation.

The major advantage of this set-up is that simultaneous measurements at both wavelengths are possible, thus eliminating all problems asso-ciated with rapidly changing atmospheric conditions (turbulences). Owing to the high repetition rate and pulse energies of the XeCl laser the apparatus is expected to give stratospheric ozone profiles every two minutes with an accuracy of 30 %.

One disadvantage of all ground-based stratospheric LIDAR measurements is the rather strong absorption of the light in the first few kilo-meters primarily due to Mie scattering, and in the UV spectral region also due to Rayleigh scattering. First measurements with the ozone LIDAR are therefore being made from the summit of the Zugspitze in the Alps (about 3 km above sea-level). This difference in altitude gives rise to an increase of the intensity backscattered from the strato-sphere by a factor of about three for clean air, and of twenty for hazy air.

CONCLUSION

The application of laser methods has led to considerable progress in the field of monitoring atmospheric constituents. For many appli-cations laser techniques prove to be superior to conventional methods with respect to sensitivity and speed of response. The usefulness of fixed-frequency lasers for monitoring purposes has been successfully demonstrated. Further measurements have to be performed and compared with conventional methods in order to establish the laser technique as a very helpful instrument in sensitive, remote, and on-line moni-toring of atmospheric constituents and properties.

ACKNOWLEDGEMENTS

The financial support of the Deutsche Forschungsgemeinschaft is grate-fully acknowledged.

REFERENCES

1. F. Pasquill: Atmospheric Diffusion, N. Y., Van Nostrand (1974)
2. R.A. McClatchey, W.S. Benedict, S.A. Clough, D.E. Burch, R.E. Calfee, K. Fox, L.S. Rothman, J.S. Garing: AFCRL-TR-73-0096 (1973)
3. J.M. Flaud, C. Camy-Peyret: J. Mol. Spectr. 55, 278 (1975)
4. W.R. Watkins, R.L. Spellicy, K.O. White, B.Z. Sojka, L.R. Bower: Appl. Opt. 18, 1582 (1979)
5. A.J. Gibson, L. Thomas: Nature 256, 561 (1975)
6. G. Megie, J.Y. Allain, M.L. Chanin, J.E. Blamont: Nature 270, 329 (1977)
7. O. Uchino, M. Maeda, M. Hirono: IEEE, QE-15,10, 1094 (1979)

1.3 Remote Sensing of Hydrazine Compounds Using a Dual Mini-TEA CO$_2$ Laser DIAL System

N. Menyuk, D.K. Killinger, and W.E. DeFeo

Lincoln Laboratory, Massachusetts Institute of Technology,
Lexington, MA 02173, USA

I. Introduction

In this paper we describe our direct-detection, pulsed, dual-CO$_2$ laser differential-absorption LIDAR (DIAL) system and report the results of measurements using this system to demonstrate the practicability of remote sensing of hydrazine, unsymmetrical dimethylhydrazine (UDMH) and monomethylhydrazine (MMH). The measurements involved LIDAR returns from a topographic target located 2.7 km from the laboratory and, to our knowledge, represent the first test of laser remote sensing of highly toxic hydrocarbon species having broad band as opposed to simple line spectra. It was found during these experiments that atmospheric fluctuations were the major factor limiting the sensitivity of these measurements. Additional experiments were carried out to achieve a better understanding of these fluctuations and of the improvements achievable through signal averaging. The results of these experiments are also discussed.

II. Experimental Apparatus

The dual-laser DIAL system[1,2] used in these experiments is shown in Fig. 1. Two line-tunable, high PRF mini-TEA CO$_2$ lasers[3] provided the

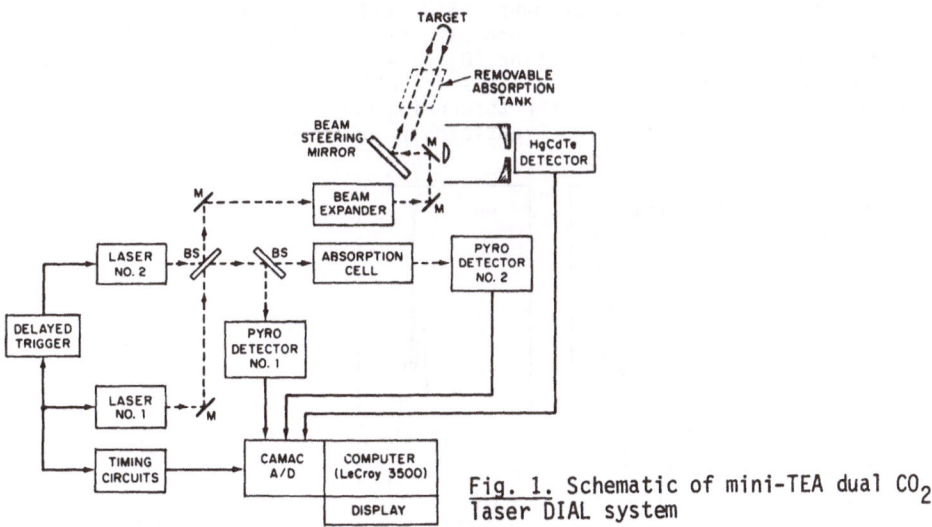

Fig. 1. Schematic of mini-TEA dual CO$_2$ laser DIAL system

*This work was supported by the Air Force Engineering and Services Center.

pulsed transmitted radiation. Laser 1 was tuned to the low-absorption transition frequency and laser 2 was tuned to the high-absorption transition frequency of the hydrazine, UDMH and MMH molecules. The time delay between the firing of the two lasers was maintained at 50 μs; for this short a delay time, the atmosphere is essentially "frozen" between firings.[1] The optical beams from the two lasers were joined at a 50/50 beam splitter. A portion of each beam was then sent to a pyroelectric detector in order to normalize the LIDAR returns to the output energy of each laser pulse, and another portion was sent through a Pyrex absorption cell to a second pyroelectric detector. The cell served to calibrate absorption levels of the hydrazine compounds at the frequencies emitted by our lasers. The major part of the laser radiation, after going through a 10X beam expander, was directed toward a 1 m x 1 m flame-sprayed aluminum plate which was located at a range of 2.7 km and served as a diffusely-reflecting target.

The LIDAR returns from this target and the pyroelectric signals were recorded and analyzed in a computerized data acquisition system which calculated the normalized differential absorption of both the LIDAR returns and the absorption cell data.

A large, enclosed, chemically inert polypropylene tank[4] (104 cm long x 62.5 cm dia. with 15° slanted polypropylene windows) was placed outside the laboratory in the LIDAR path for absorption measurements of the hydrazine compounds. The enclosed tank was essential in view of the known toxicity of the hydrazine compounds.[5]

III. Experimental Results

a) Choice of transition lines for differential absorption

The moderately strong absorption bands[6,7] of the hydrazine compounds in the 9 to 12 μm wavelength region are shown in Fig. 2. Also shown in the figure is the spectral coverage available with line-tunable CO_2 lasers. The absorption is seen to occur over broad bands rather than as individual absorption lines, such as is obtained with simple molecules. It is therefore difficult to obtain neighboring CO_2 laser transition lines with large differences in absorption levels. This is reflected in the results obtained by Loper et al.,[8] who measured the absorption cross section of these compounds at a large number of CO_2 transition frequencies using a low pressure cw laser.

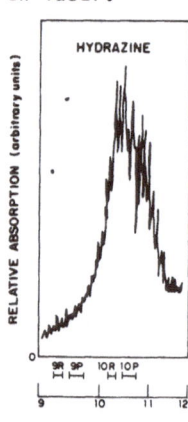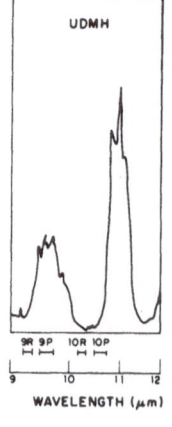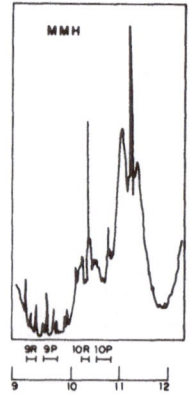

Fig. 2. Absorption spectra of hydrazine, UDMH and MMH near the CO_2 laser transitions. (D.A. Stone, Refs. 6,7)

The choice of CO_2 laser frequencies to be used for remote sensing of the hydrazine compounds included consideration of (1) the differential-absorption of the hydrazine compounds, as obtained from the measurements of Loper et al.,[8] (2) the transmission spectrum of the atmosphere, and (3) interference effects due to the potential simultaneous presence of other atmospheric species such as ethylene and ammonia.

On the basis of the above considerations, CO_2 laser frequency pairs were chosen which yielded as large a differential absorption as possible consistent with minimal interference effects from either ammonia or ethylene, and for which atmospheric attenuation levels were low. While it is desirable to choose frequency pairs close together to maximize the mutual coherence of the two laser beams, this proved impractical for both UDMH and MMH. The explicit frequency choices and the pertinent absorption parameters are given in Table I, where we have used the absorption coefficients of ammonia and ethylene obtained by Patty et al.[9] The background atmospheric attenuation values, β, are for a U. S. Standard Atmosphere.[10]

b)   Remote sensing measurements

Prior to carrying out the remote sensing measurements, both the tank and the laboratory absorption cell were filled with nitrogen gas to minimize oxidation effects. Simultaneous measurements were then made of the normalized LIDAR returns and of the laser radiation transmitted through the cell. After the 100% transmittance level was established, a known amount of a hydrazine compound in liquid form was inserted by hypodermic syringe into the cell and average relative transmittance levels were taken after a preset number of pulses from each laser. Since these measurements were carried out with the hydrazine compounds in an inert nitrogen atmosphere, the results were used to establish the absorption coefficients of the compounds at the measurement frequencies of the mini-TEA lasers. The resulting values are given in Table I and are in reasonable accord with the values obtained by Loper et al.[8] using a low-pressure cw CO_2 laser.

Shortly after the hydrazine compound was inserted into the cell, a larger amount was inserted into the nitrogen-filled tank. Simultaneous cell transmittance and LIDAR return measurements were obtained for hydrazine, UDMH and MMH. The results were qualitatively similar for all three species.[11]

CO_2 LASER TRANSITION	WAVELENGTH (μm)	ABSORPTION COEFFICIENTS (cm-atm)$^{-1}$			ATMOSPHERIC ATTENUATION β(km)$^{-1}$
		(a) Hydrazine (b)	(c) NH_3	(c) C_2H_4	(d)
P(22) P(28)	10.611 10.675	4.77 5.41 2.06 2.17	0.046 0.36	1.09 1.30	0.1142 0.0976
DIFFERENTIAL ABSORPTION		2.71 3.24	-0.316	-0.21	
		UDMH	NH_3	C_2H_4	
P(30) R(10)	10.696 10.318	2.22 1.46 0.18 0.06	0.86 0.78	1.63 1.51	0.0907 0.1142
DIFFERENTIAL ABSORPTION		2.04 1.40	0.06	0.12	
		MMH	NH_3	C_2H_4	
R(30) R(18)	10.182 9.282	1.69 1.36 0.31 0.13	0.029 0.13	0.56 0.61	0.1137 0.1418
DIFFERENTIAL ABSORPTION		1.38 1.13	-0.10	-0.05	

Table 1 Relevant absorption parameters for the remote sensing of the hydrazines

(a) This work
(b) Ref. 8
(c) Ref. 7
(d) Values given for U.S. Standard Atmosphere

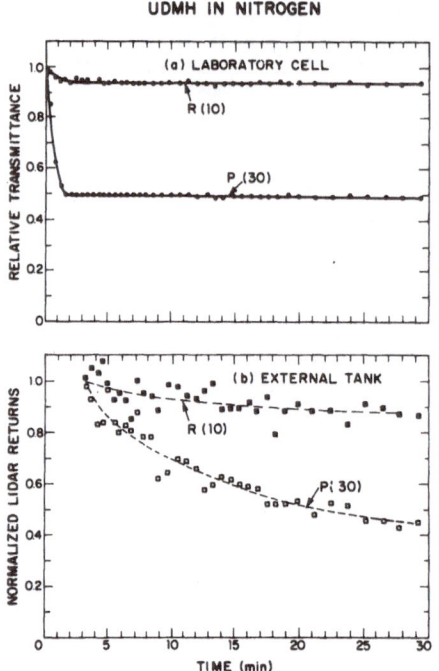

UDMH IN NITROGEN

(a) LABORATORY CELL

R (10)

P (30)

(b) EXTERNAL TANK

R (10)

P (30)

RELATIVE TRANSMITTANCE

NORMALIZED LIDAR RETURNS

TIME (min)

PULSE AVERAGES

200   500   1000

◄Fig. 3. Simultaneous measurements of UDMH in nitrogen-filled laboratory absorption cell and tank. (a) Time variation of relative transmittance through cell. (b) LIDAR returns from laser pulse passing through tank and refelcted from target

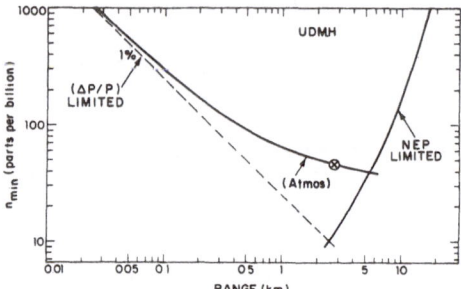

UDMH

1%

(ΔP/P) LIMITED

n_{min} (parts per billion)

NEP LIMITED

(Atmos)

RANGE (km)

Fig. 4. Minimum detectable path-averaged UDMH concentration as a function of LIDAR range. Normalization point corresponds to the experimental value $\Delta P/P = 0.05$ at a range of 2.7 km

The results of the measurements for UDMH, obtained after inserting 5.55 μℓ UDMH into the cell and 1.95 mℓ into the tank are given in Figs. 3(a) and 3(b). The initial set of points are based on 200-pulse averages, followed by 500- and 1000-pulse averages as indicated. As seen in these figures, the differential-absorption of the two laser frequencies is evident in both the laboratory absorption cell and the external tank, and demonstrates the ability of the dual-laser DIAL system to monitor the presence of UDMH over a range of 2.7 km.

A notable feature of the experimental results shown in Fig. 3 is the marked contrast between the smooth variation of transmitted radiation through the laboratory cell and the relatively large scatter in the LIDAR returns. This effect was observed with all the hydrazine compounds. The deviations from a smooth line of the points shown in Fig. 3(a) were generally less than 1% and approached the 0.1% digital quantization error of the data acquisition system. However, the fluctuations of the backscattered LIDAR returns around the smooth lines drawn in Fig. 3(b) are seen to be large, with normalized standard deviations of the order of 5%. These deviations are the major factor limiting the sensitivity of our concentration measurements.

On the basis of the absorption coefficients obtained from the laboratory absorption cell measurements shown in Fig. 3(a), the LIDAR results indicated a maximum UDMH vapor level in the tank of approximately 1500 ppm, corresponding to a path-averaged concentration of slightly under 600 ppb over the 2.7 km range. For the fluctuations and differential-absorption levels observed, our measurements established a path-averaged sensitivity of the DIAL system of 35 ppb, 45 ppb and 70 ppb for hydrazine, UDMH and MMH respectively.

The results shown, along with similar results obtained for hydrazine and MMH, established the ability of the dual-laser mini-TEA DIAL system to monitor the presence of these compounds by remote sensing over a range of 2.7 km. They can also be analyzed to determine the limits of sensitivity with which such measurements can be made at other detection ranges; that is, to establish the minimum detectable concentration, n_{min}, of a trace species as a function of range R.[11] The results of the analysis for UDMH are shown in Fig. 4. At long ranges n_{min} is determined when the difference in the backscattered LIDAR return is less than or equal to the noise equivalent power of the detector. For our system, this corresponds to the curve labeled NEP in Fig. 4. At shorter ranges a more restrictive limitation may be due to the minimum fractional change in the backscattered return ($\Delta P/P$) that can be measured accurately. Normally, ($\Delta P/P$) represents the limitation due to the equipment (i.e., $\leq$ 1%) and is independent of range, such as is observed in our laboratory cell measurements. For the conservative assumption of ($\Delta P/P$) equal to 0.01, this limitation leads to the dashed line labeled ($\Delta P/P$) limited in Fig. 4.

The effect of atmospheric turbulence will be to increase the level of the observed fluctuations beyond the level indicated above. This increase in fluctuations will be range-dependent according to the Rytov theory of atmospheric turbulence. Using this theory, one obtains[11] the curve for UDMH, labeled (Atmos) in Fig. 4, where the normalization point shown corresponds to the experimentally determined path-average concentration sensitivities of 45 ppb for a range of 2.7 km with $\Delta P/P$ = 0.05. Figure 4 indicates that a sensitivity of 100 ppb or better should be achievable between 0.5 and 5 km. Similar curves were obtained for hydrazine and MMH.

c) Atmospheric fluctuations and signal-averaging experiments

It is seen in Fig. 4 that atmospheric effects play the dominant role in limiting the minimum distinguishable pollutant concentration at intermediate ranges. In addition, it was noted that increasing the number of pulses averaged above a few hundred yielded relatively small improvement in the fluctuation level.

In view of the importance of the atmospheric fluctuations in limiting the sensitivity of our measurements a study was made to ensure that these effects were not an artifact due to the presence of the tank in the laser beam path and to directly measure the improvement achievable by signal averaging over increasing numbers of pulses.

For this purpose, the hydrazine LIDAR experiment was repeated, but with the large tank removed, so only atmospheric absorption was involved in the measurements. Over 12,000 normalized LIDAR return pulses from each laser were recorded for later statistical analysis. The process, including the time required to print out a preliminary real-time analysis after every 2048 pulses, took about 22 minutes, corresponding to an overall pulse repetition rate of slightly under 10 Hz.

The analysis included determination of the average value of the full block of 12,288 pulses and of smaller segments of that block as well as a determination of the normalized standard deviation, σ, of the LIDAR returns as a function of n, the number of pulses being averaged in each segment. The variation of the standard deviation of the returns from lasers 1 and 2 and of the pulse-pair ratio L1/L2 as a function of n is shown graphically in Fig. 5(a). It is seen that the decrease of σ with increasing n is much slower than the $n^{-1/2}$ behavior expected for a random process even for small

21

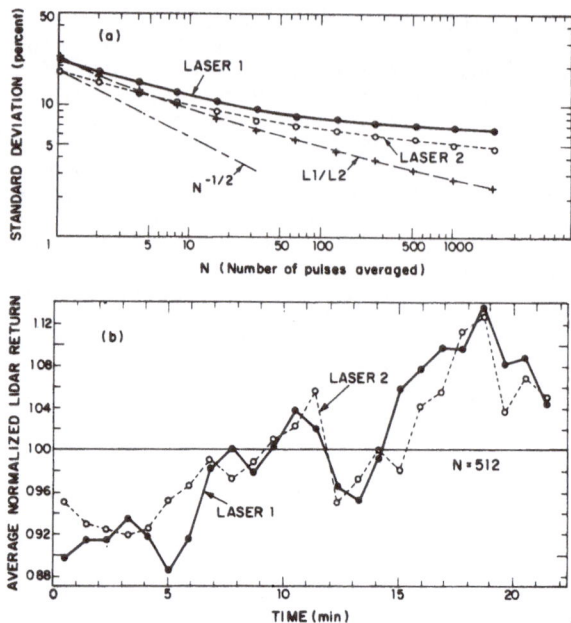

Fig. 5. Analysis of segmentally averaged block of 12288 normalized pulses from lasers 1 and 2 and their ratio. (a) Standard deviation of the segment averages as a function of N, the number of pulses averaged per segment, (b) time variation of the average value of the individual segments for N = 512

values of n and becomes almost flat for the individual lasers for n > 256. This is presumably because the atmospheric absorption being measured is varying in a nonrandom manner over time periods of the order of the measurement period. The slow decrease of the standard deviation with increasing n can be related directly to the small but persistent long-term temporal correlations of the LIDAR returns from a single laser which were observed over time periods encompassing a large number of pulses.[12] The existence of this long-term correlation severely limits the improvement attainable by pulse averaging over a reasonable number of pulses.

A notable feature of the results shown in Fig. 5(a) is that the standard deviation of the ratios of the pulse pairs, $\sigma_{L1/L2}$, while also decreasing more slowly with n than $n^{-1/2}$, yields values which are significantly lower than either σ_{L1} or σ_{L2} for large n. The source of this reduction can be understood by considering the normalized segmental average values for the entire block, as shown in Fig. 5(b) for n = 512. It is seen that there is a slow but almost constant increase in the average value of the normalized return segments from both lasers throughout the entire period, corresponding to a decreasing atmospheric absorption at both laser frequencies. Since the decrease in the atmospheric background absorption is slow compared to the measurement time of the individual segments (~ 1 min for n = 512), one can deal with the temporal cross-correlation of the segmental measurements of the two lasers in the same way as was used for pulse-to-pulse cross-correlations to relate $\sigma_{L1/L2}$ with σ_{L1} and σ_{L2}.[13] For the values shown in Fig. 5(b), the segmental cross-correlation coefficient, ρ_s, is equal to 0.9, which is consistent with the observed value of $\sigma_{L1/L2}$.[11,13] This

22

result indicates that increased accuracy can be achieved using LIDAR return ratios when long-term atmospheric shifts are superimposed on short-term effects. Under the conditions of the experiment described in this section, the use of ratios led to a reduction of standard deviation from over 6% to approximately 2.5% after averaging over the order of 2000 pulses in a 3.5 minute interval.

IV. Conclusions

We have experimentally demonstrated the capability of the CO_2 mini-TEA DIAL system to detect hydrazine, UDMH and MMH over distances approaching 3 km. Our results indicate that hydrazine, UDMH and MMH can be remotely detected with a path-averaged sensitivity on the order of 100 ppb. The concentration sensitivity of these measurements was found to be limited by atmospheric background fluctuations.

The direct detection system described in this paper has recently been modified to permit heterodyne detection of the return signals in conjunction with direct detection. This dual-laser heterodyne DIAL system will be used for direct experimental comparison of the two detection techniques under differing experimental conditions.

We would like to express our appreciation to Dr. Daniel A. Stone for permission to print the band spectra of the hydrazine compounds shown in Fig. 2.

REFERENCES

1. N. Menyuk and D. K. Killinger, "Temporal Correlation Measurements of Pulsed Dual CO_2 LIDAR Returns," Opt. Lett. 6, 301 (1981).

2. D. K. Killinger and N. Menyuk, "Remote Probing of the Atmosphere Using a CO_2 DIAL System," IEEE J. Quantum Electron. QE-17, 1917 (1981).

3. N. Menyuk and P. F. Moulton, "Development of a High-Repetition-Rate Mini-TEA CO_2 Laser," Rev. Sci. Instrum. 51, 216 (1980).

4. To avoid confusion, the large container used for remote sensing will be referred to as a tank. The term cell will be reserved for the Pyrex container used in the laboratoy absorption measurements.

5. H. W. Schiessl, "Hydrazine and its Derivatives" in Kirk-Othmer: Encyclopedia of Chemical Technology, Vol. 12, 3rd Edition (John Wiley & Sons, Inc., New York, 1980).

6. D. A. Stone, "The Autoxidation of Hydrazine Vapor," CEEDO-TR-78-17, AFESC, Tyndall Air Force Base (1978), and private communication.

7. D. A. Stone, "Autoxidation of Hydrazine, Monomethylhydrazine and Unsymmetrical Dimethylhydrazine," in SPIE Proceedings, Vol. 289, 1981 International Conference on Fourier Transform Infrared Spectroscopy, H. Sakai, Editor, (1981), p. 45.

8. G. L. Loper, A. R. Calloway, M. A. Stamps and J. A. Gelbwachs, "Carbon Dioxide Laser Absorption Spectra and Low PPB Photoacoustic Detection of Hydrazine Fuels," Appl. Opt. 19, 2726 (1980).

9. R. R. Patty, G. M. Russwurm, W. A. McClenny and D. R. Morgan, "CO_2 Laser Absorption Coefficients for Determining Ambient Levels of O_3, NH_3 and C_2H_4," Appl. Opt. 13, 2850 (1974).

10. R. A. McClatchey, R. W. Fenn, J. E. A. Selby, F. E. Volz and G. S. Garing, "Optical Properties of the Atmosphere (Third Edition)," Report AFCRL-72-0497, Environmental Research Paper No. 411 (1972).

11. N. Menyuk, D. K. Killinger and W. E. DeFeo, "Laser Remote Sensing of Hydrazine, MMH and UDMH Using a Differential-Absorption CO_2 LIDAR," (to be published).

12. C. R. Menyuk, N. Menyuk and D. K. Killinger (to be published).

13. D. K. Killinger and N. Menyuk, "Effect of Turbulence-Induced Correlation on Laser Remote Sensing Errors," Appl. Phys. Lett. 38, 968 (1981).

1.4 The Hull Coherent DIAL Programme

B.J. Rye and E.L. Thomas
Department of Applied Physics, University of Hull, Hull, HU6 7RX, U.K.

1. Introduction

An experimental programme to exploit the advantages of coherent laser radars has been in progress at Hull since January, 1976. A 10 µm, coherent laser radar system has a much greater range capability along a near horizontal tropospheric path than either a uv-visible system or an infrared system employing direct detection. Further advantages of a coherent system operating at 10 µm include eye-safety, immunity to solar background radiation, good all-weather performance and the need to transmit relatively small amounts of energy. The inherent capability for high spectral resolution enables Doppler measurements to be made. Here we consider some problems in the application of these systems to differential absorption lidar (DIAL) and discuss briefly work in our laboratory.

2. Measurement sensitivity

In DIAL the concentration γ of a species is inferred from (1)

$$\gamma = \frac{1}{2\alpha_D \Delta r} \log_e \frac{<P_{12}><P_{21}>}{<P_{11}><P_{22}>} \quad , \tag{1}$$

where the $<P_{ij}>$ are the mean return powers observed from aerosol backscatter at wavelength λ_j and range r_i, the range resolution $\Delta r = r_1 - r_2$ and α_D is the differential absorption coefficient (eg. in $atm^{-1} m^{-1}$).[1] The sensitivity of the measurement is given by the variance

$$<(\Delta\gamma)^2> = \frac{1}{(2\alpha_D \Delta r)^2} \sum_{i,j=1}^{2} \sigma_{Pij}^2 \quad , \tag{2}$$

where the normalised variances $\sigma_p^2 = <(\Delta P)^2>/<P>^2$. The P are obtained from the lidar equation

$$P = (\tfrac{1}{2} A\beta c T_A/r^2) E_o \quad , \tag{3}$$

where E_o is the laser energy output, T_A the atmosphere attenuation ($\exp(-2\alpha_A r)$), β the backscatter coefficient and A the system antenna area.

In a lidar context incoherent backscatter from a deep distributed target leads to speckle fluctuations in the return. These are squared and averaged at the photodetector in a direct detection system, the resulting fluctuation usually being dominated by background noise, in which case

$$\sigma_p^2 = (1/SNR)^2, \quad SNR = <P>/(DNEP \sqrt{B_{\Delta r}}) \quad , \tag{4}$$

where DNEP is the noise equivalent power and the bandwidth $B_{\Delta r} \sim c/(2\Delta r)$ is determined by the range resolution Δr. In a coherent lidar, detection and therefore speckle averaging is achieved electronically. It is also necessary to take account of excess noise arising from variations in the lidar equation coefficients T_A and β, the latter depending on the optical properties of the aerosol which are humidity and wavelength dependent (2) and on its concentration (3). While these sources affect direct detection systems, for coherent lidar there are also fluctuations in the effective area A (4) arising from refractive turbulence (5). Whereas the speckle bandwidth is at least that of the transmitted pulse (6), atmospheric excess noise is narrowband so long as averaging times are needed to obtain $<P>$. To obtain expressions for the sensitivity of a coherent DIAL system we therefore suppose that m independent observation of a Rayleigh distributed signal (return + background noise) - or 2 m such values if the latter is Gaussian - are squared, averaged and reduced by subtraction of a known mean noise level to obtain a sample value P within the return from each individual transmitted pulse (7); samples from successive pulses separated in time by more than the correlation period of the atmospheric lidar equation parameters are independent yielding

$$\sigma_p^2 = \sigma_{speckle}^2 + \sigma_{ex}^2 \tag{5}$$

where

$$\sigma_{speckle}^2 = 1/m(1+1/CNR)^2, \quad CNR = <P>/(m(CNEP)B_{\Delta r}) \tag{6}$$

$$\sigma_{ex}^2 = (1 + 1/m)(\sigma_A^2 + \sigma_\beta^2 + \sigma_{TA}^2) \quad . \tag{7}$$

The σ^2 in equation (7) are normalised variances in the subscripted variable and CNEP (in W/Hz) is the heterodyne system NEP. For a given long term average $<P>$, $\sigma_{speckle}^2$ is minimised by making $CNR = 1$ (8-10); for a given CNR advantage can only be taken of increased $<P>$ (eg. using higher E_o) by increasing m.

3. Intrapulse averaging

Advantage might be taken of the (doubly) differential nature of DIAL to remove or at least minimise the excess noise contribution in eqn (5) and reduce the integration times or laser pulse repetition frequencies otherwise required. Suppose samples P_a and P_b are taken within the atmospheric coherent time. In a DIAL context we might sample the returns from different ranges within the return from a single pulse, making $P_a = P_{i1}$, $P_b = P_{i2}$ or sample returns from a given range at two wavelengths making $P_a' = P_{1j}$, $P_b = P_{2j}$. Then a suitable statistic based on the ratio P_a/P_b can be used to estimate functions of the ratio R_{ST} of the short term means (STM) $<P_a>_{ST}$ and $<P_b>_{ST}$, ie. the values of $<P_a>$ and $<P_b>$ that would obtain if the atmospheric lidar parameters were to be frozen at the values they had during transmission of the pulse. Two such statistics are $z^1 = \frac{1}{2}\log_e(P_a/P_b)$ and $F^1 = P_a/P_b$ which can be used to estimate $\chi_{ST} = \log_e R_{ST}$ by forming either

$$\chi_1 = <2z^1> = <\log_e P_a> - <\log_e P_b> \tag{8}$$

or

$$\chi_2 = \log_e((m-1)/m <F^1>) = \log_e((m-1)/m < P_a/P_b>) \quad . \tag{9}$$

The fractional bias $\chi/\chi_{ST}-1$ is plotted in Fig. 1 for each of χ_1 and χ_2. While these are unbiased in the zero noise limit it is seen that bias depending on both CNR and R_{ST} increases rapidly with increasing noise. The effect of atmospheric fluctuations being to vary CNR and (possibly) R_{ST}, the bias itself can be regarded as having a fluctuating component. For

26

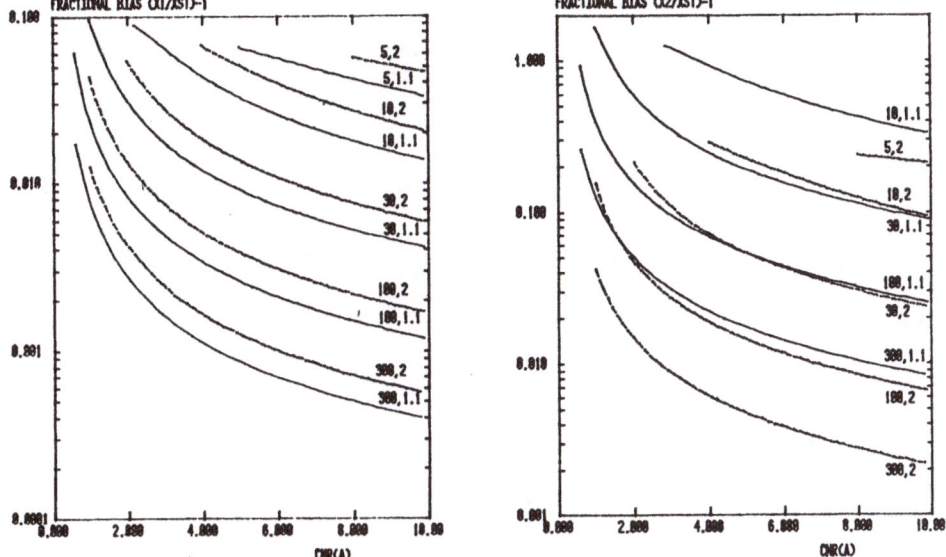

Fig. 1. Fractional bias in estimation of the ratio of the short term mean return powers (see text). The numbers on each curve are m, R_{ST}. CNR is quoted for the numerator in R_{ST}

given m and R_{ST} Fig 1 shows that it is preferable to use the estimator $2z^1$ and to operate at CNR > 1; the bias problem can probably be eliminated in practice in this way.

Fluctuations in R_{ST} have the second consequence that they need to be averaged over several atmospheric coherence times. This is a less serious problem than averaging over the values of P_a and P_b separately, since any correlation in the joint density of the STMs reduces the variance of their ratio. The general requirement is that the latter should become less than the single pulse speckle noise in the ratio found from eqn (6). Formation on an intrapulse basis of the estimator $\log_e (P_{12} P_{21} / (P_{11} P_{22}))$ might be expected to reduce this contribution further.

4. Sensitivity comparisons

Comparison is now made between the sensitivities obtained using direct detection and coherent systems. The basis of the comparison is that the laser output E_O is the same for both systems and that in view of the above discussion the coherent lidar CNR is a fixed parameter and σ_{ex}^2 can be neglected; it further supposed conservatively that all the P_{ij} are independent (8-11) so we need only compare the variances in estimation of the return from a single range-wavelength element. In eqn (3) direct detection gives $A=A_R$ while for the coherent case in the absence of turbulence $A = \eta_a A_T$ where A_R and A_T are receiver and transmitter antenna areas respectively. The efficiency factor η_a for the far field (focused) diffraction limited return $\simeq 3.5$dB for an optimum Gaussian system (4) and values for an annular uniform transmitter (a rough and ready model for systems using unstable resonator lasers) are indicated in Fig 2. In both cases $<A> \gtrsim \pi \rho_o^2$ in strong turbulence (5) where ρ_o is the lateral coherence length. Here we put $A_T = A_R$ so combining eqns (3) (4) and (6) leads to

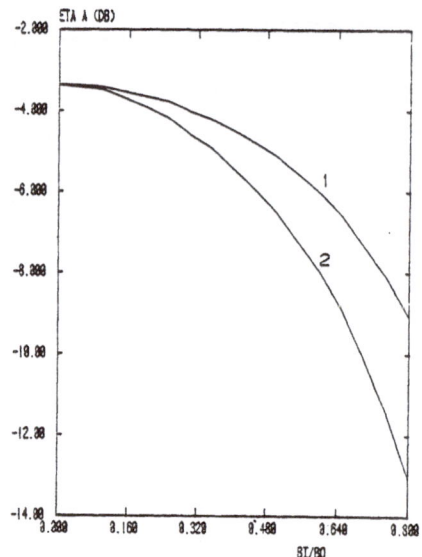

<cf>Fig. 2.** Far field diffraction limited antenna efficiency η_a for uniformly illuminated annular transmitter of inner and outer radii b_i and b_o. For curve (1) the receiver is circular of radius b_o and for (2) it is a matching annulus. In all cases A_T is taken to be $\pi b_o{}^2$

$$\frac{(\sigma_p{}^2)\ \text{direct}}{(\sigma_p{}^2)\text{coherent}} = \frac{\eta_a\,(\text{CNR})}{A_R\beta cT_A E_o\,(\text{CNEP})}\left(\frac{r(\text{DNEP})}{1+\text{CNR}}\right)^2 \qquad . \tag{10}$$

As a numerical example using DNEP = 10^{-12} WHz$^{-\frac{1}{2}}$, $r = 10^4$ km, exp $(\alpha_A r) = 3$, $\eta_a = 0.1$, $A_R = 0.1$ m^2, $\beta = 5 \times 10^{-8}$ m^{-1}, CNEP = 8×10^{-20} W/Hz, laser output $E_o = 0.1$J, and CNR = 1 (CNR = 5 in parentheses) we find

$$\frac{(\sigma_p{}^2)\ \text{direct}}{(\sigma_p{}^2)\ \text{coherent}} \sim 1600\ (900) \qquad .$$

The same numbers entail $P_R \sim 8 \times 10^{-11}$W allowing m = 200 (40) at $B_{\Delta r}$ = 2.5 MHz or Δr = 60 m. The direct detection system has been assumed loss and bias free but for the coherent lidar some allowance for turbulence has been made in η_a. Even with CNR > 1 coherent systems should have advantages over direct detection in infrared DIAL especially for low level returns (large r, α_A, small β) or where it is desirable to make the system compact (small A_R, E_o).

5. The experimental programme

After demonstrating the use of a coherent system using a single, TEA, CO_2 laser for range resolved anemometry (12), the Hull system has been completely rebuilt. Efforts are now directed towards making DIAL measurements. The present system employs two hybrid TEA, CO_2 lasers, one laser being tuned to the on-resonance frequency and the other to the off-resonance frequency of the target gas. Each laser is capable of emitting 200 mJ in a single mode at 100 Hz when operated separately or 50 Hz when operated together. The gain sections can be operated as two oscillators, an oscillator-amplifier combination and in stable and unstable resonator configurations. When operating as oscillators, each oscillator can be on the same or different frequencies and the delay between the laser pulses is variable. Each transmitting laser has its own local oscillator laser and the fixed transmitting and receiving

28

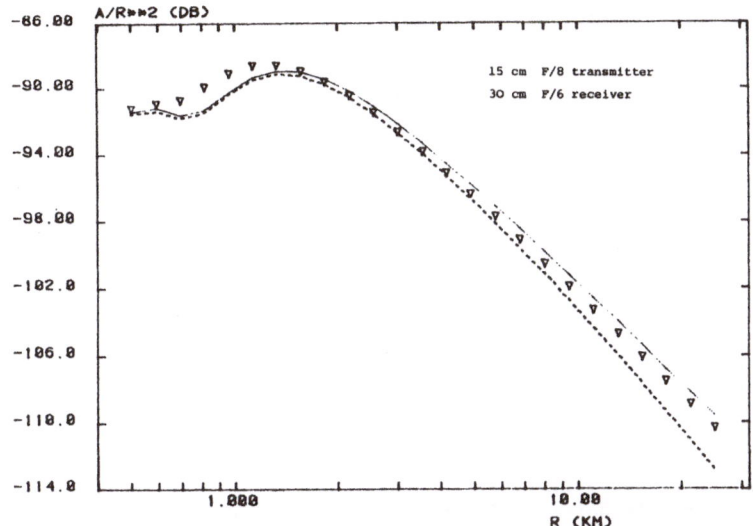

Fig. 3. Computed mean return for the Hull coaxial telescope. The Gaussian transmitted beam has $1/e^2$ radius = 0.825 x transmitter diameter and the receiver is uniformly filled. The curves are (-) diffraction limited collimated beams (....) same + uniform turbulence with $C_n^2 = 10^{-14}$ $m^{-2/3}$ and (Δ) with parabolic mirror aberrations for 10mrad misalignment on transmitter and receiver; the tangential planes are coplanar but the misalignments are on opposite sides of the mirror axes. At short range the return falls because of the reduction in beam overlap and at long range the return is similar to that of a 15cm transceiver

telescopes are in a coaxial configuration. An indication of the range dependence of the return to be expected using this geometry is given in Fig 3. The design philosophy was to make the system as versatile as possible.

Whilst the coherent DIAL work is the main project there are associated projects involving atmospheric modelling, and development of multi-atmosphere CO_2 lasers, narrow bandgap diode lasers and fast, photovoltaic detectors. It is intended to incorporate the multi-atmosphere laser within the system in the near future.

References and Footnotes

(1) R. M. Schottland, Errors in the lidar measurement of atmospheric gases by differential absorption, J. Appl. Met. 13, 71 (1974).

(2) J. C. Petheram, Differential backscatter from the atmospheric aerosol: the implications for IR differential absorption lidar, Appl. Opt. 20, 3941 (1981).

(3) M. J. Post, Experiemntal measurements of atmospheric aerosol inhomogeneities, Opt. Lett., 2, 166 (1978).

(4) B. J. Rye, Primary aberration contribution to incoherent backscatter heterodyne lidar returns, Appl. Opt., 21, 839 (1982).

(5) B. J. Rye, Refractive turbulence contribution to incoherent backscatter heterodyne lidar returns, J. Opt. Soc. Am., 71, 687 (1981).

(6)   Where the bandwidth of the pulse is large compared with atmospheric Doppler shifts the frequency-time correlation properties of the return are described by the radar ambiguity function.   For simple pulses the correlation time is about the pulse duration; unfortunately except for chirp-free SLM operation laser pulses are not simple.

(7)   Ambiguity considerations suggest that the notion of obtaining temporally independent returns from essentially the same scatter volume is possible in principle provided the pulse bandwidth is much greater than the reciprocal pulse duration and pulse compression (matched) filters are not used.   How closely the ideal can be realised in practice is not considered here.   An alternative approach would of course be aperture (spatial) averaging which is demanding technically as use is made of a photodetector array; the conclusions here would be essentially unaffected.

(8)   R. M. Hardesty, A comparison of heterodyne and direct detection CO_2 DIAL systems for ground based humidity profiling, NOAA Tech. Memo, ERL/WPL-64 (1980).

(9)   P. Brockman, R. V. Hess, L. D. Staton and C. H. Blair, DIAL with heterodyne detection including speckle noise, NASA Tech. Paper 1725 (1980).

(10)   G. Megie and R. T. Menzies, Complementarity of UV and IR differential absorption lidar for global measurements of atmospheric species, Appl. Opt. 19, 1173 (1980).

(11)   B. J. Rye, DIAL system sensitivity with heterodyne reception, Appl. Opt. 17 3862 (1978).

(12)   E. L. Thomas, A coherent laser radar for trace gas and meteorological measurements, Topical Meeting on Coherent Laser Radar, Aspen, 1980.

1.5 Remote Measurement of Trace Gases With the JPL Laser Absorption Spectrometer[1]

M.S. Shumate, W.B. Grant, and R.T. Menzies

California Institute of Technology, Jet Propulsion Laboratory,
4800 Oak Grove Drive, Pasadena, CA 91109, USA

Introduction

Measurement of anthropogenic trace gases in the troposphere has become an increasingly difficult problem due to the need for extensive data from large regions. Atmospheric models for the study of air pollution are becoming more sophisticated and require large amounts of input data collected over extended volumes of the troposphere. The need for an instrument capable of rapid monitoring the atmosphere in a region is obvious. The work reported in this paper represents the results of efforts to develop an instrument which is capable of remotely measuring gases in the troposphere.

The Laser Absorption Spectrometer (LAS) is a remote sensing instrument, designed and built for use in aircraft, and has been used since 1977 to measure tropospheric ozone distributions[1-6]. It is an active, nadir-directed instrument which measures the vertical column abundance of ozone between ground-level and the aircraft altitude. The basis of the measurement is differential absorption of a pair of transmitted wavelengths which are selected to interact with a sharp spectral feature of the ozone ν_3 band, near a wavelength of 9.5 μm. Two grating-tunable tunable waveguide CO_2 lasers provide the transmitted radiation, and two heterodyne receiver channels in the instrument respond to a small portion of the laser radiation which is backscattered off the earth's surface below the aircraft and propagated back to the collecting telescope. The LAS has the capability of easily measuring ozone, ethylene, ammonia and water vapor. Utilization of other isotopes of carbon dioxide will extend the capability of the instrument.

Ozone was selected for study during demonstration measurements for the following reasons: (1) ozone is an extremely important tropospheric constituent, being the chief precursor of tropospheric photochemistry on a global scale and the principal product of photochemical smog near urban areas; (2) ozone has a strong vibration-rotation band centered at 9.5 μm; and (3) ozone can be measured with relatively high accuracy using a number of in-situ measurement techniques, which provides the potential for calibrating the operational accuracy of the LAS under a variety of conditions.

A series of correlative ozone measurements, involving the LAS in one aircraft and an in-situ ozone measuring instrument in another aircraft, have been conducted over the past three years. The airborne in-situ data, along with ozone data from ground-based monitoring instruments, were used to assess the accuracy of the ozone column abundance data collected by the LAS.

Instrument Description

The design considerations for the LAS instrument were heavily influenced by aircraft environmental factors. The NASA/JPL Beechcraft twin-engine airplane was

[1] This paper presents one phase of research carried out at the Jet Propulsion Laboratory, California Institute of Technology, for the National Aeronautics and Space Administration under contracts NAS7-100.

selected to carry the LAS on its first flight series, and the instrument as a consequence was required to operate in a limited space and in a noisy environment. Two slightly modified Hughes Research Labs carbon dioxide waveguide lasers were chosen as the transmitters, with compactness and ruggedness being major considerations. The waveguide lasers were tested for amplitude and frequency stability on a shake table, which was operated at a vibration level which would be encountered in the Beechcraft. The laser is a flowing gas type designed to emit at a one-watt power level on a large number of lines which can be selectively tuned by tilting a Littrow-mounted grating and adjusting the voltage on a piezoelectric bimorph. The LAS telescope and other optical components were placed on an aluminum honeycomb plate to provide additional vibration damping, and novel mirror mounts were used which have a special clamping feature which can be activated after alignment adjustments are made.

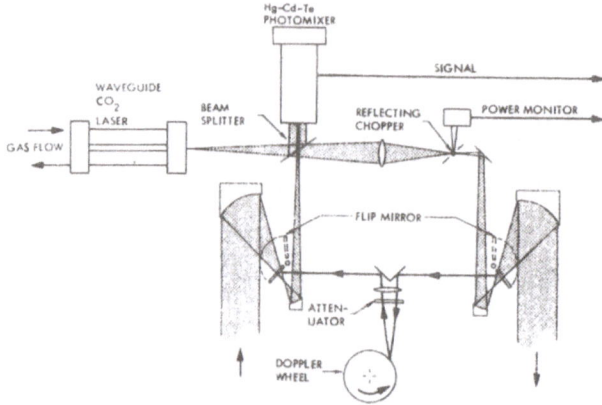

Fig. 1. Optical Schematic of the LAS

An optical schematic for the Laser Absorption Spectrometer is shown in Figure 1. Only one of the two (signal wavelength and reference wavelength) channels is shown, since they are identical. A small portion of the laser output is reflected off the beamsplitter into the photomixer to serve as a local oscillator. The remainder of the laser radiation is transmitted through a lens which provides optimum coupling to the telescope for a collimated output. At the lens focus, a 1000 Hz tuning fork chopper with a reflective tine modulates the beam with a 50% duty cycle and directs the remainder of the power into a calibrated power monitor. The received signal is collected by the telescope and focused onto a liquid nitrogen cooled mercury-cadmium-telluride photomixer. A frequency offset between the local oscillator and the received signal is accomplished by tilting the transmit/receive telescope so that the transmitted beam is pointed a few degrees ahead of nadir. The scattered return radiation is Doppler-shifted a few MHz by the aircraft's motion. A single telescope, with a 15 cm diameter primary mirror, is divided into four non overlapping 5 cm subaperatures, two for the transmitted wavelengths and two for the independent heterodyne receivers. An open port is provided in the belly of the airplane to allow the instrument an unobstructed view of the ground.

The receiver electronics block diagram is shown in Figure 2 for both channels. The received signals are amplified in a 2-30 MHz passband, and then envelope-detected with a square-law detector. Synchronous demodulation at the chopper frequency is followed by an electronic ratiometer and data recording system. Care was taken in the electronics to make certain that both channels had well-matched frequency response and amplitude-response characteristics. During flight operations, the IF amplifier gains were adjusted to limit the signal amplitude into the square-law detectors, in order to keep from driving them into a range of non-square-law operation.

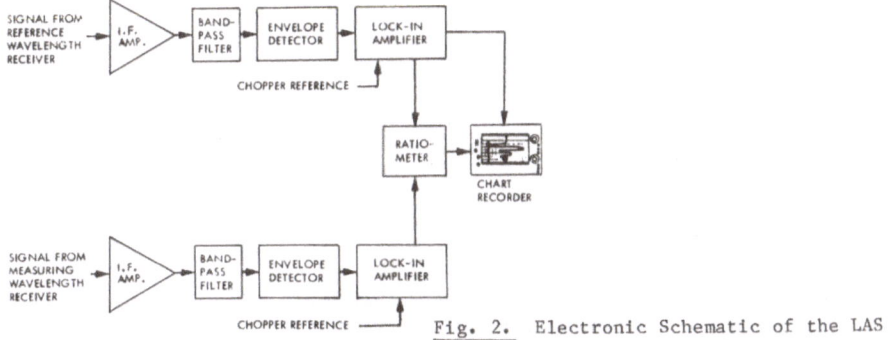

Fig. 2. Electronic Schematic of the LAS

The LAS has an internal reference feature. In order for the system to function properly, the net gain in the two channels must be checked frequently and, ideally, adjusted to be equal. The instrument is placed in the reference mode by energizing a solenoid-actuated motor which inserts a flip mirror into the two transmit beam paths and deflects the transmitted signals to an internal Doppler shifter. The Doppler-shifted radiation is attenuated to the desired level and directed into the receiver optics to produce fixed receiver output signals. The gains of the two channels are adjusted to assure that the system is balanced and can thus accurately measure the differential absorption due to the atmospheric path when the signals are transmitted. During actual flight operations, the LAS is switched to the internal reference mode about 10% of the time, in order to check for slow drifts in the system gains.

Theory

The equation used to calculate the atmosheric ozone concentration for the LAS is derived here. Using the Beer-Lambert law we obtain

$$s = kPTS^2R\frac{A}{h^2} \exp\left\{ -2\int_0^h \alpha(x)\rho(x)dx\right\}\qquad(1)$$

where

s	=	received signal strength
k	=	receiver conversion gain
P	=	transmitter intensity
T	=	total optical system losses
S	=	atmospheric scattering losses
R	=	surface diffuse reflection coefficient
$\alpha(x)$	=	ozone absorption coefficient expressed as a function of altitude
h	=	aircraft altitude above ground
$\rho(x)$	=	ozone concentration expressed as a function of altitude
A	=	receiver area

The expression can be simplified by assuming that there is no variation in ozone concentration with altitude. The LAS has been operated mostly at an altitude of 1 km, which is below the inversion layer height where vertical mixing keeps the ozone concentration quite constant. We let $\rho(x) = \rho$ and $\alpha(x) = \alpha$. Since the LAS has two completely separate systems operating at two different wavelengths [1], we rewrite Eq. (1) with the subscripts 1 and 2 to denote the two wavelengths. The system signal processing produces a ratio of the received signals, r_s, which can be written:

33

$$r_s = \frac{k_1 P_1 T_1 S_1{}^2 R_1}{k_2 P_2 T_2 S_2{}^2 R_2} \, e^{-2(\alpha_1-\alpha_2)h\rho} \qquad . \tag{2}$$

The optical system also contains an internal reference device, which produces a constant output ratio r_r

$$r_r = \frac{k_1 P_1 T_1' r_1}{k_2 P_2 T_2' r_2} \tag{3}$$

where r_1, r_2 are the amplitude of the optical reference signals at the two wavelengths, and T_1', T_2' are the transmission of the modified optical paths. The signal-processing system then compares the two ratios r_s and r_r,

$$\Psi = \frac{r_s}{r_r} \qquad , \tag{4}$$

Cancelling terms and rewriting Eq. (4), we obtain

$$\Psi = \left[\frac{(T_1/T_2)(S_1^2/S_2^2)}{(T_1'/T_2')(r_1/r_2)}\right] \frac{R_1}{R_2} e^{-2(\alpha_1-\alpha_2)h\rho} \qquad . \tag{5}$$

All of the terms in the braces [] are assumed to be constant over the measurement period so these are expressed together as a single constant Ψ_o. Equation (5) can now be solved for the ozone concentration ρ

$$\rho = \frac{1}{2(\alpha_1-\alpha_2)h} \, [\ln(\Psi_o/\Psi) + \ln(R_1/R_2)] \qquad . \tag{6}$$

Each of the terms in Eq. (6) must be carefully considered. The values of α_1, α_2, and h are all known in advance, with α_1 and α_2 having been determined from laboratory measurements, and the system operating altitude, h, having been obtained from the aircraft's flight instruments. The value of Ψ is the primary measured quantity and is determined by the instrument's signal-processing system. The value of Ψ_o is the instrument's calibration constant and must be determined by comparison of ozone measurements with other ozone-monitoring instruments. The remaining term, R_1/R_2, is the ratio of the reflectivity of the surface material at the two wavelengths.

LAS Measurement Programs

The Laser Absorption Spectrometer system has been involved in several different measurement programs since it was constructed in 1976. These are listed in the table on the next page. During the first project the LAS was installed in NASA's Convair 990, and for all subsequent projects it was installed in a twin-engine Beechcraft Queen Air. The most recent set of measurements was made on the transport of ozone from the Los Angeles Basin during July 1981. The goal of a large program sponsored by the California Air Resources Board was to determine how ozone gets past the mountains to the desert regions north and east of the Basin. For this set of measurements, the LAS and a Dasibi ozone monitor were flown near 10,000 feet above mean sea level, first in an eastward direction through the basin starting at the Burbank Airport, then through the Banning Pass, then north, then west just north of the San Gabriel mountains to the end of the flat desert region about 20 miles west of Edwards Air Force Base. A typical set of data is shown in Figure 3. Elevation is indicated by parallel lines--the taller mountains (above 5000') are indicated by lines ascending to the right; the region from 2500' to 5000' is indicated by lines in the perpendicular direction. Regions below 2500' are not cross-hatched. The figures below the line indicate the ozone values measured at the air-craft altitude by the Dasibi ozone monitor; those figures above the line indicate the values determined from the LAS data. Note that the LAS-determined values peak near Riverside in the basin and north of Big Bear Lake in the desert. This can be

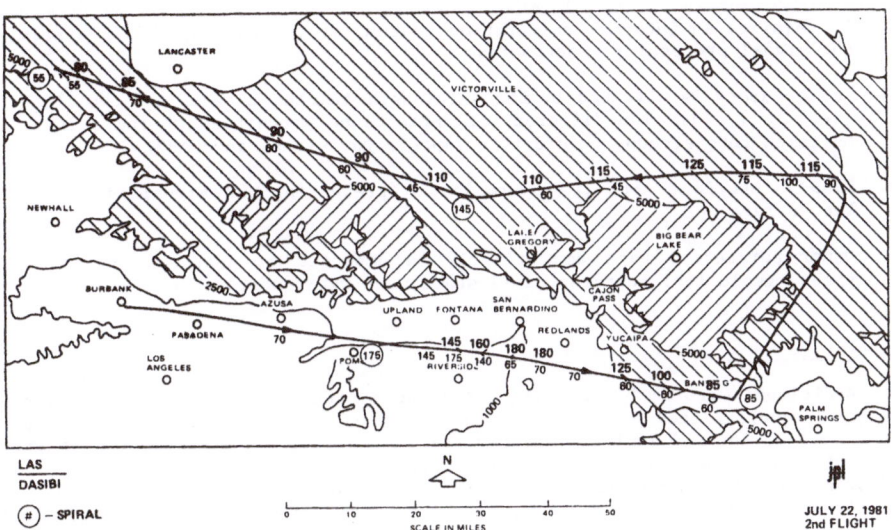

Fig. 3. LAS Measurements of Ozone Made During the 1981 LA Basin Pollutant Transport Study. The numbers above the flight path line are LAS ozone measurements and the numbers below the line are ozone measurements from an in-situ monitor in the aircraft carrying the LAS

MEASUREMENT PROGRAMS INVOLVING THE JPL LASER ABSORPTION SPECTROMETER

Name	Year	Molecule	CO_2 Laser Lines* Sig. λ	Ref. λ	Ref.
NASA-ESA Space Shuttle Simulation	1977	O_3	P(12)	P(24)	2
SE Virginia Urban Plume Study	1978	O_3	P(12)	P(24)	1
SE Virginia Urban Plume Study	1979	O_3	P(14)	P(22)	4
L.A. Basin Study	1979	O_3	P(14)	P(22)	1
NE Regional Oxidant Study	1980	O_3	P(14)	P(22)	5
Ammonia Spill Test	1981	NH_3	P(18)	P(20)	-
L.A. Basin Pollutant Transport Study	1981	O_3	P(26)	P(24)	6

*All measurements done in the 9.4 µm band of the CO_2 laser.

explained by winds blowing from the southwest. Data on other days suggested that the ozone was transported through the pass on one day and over the mountains near Azusa on another day.

Results of the LAS Development Program

After five years of flight experience, the LAS has been well characterized in its present form. The general characteristics are listed below.

1. The LAS is capable of measuring ozone concentrations to an accuracy of ± 20 ppb-km. The horizontal resolution is 80 m along the flight track.

2. It requires an internal calibration system for the optics, the IF system, and the signal-processing electronics.

3. The accuracy is affected by the wavelength spacing between the two lines used [Ref. 1, 7, 8]. The differential reflectance properties of the surface must be taken into account in order to improve the system accuracy.

4. Clouds interfere with its operation by preventing the signals from reaching the surface.

5. Other parameters that can probably be detected or measured include: other gases, geological features, soil state, and water surface state.

IV.   Future Work

Additional work is being planned for the system in the near future. Improvements being contemplated include: 1) installation of a digital processing system for data collection and analysis and for system control, 2) installation of grating angle controllers for rapid tuning of the lasers' wavelengths, and 3) replacement of the flowing-gas CO_2 waveguide lasers with sealed-off CO_2 waveguide lasers. This would allow the use of CO_2 isotopes and reduce the system weight and complexity.

In addition, the design of a improved version of the LAS is being considered. One desirable feature beyond those mentioned above is increasing the number of lasers to three or four. This allows more than one species to be measured at a time and can be used to reduce measurement error.

New measurement areas that are being considered include:

1) other gases (NH_3, H_2O, vinyl chloride),

2) geological features, especially different rock types,

3) soil conditions (freshly disturbed, soil moisture),

4) water surface state (effect of winds, ship wakes).

REFERENCES

1. M. S. Shumate, R. T. Menzies, W. B. Grant, and D. S. McDougal, "Laser Absorption Spectrometer: Remote Measurement of Tropospheric Ozone," Applied Optics 20, p. 545 (15 Feb. 1981).

2. R. T. Menzies and M. S. Shumate, "Torposhperic Ozone Distributions Measured with an Airborne Laser Absorption Spectrometer," J. Geophys. Res. 83, p. 4039 (Aug. 20, 1978).

3. M. S. Shumate and R. T. Menzies, "Airborne Laser Absorption Spectrometer: A New Instrument for Remote Measurement of Atmospheric Trace Gases," in <u>Proc. of the Fourth Conference on Sensing of Environmental Pollutants</u> (American Chemical Society, Washington, D. C.), p. 114 (1978).

4. M. S. Shumate, W. B. Grant, and R. T. Menzies, "Participation of the Laser Absorption Spectrometer in the 1979 Southeast Virginia Urban Plume Study," Jet Propulsion Laboratory Technical Report 715-26 to NASA Langley Research Center (1980).

5. M. S. Shumate, "Participation of the JPL Laser Absorption Spectrometer in the 1980 PEPE/NEROS Program in Columbus, Ohio," Jet Propulsion Laboratory Technical Report 715-84 to NASA Langley Research Center (1980).

6. W. B. Grant, "Measurement of Ozone Transport from the Los Angeles Basin Using the Airborne Laser Absorption Spectrometer and a Dasibi Ozone Monitor," Jet Propulsion Laboratory Technical Report 5030-512 to the California Air Resources Board (1981).

7. J. Boscher, W. Englisch, and W. Wiesemann, "Foreign Gas and Differential Albedo Effects in CO_2 Laser Long Path Absorption Monitoring," paper presented at the Topical Meeting on Coherent Laser Radar for Atmospheric Sensing, Aspen, Colorado, July 15-17, 1980. Sponsored by the National Oceanic and Atmospheric Administration and the Optical Society of America.

8. M. S. Shumate, S. Lundqvist, U. Persson, and S. T. Eng, "Measurements of the Diffuse Reflectance of Several Natural and Man-Made Materials and CO_2 Laser Wavelengths," Technical Report No. 8193, Dept. Of Electrical Measurements, Chalmers University of Technology, Göteborg, Sweden (1981).

1.6 Laser Remote Sensing Measurements of Atmospheric Species and Natural Target Reflectivities*

W. Englisch, W. Wiesemann, J. Boscher, M. Rother

Battelle Institut, D-6000 Frankfurt a.M., Fed. Rep. of Germany, and

F. Lehmann, ZGF München, D-8000 München, Fed. Rep. of Germany

1. Introduction

This paper shall give an overview of our CO_2 laser remote sensing program based on the airborne DIALEX instrument. The objectives of this program are twofold:

o long path monitoring of atmospheric species by differential absorption spectroscopy, and
o petrographic mapping of the uncovered earth's surface (minerals and soil moisture) by differential reflexion spectroscopy.

The program comprises the following R & D activities:

o spectroscopic investigations
  - molecular absorption cross section
  - natural target reflectance
o technology development
o development of the DIALEX instrument and application in flight campaigns.

The operational principles of the DIALEX instrument are indicated in fig. 1. It is a single ended, dual wavelength system with heterodyne detection which operates from flying platforms. The return signal is provided by scattering and reflection at the earth's surface. Depending on the selected CO_2 laser lines various atmospheric species like O_3, H_2O, C_2H_4, SF_6 and others can be measured. The system yields mean concentration values within the wedgeshaped probe volumes. The width of the wedge is given by the scan angle, while its length is determined by the integration time which can be adjusted according to the special measurement needs. E.g. sensitivity may be improved by choosing a longer integration time constant however at the expense of reduced horizontal resolution.

With that kind of instrument two major types of interaction have to be considered:

o molecular absorption (due to the species of interest as well as to interfering gases), and
o ground surface reflexion and scattering (spectral albedo)

Both effects are more of less strongly dependent on wavelength and may mask each other as illustrated by the mathematical expression

---
*) Work supported by the German Federal Ministry for Research and Technology, represented by DFVLR-BPT

PRINCIPLE OF OPERATION

- single ended
- airborne

- two (or more) wavelengths
- heterodyne detection (Doppler)
- mean concentration
  of a single species, e.g.
  H_2O vapor
  O_3
  NH_3
  C_2H_4
  SF_6

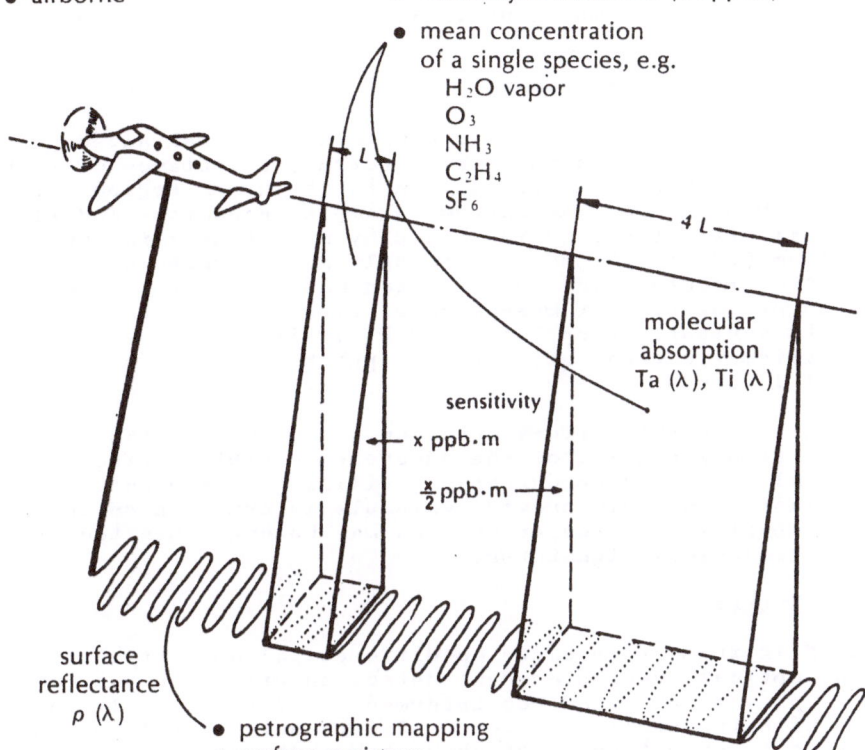

molecular
absorption
$Ta (\lambda)$, $Ti (\lambda)$

sensitivity

x ppb·m

$\frac{x}{2}$ ppb·m

surface
reflectance
$\rho (\lambda)$

- petrographic mapping
- surface moisture

processed signal:

$$Q = \left[\lg \frac{c (\lambda_1) \, P_0 (\lambda_1)}{c (\lambda_2) \, P_0 (\lambda_2)} + \lg \frac{Ta (\lambda_1)}{Ta (\lambda_2)} + \lg \frac{Ti (\lambda_1)}{Ti (\lambda_2)} + \frac{1}{2} \lg \frac{\rho (\lambda_1)}{\rho (\lambda_2)} \right]$$

Fig.1

| internal | measured | foreign gas | differential |
| calibration | species | interference | albedo |

given in fig. 1: The first term comprises all system dependent
parameters and can be made zero by internal calibration. The
second and third term describe molecular absorption by the species
to be measured and by interfering other gases respectively. The
fourth term is due to differential albedo effects. In order to
find the most appropriate wavelength combinations extensive
spectroscopic measurements have been performed in the laboratory.

2. Molecular Absorption Measurements

For 15 interesting gases the absorption cross sections, their
pressure dependence, and in some cases also the variation with

temperature have been determined experimentally for lines of the
"normal" and the "isotopic" CO_2 laser. About 5000 data have been
collected and stored on magnetic tape. This data base has proved
to be a valuable tool for optimum wavelength selection according
to the criteria of maximum sensitivity and minimum foreign gas
interference /1/.

3. Spectral Albedo Measurements

The other kind of interference is caused by spectral albedo
effects. It is well known that the spectral reflectance of rocks
and minerals can vary considerably in the 8-12 /um region where
Reststrahlenbands occur. An extreme example is quartz-sand /1/.
Its spectral reflectance is increased by more than a factor of
5 going from 9.7 to 9.3 /um. Accordingly a pronounced differential
albedo effect occurs which entirely can mask differential molecular absorption. E.g. if measuring O_3 over an open surface with
high quartz content, using P(12) and P(24) lines in the 9 /um
band, a systematic error of up to 170 ppb·km may be caused by
differential albedo.

On the other hand this effect opens up a new and exciting possibility for remote sensing of the uncovered earth's surface. Active
differential reflexion spectroscopy utilizing appropriately
selected laser lines can provide valuable information an the minera-
logical composition of the ground and may become a new tool to
recognize geological signatures.

4. DIALEX-Instrument

The DIALEX instrument is based on differential absorption spectroscopy and optical superheterodyne detection utilizing the Doppler
effect to provide the required intermediate frequency. It is
equipped with two sealed-off cw CO_2 lasers which are line tunable
and provide an output power of about 5 W each. An off-axis
telescope with an aperture of 7 cm serve simultaneously for transmission and reception and is common to both spectral channels.
However two heterodyne detectors are used to avoid channel cross-
talk. Furthermore a rotating disc is incorporated which simulates
the moving earth's surface and is used for internal calibration
and test runs in the laboratory.

Fig. 2 illustrates the optical arrangement on the mounting plate
(on scale) which carries on top all optical and control elements
like various mirrors, beamsplitters (BS) for generating the local
oscillator beams which can be scaled in power independently of the
laser output power by means of attenuating cells (AC), the
Brewster-plate beam combiner (BC), the rotating disc (RD) with
separate attenuator (A), the two heterodyne detectors (D) and
various optics for alignment and monitoring (monitor detector MD,
monitor points MP). The two CO_2 lasers are mounted underneath
the table. The overall optical subsystem has dimensions of about
$40\times50\times120$ cm^3 and is shock—proof mounted in an aluminum frame.

Fig. 3 shows a block diagram of the signal processing electronics.
For each spectral channel the circuitry is nearly identical
comprising various amplifiers (A),a calibrated attenuator (CA),
a band-pass filter (B), an amplitude demodulator (AD), a lock-in

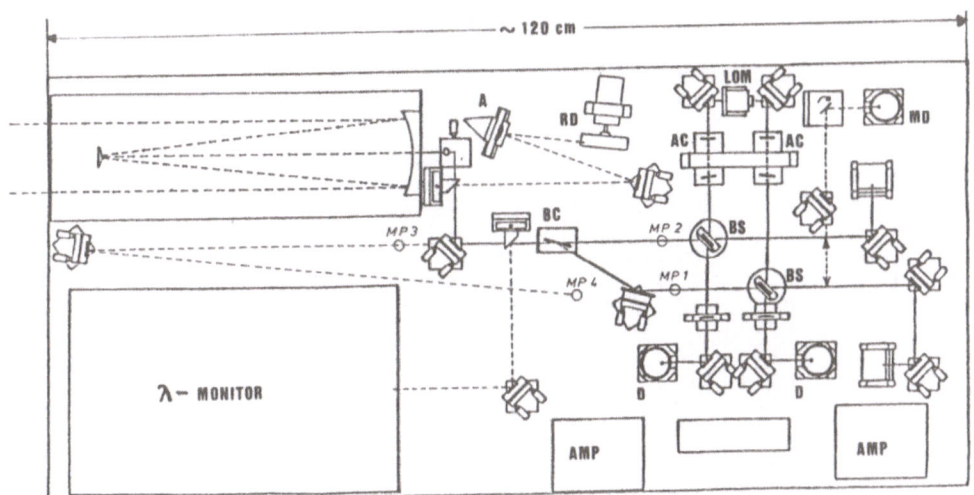

Fig.2: Optical arrangement on mounting plate (on scale)

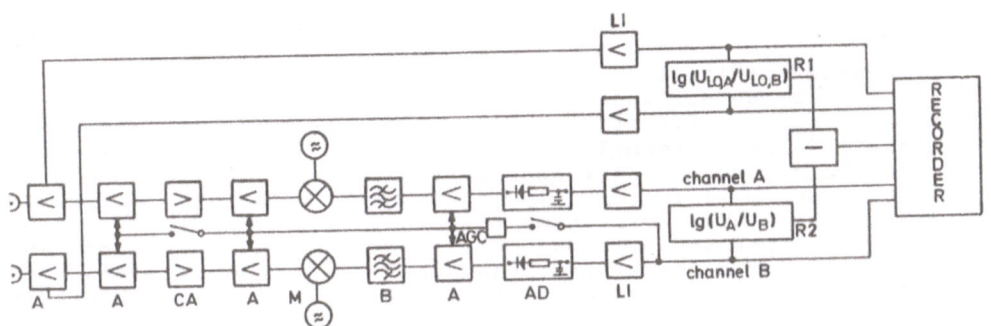

Fig.3. Signal processing electronics

voltmeter (LI) and a logarithmic ratiometer (R). A second electronic mixing stage (M) is included to compensate for long term
drifts in the intermediate frequency due to the varying inclina-
tion of the aircraft. An important feature is the use of an automatic gain control circuit (AGC) which keeps the reference signal
at a constant voltage level and applies the same amplification
factor to the signal channel. This is essential to compensate for
the large dynamic range of the backscattered signal power which
otherwise may cause overload or insufficient voltage levels at the
lock-ins.

5. Airborne Measurements

The DIALEX instrument has proven its performance and versatility
during various flight experiments using the twin-engine Do 28
Skyservant operated by DFVLR, Oberpfaffenhofen.

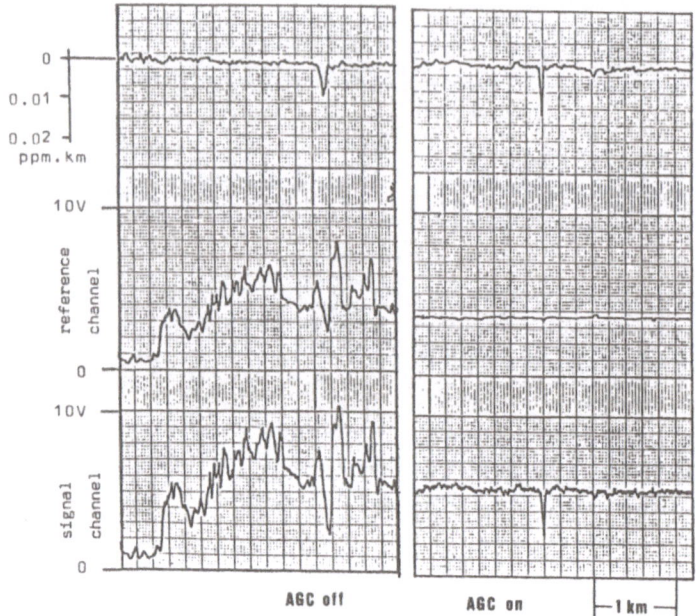

Fig.4. Tracing of a small SF$_6$ plume

One example is its application in a tracer gas experiment with the objective to detect and measure an artificially generated very small SF$_6$ plume. Figure 4 shows the original traces of the signals in the reference channel, the signal channel, and the logarithmic ratio channel versus distance for two crossings. The ratio trace exhibits sharp minima exactly at the points where the flight tracks crossed the plume. Horizontal resolution is about 40 meters and can be further improved by choosing a shorter integration time. This experiment demonstrates the feasibility of measuring small plumes of a tracer gas with good horizontal resolution and high sensitivity.

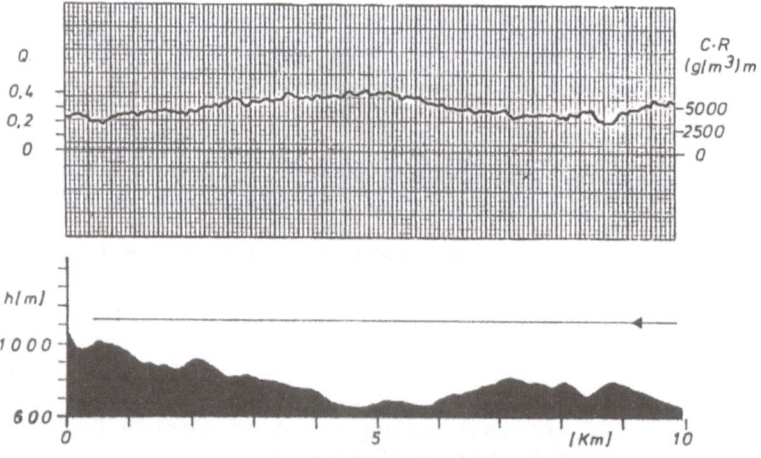

Fig.5. Water vapor column content measurement over a hilly area

Another application is the measurement of atmospheric species which are widely distributed and show only slow changes in spatial concentration. As an example fig. 5 demonstrates water vapor measurements while flying over a hilly area at constant pressure altitude. The signal indicating the water vapor column content runs approximately invers to the altitude profile above ground. This is to be expected if the spatial water vapor concentration is nearly constant over this area. Horizontal resolution is about 120 m.

Furthermore the concept of remote sensing of the uncovered ground by differential reflexion spectroscopy was pursued by some preliminary experiments /2/ with the objective to "see" quartz. The results are quite encouraging and demonstrate the principal feasibility of this new technique. More detailed experiments are scheduled for spring 1982.

Literature

/1/ J. Boscher, W. Englisch, W. Wiesemann: Foreign gas interference and differential albedo effects in CO_2 laser long path absorption monitoring.
Top. Meeting on Coherent Laser Radar for Atmospheric Sensing, July 15-17, 1980, Aspen Col.

/2/ W. Englisch, W. Wiesemann: Remote sensing of atmospheric trace gases by differential absorption spectroscopy.
Proc. OST Conf., March 6-11, 1978, Toulouse, France.
ESA special publication ESA-SP-134, p. 465-473

1.7 Airborne CO_2 Laser Heterodyne Sensor for Monitoring Regional Ozone Distributions

Kazuhiro Asai, Toshikazu Itabe, and Takashi Igarashi

Radio Research Laboratories, Ministry of Posts and Telecommunications
Koganei, Tokyo 184, Japan

1.INTRODUCTION

It is said that urban ozone molecules generated in big cities increase in number, being diffused and transported toward suburbs and sea by atmospheric motions. Therefore, it is very important to measure regional distributions of ozone concentrations in a short time for resolving a mechanism of photo-chemical smogs. The airborne laser remote sensing should be featured as a powerful and useful technique for this purpose. Several groups have been working on developments of airborne laser remote sensing systems with cw CO_2 lasers and pulsed CO_2 lasers.[1-4]

This paper will provide a discussion of measurement error in the airborne CO_2 laser heterodyne sensor for ozone measurements and albedo spectra for natural and artificial materials obtained in laboratory measurements. A brief description of our airborne system and flight experiments for albedo of various ground surfaces are also described.

2.ANALYSIS OF MEASUREMENT ERROR

We assume that there is no interference from other gases. When the aircraft is flying toward the x direction, ozone column density between the aircraft and the ground at point x is estimated by using the differential absorption technique as follows:

$$N(x) = \frac{1}{2 \Delta \sigma H} \{ \ln R_I(x) + \ln R_A(x) \} \qquad (1)$$

$$; \quad R_I(x) = \frac{I_{if_2}(x) / I_2}{I_{if_1}(x) / I_1}$$

$$R_A(x) = \frac{\gamma_1(x)}{\gamma_2(x)}$$

where $N(x)$ is a column density of ozone at point x, $\Delta \sigma$ is the differential absorption cross section of ozone molecule, H is an aircraft altitude, $I_{if}(x)$ is an intensity of if beat signal in the sky at the point x, I is the laser power, γ is the albedo of the ground surface at the point x, and subscript 1,2 correspond to on-resonance line and off-resonance line, respectively. R_A defined as the ratio of albedo should strongly depend on materials of ground surfaces. Knowledge about R_A at various ground surfaces, e.g., town, suburbs, country, sea, etc., is required in advance for calculation of the ozone density from eq. (1).

According to "a law of error propagations", a square of the relative standard error of ozone number density ($\Delta N / \bar{N}$)^2 is given by

$$(\frac{\Delta N}{\bar{N}})^2 = (\frac{\Delta H}{\bar{H}})^2 + \frac{1}{(2 \Delta \sigma \bar{N} \bar{H})^2} \{ (\frac{\Delta R_I}{R_I})^2 + (\frac{\Delta R_A}{R_A})^2 \} \quad (2)$$

where ΔN, ΔH, ΔR_I, ΔR_A are standard deviations of N, H, R_I, R_A and $\bar{N}$, $\bar{H}$, $\bar{R_I}$, $\bar{R_A}$ are averages of N, H, R_I, R_A, respectively. We will discuss on each term in the following;

(i) Term of $\Delta H / \bar{H}$; this term is resulting from random changes of pitch and roll angles of the aircraft during flights. However, even if these angles randomly change ±5 degrees, $\Delta H / \bar{H}$ gives only an error of 0.8 % in $\Delta N / \bar{N}$.

(ii) Term of $\Delta R_I / \bar{R_I}$; this $\Delta R_I / \bar{R_I}$ is divided into two factors. One of these is due to changes of optical axes of transmitter, heterodyne receiver optics by mechanical vibrations and heating in the aircraft. The other factor is resulting from instabilities of signal processing

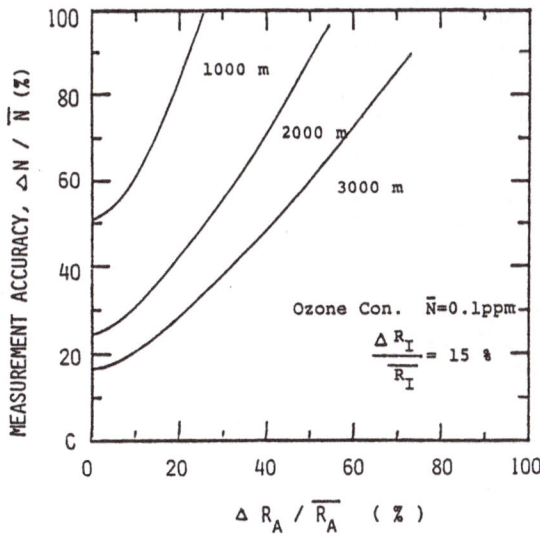

Fig. 1. Estimated curves for measurement error, $\Delta N \sqrt{N}$ vs $\Delta R_A \sqrt{R_A}$ with parameters of aircraft altitude

and recording electronics instruments. The error caused by these factors is called the systematic error. This error should be reduced by careful designs and alignments for the system.

(iii) term of $\Delta R_A / \overline{R_A}$; this term is due to that the albedo of the ground surface depends on laser lines. The error caused by this frequency-dependence of ground surface's albedo is an uncontrollable factor. Measurement error of this airborne system might be eventually determined by this term.

Fig.1 shows a calculated result for the relative standard error of R_A vs $\Delta N / \overline{N}$. As shown in Fig.1, the fluctuations of R_A will be serious problems in the flight operations for measuring regional ozone concentrations.

3. MEASUREMENT of R_A [5-6)]
3.1 Laboratory Experiment
Fig.2 shows a block-diagram of an experimental apparatus we used . Iris 1 and 2 are put in an optical path to assure that each laser beam of laser lines in P and R branch of 9 um band accurately illuminates on the same area of samples. Laser power is 15 mW and a diameter of the beam on the sample surface is approximately 10 cm. Scattered ir is collected by receiving optics with an aperture of 30 cm. Measurements of R_A in the laboratory were made for seven natural and four artificial materials. Measured $\overline{R_A}$ for each material are listed in Table 1.

3.2 Flight Experiment
A block-diagram of our airborne laser heterodyne sensor is shown in Fig.3. The system under development is divided into three sections; i.e., transmitting- receiving optics including two waveguide CO_2 lasers, the heterodyne receiver, data acquisition electronics and recording

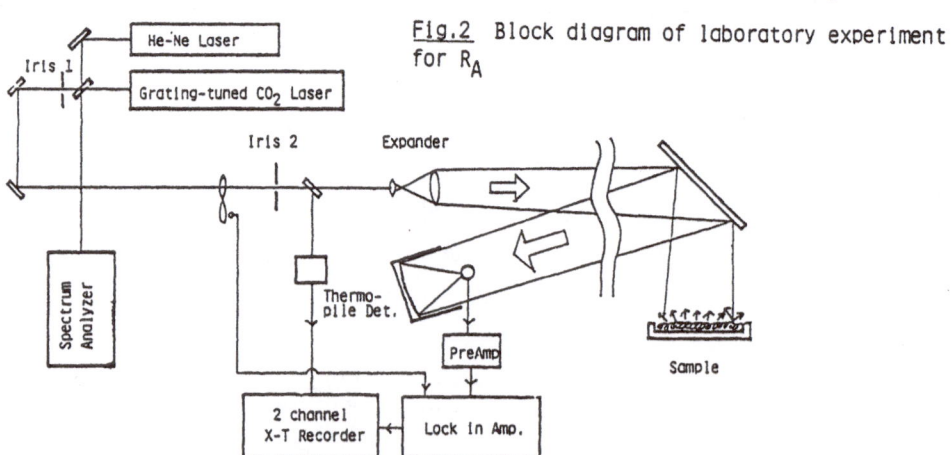

Fig.2 Block diagram of laboratory experiment for R_A

He-Ne Laser

Iris 1

Grating-tuned CO_2 Laser

Iris 2   Expander

Spectrum Analyzer

Thermo-pile Det.

PreAmp

Sample

2 channel X-T Recorder   Lock in Amp.

Materials	Albedo		$R_A = \gamma_{14} / \gamma_{24}$
	γ_{14}	γ_{24}	
lawn	0.2	0.21	0.95
needle-leaves	0.31	0.35	0.89
broad-leaves	0.62	0.69	0.9
dead-leaves	0.48	0.52	0.92
plywood	2.84	2.98	0.95
painted matter	2.08	2.1	0.99
concrete block	0.66	0.68	0.97
brick	0.94	0.88	0.83
sand	1.78	1.68	1.06
soil	0.5	0.48	1.04
kaolin	0.97	0.85	1.14

Table 1. Experimental results for R_A of natural and artificial materials

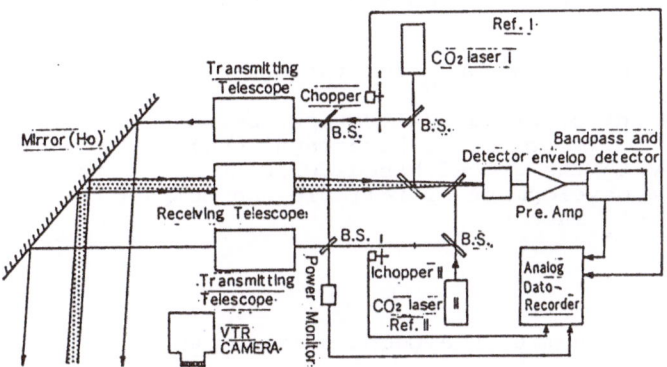

Fig. 3. Block diagram of airborne CO_2 laser heterodyne sensor for monitoring regional ozone concentrations

instruments. P(14) is used for on-resonance line, P(24) for off-resonance line. Both outputs of each laser are about 500 mW. Two laser beams are pointed the direction of 7 degrees back of nadir. Frequency of if beat signal from the HgCdTe mixer was 1.5-2.5 MHz at the flying speed of 250 km/hour. The if beat signals were detected by an envelope detector through a preamplifier and an if bandpass filter, then recorded in an analog data recorder. Off-line data analysis were made by a computer in the laboratory for all data recorded in the aircraft during flights. The experiments were made over 20 hours observing various ground surfaces; i.e., big cities, suburbs, countries, forest, sea and ground covered with snow. The aircraft used in the flight was Grand Commander 685. Table 2 shows $\overline{R_A}$ normalized by $\overline{R_A}$ of a Doppler shifter in aluminum for big cities and countries.

$\overline{R_A}$ normalized by aluminum doppler shifter	
big cities	0.9
country	1.26

Table 2. Measured R_A in the flight experiment

4. SUMMARY

We discussed the importance of preliminary knowledge about the ratio of the albedo, $\overline{R_A}$ and ΔR_A, of various places for minimizing the measurement errors. We do not think that our experimental data are enough to determine $\overline{R_A}$ and ΔR_A for the above places (big cities and countries). It is still necessary to fly for the determinations of this R_A at various ground surfaces.

For a goal of this program, the subjects which we have to investigate are ;

1. collection of $\overline{R_A}$ and ΔR_A for various ground surfaces,
2. establishment of cross calibrations between data with airborne laser heterodyne sensor and that with the ground-based CO_2 DIAL,
3. research & development of reliable and compact airborne CO_2 lasers.

REFERENCES

1) R.T.Menzies and M.S.Shumate ; J.Geophys. Res., 89 4039(1978)
   M.S.Shumate,R.T.Menzies,W.B.Grant and D.S.McDougal ; Appl. optics, 20, 545(1981)
2) R.B.Wiesemann,W.English and K.Gurs ; Appl. Phys. 15, 257(1978)
3) R.W.Stewart and J.Bufton ; Optical Eng., 19, 503 (1980)
4) P.Brockman, R.V.Hess, L.D.Staton and C.H.Bais ; Proceeding of an international conference at Williams- burg,557(1980)
5) J.Boscher and F.Lehmann ; Battelle Institute e.V., D-6000 Frankfurt am Main, West Germany (1979)
6) K.Asai, T.Itabe, T.Matsui and T.Igarashi ; Symposium on Laser Radar in Japan at Hamana, Feb.(1981)

Part 2

Spectrometric Techniques

2.1 Tunable Laser Heterodyne Spectrometer Measurements of Atmospheric Species

Frank Allario, S.J. Katzberg, and J.M. Hoell

NASA Langley Research Center, Hampton, VA 23665, USA

INTRODUCTION

Spectroscopic measurements using tunable laser heterodyne spectrometers in the 3-30 micron range of the spectrum have the potential to measure the vertical profiles of tenuous gas molecules in the atmosphere with ultra high spectral resolution ($\Delta v < 0.001$ cm^{-1}) and high sensitivity [1]. At the NASA Langley Research Center (LaRC), the technology and system level development for demonstrating the "technology readiness" of this technique has been pursued for some time, and a major activity has included technology development of reliable tunable semiconductor lasers [2], laboratory research in the fundamental noise sources of the heterodyne system [3], laboratory measurements of spectroscopic parameters with ultra high spectral resolution [4, 5], sensitivity analyses for potential applications [6] from space and balloon platforms, and development of a Laser Heterodyne Spectrometer (LHS) experiment to measure trace species in the atmosphere from the NASA CV-990 Airborne Laboratory [7]. The experimental flight demonstration is currently scheduled for implementation in the first quarter of fisical year 1983.

The purpose of this paper is to provide an overview of the LHS concept for measuring the vertical profiles of tenuous gas molecules in the upper atmosphere from space and airborne platforms, and to discuss the sensitivity ranges for this technique based upon a series of mathematical simulations, laboratory experiments, and ground-based solar viewing experiments performed in our laboratory. The status of the LHS aircraft flight experiment will also be reviewed with particular emphasis on specified design criteria and performance based upon laboratory measurements of the various subsystems. Comments on future technological thrusts to reduce size, weight, and volume for autonomous operation from free-flyer platforms will also be presented, since this technology could lead to greater operational flexibility and reliability in non-space oriented applications.

Laser Heterodyne Concept

Figure 1 presents in schematic form the instrument concept for a laser heterodyne receiver assumed to utilize solar energy as the radiation source. A conventional collimating lens or reflecting telescope focuses solar radiation attenuated by individual vibrational-rotational absorption lines of atmospheric molecules on the surface of a high-speed mercury-cadmium-telluride (Hg-Cd-Te) detector. An infrared laser in the

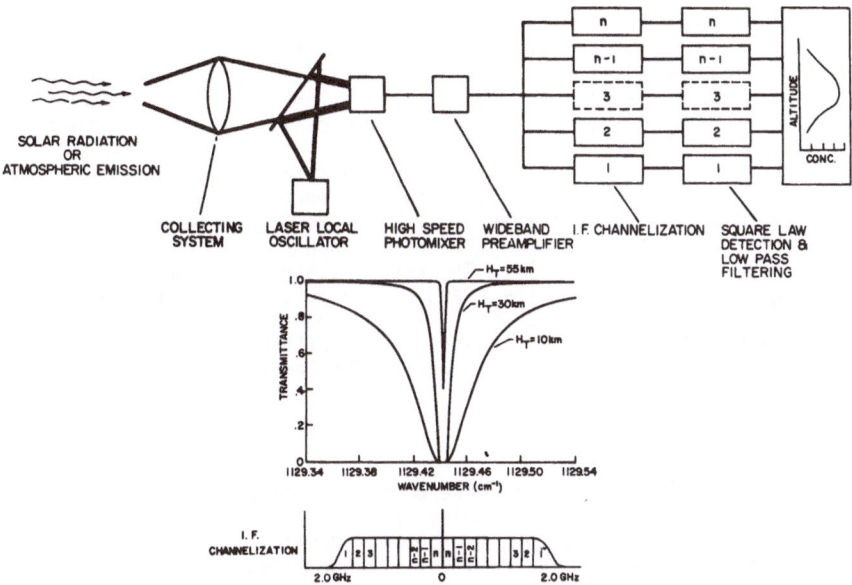

Fig. 1. Instrument concept for a passive laser heterodyne receiver

3-30 μm range is used as a local oscillator (lo) and tuned to the absorption maximum of an individual line. The lo is then mixed with incoming solar radiation on the surface of a high-speed photomixer. Phase fronts of the incoming radiometric signal and lo beam must be matched closely to generate maximum mixing efficiency. The photomixing process generates an electronic signal in the radio frequency (rf) range of the spectrum, preserving amplitude and phase properties of the solar radiation. The rf signal serves as the intermediate frequency (if) signal for electronic processing, and consists of a range of if frequencies in both upper and lower sidebands, centered about the dc photomixed if current. The upper limit of the if frequency is determined by the frequency response (i.e., the bandwidth) of the photomixer and associated preamplifier. The two sidebands are folded about the center of the absorption line so that an individual intermediate frequency filter contains contributions from both upper and lower sidebands. A multiplex advantage is gained by simultaneously sampling the if range with a set of consecutive if filters.

The channelized if signals are proportional to the atmospheric transmissivity at a particular spectral location within the absorption line, and are a measure of the distribution of spectral absorption of a molecular transition. The if signals are subsequently converted to a molecular gas concentration at the tangent altitude by an inversion algorithm utilizing Beer's law. Vertical profiles of gas concentrations in the stratosphere are obtained by either using an instrument pointing system to lock the telescope on a fixed location on the solar disc or locking the telescope at a fixed altitude above the Earth's limb while the sun's disc passes through that altitude. In either case, the lower altitude limit is determined by cloud height. The vertical resolution obtainable from the measurement is determined essentially by three parameters: (1) the field of view, (2) the post detection integration time (τ) to achieve signal-to-noise ratios defined by the desired sensitivity of the measurement, and (3) the inversion algorithm.

51

Although the foregoing basically describes any laser heterodyne system, present systems exhibit significant differences. First, some heterodyne systems employ a discretely tunable gas laser as a local oscillator, while others use solid state tunable diode lasers which can be fabricated to be tuned anywhere from 3μm - 30μm. Both systems are currently being further developed within NASA. Secondly, given either a fixed or tunable-frequency local oscillator, the number of if filters in the system constitutes a significant difference. The number of if filters necessary is in turn a function of the method used to invert the data. The "onion-peeling" inversion method requires data from only one if channel and thus use of this method implies very few if channels. The other method, a spectral inversion, requires knowledge of the shape of the absorption line. Thus, this method requires a sizeable number (5-10) of channels depending on the gas to be examined. Details of the inversion algorithms have been previously described [6, 8, 9] and will not be further discussed in this paper since the impact on the instrument design is restricted to the number and bandwidth of the if filters.

In the description of the laser heterodyne concept just given, solar radiation has been assumed as the incoming radiometric signal and information on the concentration of atmospheric consituents is obtained from the depth and shape of the absorption lines of atmospheric molecules measured. This measurement scenario is dictated by the details of the photomixing process in which high signal source temperatures are translated into high signal to noise ratios [10], especially when sensitive absorption measurements need to be obtained. However, it is worth mentioning that other "hot" sources can be used as radiation sources including backscattered laser radiation [11] or spectroscopic emissions from hot exhaust plumes. In the remaining sections of this paper, discussions will be limited to the use of solar radiation as the energy source for the laser heterodyne receiver and the long path absorption technique due to the importance of this measurement scenario for measuring tenuous gas molecule concentrations from space, balloon, aircraft and ground-based platforms with the laser heterodyne technique [12].

Modeling Studies for Laser Heterodyne Spectroscopy

In order to determine the sensitivity of laser heterodyne spectroscopy for measuring vertical profiles of low concentration molecules in the upper atmosphere, extensive modeling activities for the solar occultation measurements from space platforms have been reported [1, 6]. These modeling activities included (1) mathematical modeling of the radiative transfer process with spectral resolution, $\Delta \nu \leq 0.001$ cm^{-1}, (2) generation of an instrument transfer function (ITF) for the optical receiver, and (3) development of appropriate inversion algorithms to convert the high-frequency if singals into vertical profiles of atmospheric molecules. In addition, spectroscopic parameters for the relevant target gases have been generated using tunable diode laser experiments [13, 14] as a function of pressure and temperature, and a survey of available spectroscopic parameters from the scientific community was recently performed for those molecules of interest in upper atmospheric research.

To assess the validity of the modeling studies, laboratory measurements of several critical subsystems of the heterodyne receiver were evaluated to provide experimental values of signal to noise ratios, inherent noise effects due to laser instabilities of commercially available laser diodes, and end-to-end system performance. These system studies were used to define minimum performance specifications for tunable diode laser performance from a space platform. For application from space platforms, considerations have been given to develop an experiment package with a high degree of autonomous control to reduce the interaction of the experiment specialist with control, calibration and verification of the scientific and engineering experimental data, vis-à-vis management of the measurement procedure. To achieve this objective, the experiment has been designed with system level microprocessor control of critical subsystems. In previous publications, the modeling studies were reported and serve as the design criteria for the current fabrication of a dual gas, Laser Heterodyne Spectrometer (LHS) experiment currently being developed for a technology demonstration on the NASA CV-990.

In the following sections, the various subsystems for the LHS experiment will be briefly reviewed, and interested readers can refer to the sensitivity studies previously published [1, 6]. In figure 2, the model for

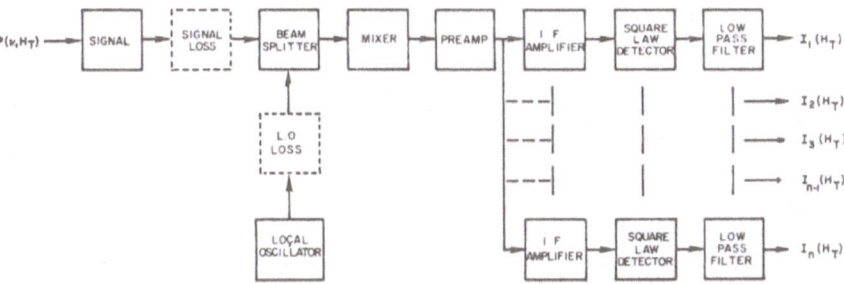

Fig. 2. Model for the LHS instrument transfer function

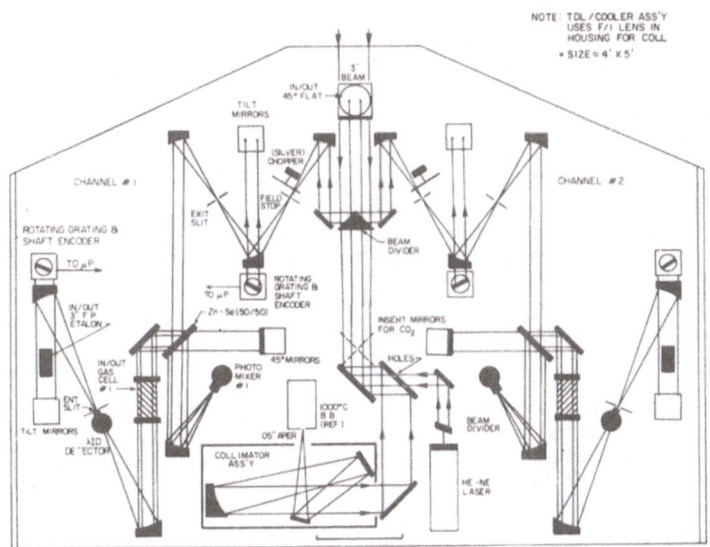

Fig. 3. Schematic of a dual gas LHS optical receiver

the ITF used in the sensitivity analyses is shown and accounts for processing of the input solar radiometric signal and laser local oscillator through the respective transfer optics, photomixer, wideband amplifier, five intermediate frequency amplifiers, square law detectors and low pass filters. In figure 3, a schematic of a dual gas LHS optical receiver is shown which was used in the sensitivity analyses to generate the final system sensitivity limits of the experiment.

Dual Gas Optical Receiver

Optical Receiver

The optical receiver consists of two optical receiver channels in an integrated package, bore-sighted to observe the same tangent altitude in the stratosphere. Functionally, the optical receiver is composed of three basic subsystems including: (1) the transfer optics to simultaneously irradiate the photomixers with the solar radiation and local oscillator (lo) radiation; (2) a wavelength identification and control system to identify the frequency of the lo to the precision required to invert the radiometric signal, and a microprocessor controlled command system to lock the lo wavelength within 5 MHz (precision) of the center of the absorption line to be measured; and (3) a calibration system with a $2000°K$ blackbody reference source to provide an eight point calibration of the instrument prior to and following measurements performed during the solar occultation measurement period.

The lo transfer optics couple the laser radiation to the optical train of the solar signal through either an f/1 germanium lens, or an f/1.5 parabolic mirror at a 50:50 zinc selenide beam splitter. The semiconductor laser is assumed to be completely polarized along the p-n junction (p-type polarization). The transfer optics for the solar radiation consist of a telescope designed to provide a 3-inch collimated beam as the input to the optical receiver. A reflecting beam divider separates the collimated beam into two beams which serve as the input optical trains to the LHS optical receiver. Each optical train is directed to a diffraction grating, blazed to the wavelength of interest and controlled by a shaft encoder with inputs from a central microprocessor system. The diffracted solar signal is focused on the photomixer through two off axis parabolas (f/3) and combined with the lo radiation at the Zn-Se beam splitter. The angle of the diffraction grating is set to center the bandpass of solar radiation at the wavelength of the absorption line of the target molecule. Laboratory experiments have been performed to demonstrate that a diffraction grating blazed at the appropriate wavelength (for NH_3, $\nu = 965$ cm^{-1}, 100 grooves/mm) and a bandpass of 1.5 cm^{-1} at the photomixer, yields a spectral rejection ratio of (1000:1) for solar radiation outside the bandpass.

An important design criterion for the optical receiver shown in figure 3 is the lack of a mode rejection system for radiation from the tunable semiconductor laser. This design is based upon results reported by Jackel and Guekos [15] in which the high-frequency noise intensity spectra from CW GaAlAs semiconductor lasers were studied, and it was generally observed that a lasing mode which is optically isolated from other axial modes of the laser exhibited much stronger intensity fluctuations when compared to the total output of the semiconductor laser, particularly in the freqeuncy range below the intrinsic natural resonance frequency of the laser. Similar noise measurements have been performed with PbSnSe lasers at the Langley Research Center and have been reported [16] confirming these conclusions for Pb-salt lasers. Although all Pb-salt lasers which have been tested do not exhibit these noise properties, the number is sufficiently large with present state-of-the-art lasers to warrant this design. Therefore, in the design shown in figure 3, the solar signal is filtered to a sufficient extent to require the local oscillator to have a mode spacing ≥ 1.5 cm^{-1} in order to confine heterodyning with the solar signal to one axial mode.

In designing the optical receiver in which several modes simultaneously irradiate the photomixer, the added shot noise induced by the extraneous axial modes must be considered in the evaluation of the ultimate signal-to-noise ratio as well as excess noise effects. To minimize effects of the added shot noise on the photomixer current, a requirement must be imposed upon the semiconductor laser which takes into account both excess noise and lack of spectral purity. To develop this requirement, a series of semiconductor laser performance criteria have been generated based upon the following analysis. The signal-to-noise ratio for the heterodyne process can be written as

$$S/N = K \frac{P_s P'_{lo}}{2e^2 \alpha P_{lo} + \dfrac{(F-1)kT_o}{R_m} + 2e^2 \alpha N_e} \tag{1}$$

where
P_{lo} = total local oscillator power
P'_{lo} = local oscillator power per mode within the bandpass of the filtered solar signal

N_e = equivalent excess noise power

K = a proportionality factor containing various system constants

P_S = blackbody power

F = noise figure

T_o = 273°K reference temperature for noise figure .

Neglecting the amplifier noise term (second term in the denominator) this equation can be rewritten as

$$S/N = K' \frac{P_s P'_{lo}/P_{lo}}{1 + N_e/P_{lo}} \tag{2}$$

and $K' = K/2e^2\alpha$. If the laser spectral purity is defined as $S_p = P'_{lo}/P_{lo}$ and the excess noise ratio as $E_{NR} = 1 + N_e/P_{lo}$, equation (1) can be written as

$$S/N = \frac{K' \, S_p}{E_{NR}} = K' \, F' \tag{3}$$

where $F' = S_p/E_{NR}$ is a figure of merit which defines minimum performance specifications for tunable semiconductor lasers. For a large number of tunable semiconductor lasers which have been tested, the parameter E_{NR}

ranged from 1.0 to 1.8 in the quiet regions of the tuning range which for a value of $S_p = 1.0$ results in a typical figure of merit, $F' = 0.55$. This represents a degradation of approximately 0.50 from the theoretically calculated S/N ratio, and is the figure of merit used in the ITF as typical of the expected performance criteria for semiconductor lasers to be used in the optical receiver shown in figure 3. For semiconductor lasers in which the spectral purity (S_p) is <1.0, E_{NR} must be less than 1.8 to maintain a

degradation of 0.5 (e.g., for $S_p = 0.8$, $E_{NR} = 1.6$; for $S_p = 0.5$, $E_{NR} = 1.0$).

Moreover, by measuring the total laser power of the semiconductor laser, the figure of merit F', and the noise figure of the preamplifier, the expected signal-to-noise ratio for the measurements can be obtained.

In specifying performance criteria for high performance but still practical tunable semiconductor lasers as lo's in the experiment, the following criteria are established at the exit window of the cryogenic cooler:

Total Power $\geq$ 700 µWatts in "P" polarization
Spectral purity $\geq$ 80 percent in central mode
Mode separation $\geq$ 1.5 cm^{-1}
E_{NR} < 1.8

These performance criteria assume the following optical efficiencies in the optical receiver design. Total lo radiation losses include the beam splitter (0.5), overfill losses of the lo beam image at the photomixer (0.10), and transmission losses of the photomixer window (0.04) for a total transmission efficiency of 43 percent. With 700 µW power at the exit window of the cryogenic cooler, 299 µW of power irradiates the photomixer element which is sufficient to produce 1425 µamps of shot noise current. In the optical train of the solar radiation, optical reflection losses are 0.85 and the loss at the beam splitter is 0.5, for a total transmission efficiency of 0.43. Furthermore, due to the inability to perfectly match the Airey pattern of the solar radiation and the laser lo beam, a heterodyne efficiency of 0.84 is assumed in the model. Calculations and trade-off studies performed to optimize the heterodyne efficiency for this design have been previously report [17].

Photomixer, Preamplifier, and if Processing Electronics

The effect of the photomixer, preamplifier, and if processing electronics on the signal is modeled using power spectral density transfer functions. The photomixer is modeled using an equivalent electrical circuit [18] with signal rolloff and shot noise effects included. In the results of the sensitivity analysis previously reported, the measured frequency response for photomixer A shown in figure 4 was used from data generated in laboratory experiments [19]. The transfer function was calculated by normalizing curves to a dc value of 1. The shot noise components included

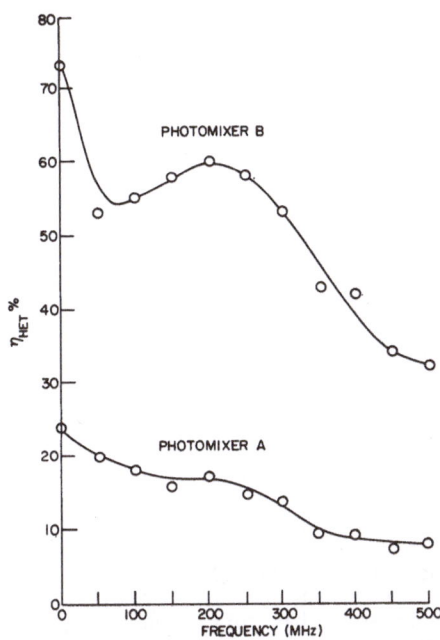

Fig. 4. Measured photomixer quantum efficiency versus frequency. Photomixer A - research photomixer: Photomixer B - commercial device

in the equivalent circuit are due to the dc current induced by the signal, the local oscillator, and the dark current. In addition, provision was made in the model for a shot noise component induced by excess noise. Noise sources in components following the photomixer are primarily Johnson noise and are combined with the Johnson noise of the photomixer.

The if processing electronics represent the straightforward application of multipole filters, power distribution amplifiers, and square law RF detection diodes for power detection. The final subsystem in the signal chain is composed of a full-wave synchronous demodulator, running mean integrator and sample and hold circuit. Each element just described was checked to determine fidelity to specification or design and, where necessary, incorporated into the ITF.

Wavelength Identification and Control Subsystem

In figure 3, a wavelength identification and control system is shown in which the laser radiation from the beam splitter is transmitted through an in/out gas cell, a diffraction grating used to isolate the axial mode of the laser which overlaps the absorption line of interest and a 2.5-cm diameter in/out Fabry-Perot etalon inserted prior to detection with an Hg-Cd-Te wavelength identification (λID) detector. The primary functions of the wavelength ID and control subsystem are to 1) tune the wavelength of the lo to the center of the absorption line to an accuracy between 5 to 20 MHz, depending upon the species of interest, (2) provide command and control logic to lock the wavelength to the required accuracy, and (3) provide command and control logic to the cryogenic cooler and laser wavelength drive to utilize a laser operating point free of "excess-noise" effects.

The wavelength ID and control system has been designed to perform identical wavelength calibration functions as used in high-resolution spectroscopic measurements of gas molecules, but the control functions in this experiment will be completely automated and controlled by a central microprocessor system. The wavelength ID system is being assembled to perform the following functions in sequence. First, a coarse wavelength selection of the laser mode is performed using a servo-driven grating monochromator at the same time that the solar isolation grating is tuned to the center of the atmospheric absorption line to be measured. A microprocessor system sets the cooler temperature and ramps the injection current to determine operational limits that locate the laser mode within the bandpass of the wavelength identification monochromator. The spectral response of the λID detector is stored by the microprocessor as well as the threshold current I_1 and the current at mode power cutoff I_2. Secondly, the gas cell is driven into the laser beam and the laser signal is ramped again. The reference gas in the cell must be a stable gas with at least two well-known vibration-rotation lines lying on either side of the atmospheric absorption line to be measured. Suitable reference gases for the 7.5 to 13.0 μm region include OCS, SO_2, and NH_3. The spectral response of the λID detector is again stored by the microprocessor and the two response functions ratioed to obtain the normalized absorption spectra of the reference gas. Third, the Fabry-Perot etalon with free spectral range of 500 MHz is inserted, with the reference gas cell removed, to generate interference fringes to establish the tuning rate for the laser mode. Following this operational step, the microprocessor has sufficient information to set the current of the laser to lock the wavelength near the center of the atmospheric absorption line.

Fine tuning of the lo wavelength is achieved by microprocessor interrogation of the five IF signal channels to insure, during the first solar occultation measurement period, that the spectral distribution of the five IF channels conforms to expected values. The microprocessor system will be designed to provide this fine tuning capability through control of the temperature and injection current of the semiconductor laser. In the event that the expected absorption line is not identified in the atmosphere, the microprocessor control unit will be placed in a scanning mode of operation to generate absorption spectra of the atmosphere over the tuning range of the TDL mode.

Calibration and Alignment Subsystem

The calibration subsystem shown in figure 3 consists of a 2000°K blackbody source and a collimator assembly unit designed to completely retrace the solar signal through the two channels of the LHS optical receiver, and provides for injection of a highly stable, ruggedized CO_2 laser as a reference lo signal to validate performance of the photomixer and the IF processing electronics. The calibration system will be used for ground and flight calibration and alignment. A He-Ne laser with a divided two-beam output will be used for ground set-up and alignment (detail not shown on the figure).

Laboratory Measurements

A series of laboratory measurements has been performed to substantiate some of the basic design criteria of the instrument shown in figure 3 with a laboratory 1-channel LHS instrument. The instrument system consisted of three radiation sources including a 1273°K blackbody source, a highly stable CO_2 laser and several tunable semiconductor lasers mounted in a standard cryogenic mechanical refrigerator. Experiments were performed in which the blackbody source could be heterodyned with either the CO_2 or semiconductor laser. Similarly, heterodyning experiments have been performed with the CO_2 and TDL lasers. Two photomixers were used in the experiment including a commercial photomixer with a measured effective heterodyne quantum efficiency η of 12 percent from 5 to 115 MHz, and a research photomixer with a measured η of 38 percent. The research photomixer was provided by the MIT Lincoln Laboratories, D. L. Spears. Preamplifier noise figures for both photomixers were 2 dB. In general, the laboratory experiments addressed three basic questions:

(1) The signal-to-noise ratio obtained with the tunable semiconductor lasers when compared to the signal-to-noise measurements obtained with a stable CO_2 laser.

(2) The phenomena of "excess noise" and expected degradations for typical tunable semiconductor lasers either procured under contract, obtained from NASA sponsored research programs,[1] or obtained on consignment from the General Motors Research Laboratories.[2]

(3) The relative magnitude of the RF noise induced by photomixing a multiplicity of axial modes from the tunable semiconductor lasers and the blackbody source and the RF noise induced by a frequency isolated mode from the semiconductor lasers.

In order to determine the projected sensitivity of the LHS experiment to measure trace gas constituents in the stratosphere, laboratory measurements have been performed to compare signal-to-signal ratios with selected semiconductor lasers and highly stable CO_2 laser local oscillators. These results are summarized in figure 5 for CO_2 lasers (solid lines) and a semiconductor laser (squares). This particular semiconductor laser exhibited nearly single mode output over the current tuning range for a fixed operating temperature. The measured signals for the CO_2 laser measurements were compared to a theoretically derived curve for a 1273°K blackbody, with appropriate losses due to the transfer optics. Comparison between theory and experiment is shown in the figure for the two photomixers. The experiments performed with the semiconductor laser generated a maximum heterodyne detector photocurrent of 400 microamperes (η = 38 percent). The higher signal-to-noise ratio for the semiconductor laser results from minor alignment differences and experimental scatter. The important point of these results is that semiconductor lasers, when properly selected for quiet operation, can be as effective as CO_2 lasers in providing signal-to-noise at equal power.

---

[1]Research lasers were obtained from Laser Analytics Incorporated under contract NAS1-15190. Partial funding for this contract was provided by the Los Alamos Laboratories, technical monitor Dr. H. Flicker.

[2]A number of research lasers were obtained from the General Motors Research Laboratories on consignment for test and evaluation in the heterodyne mode. Technical collaborators included Dr. John Hill and Dr. Wayne Lo.

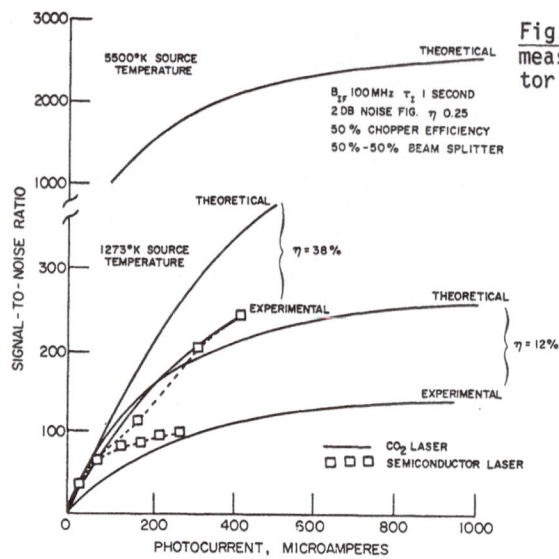

Fig. 5. Comparative signal-to-noise measurements for CO_2 and semiconductor lasers

Utilizing the experimentally measured value of 1273°K, the experimental curve was extrapolated to a solar source temperature of 5500°K for an assumed effective heterodyne quantum efficiency of 25 percent and other parameters listed in the figure. These results demonstrate that for semiconductor lasers operating with single mode output, signal-to-noise ratios in excess of 1000 can be achieved using heterodyne detection without requiring the heterodyne photocurrent to be in the shot noise limited regime.

In general, tunable semiconductor lasers operating in a single mode over desired wavelengths for the LHS experiment are difficult to obtain with current technology. The "excess noise" effects in certain regions have been discussed by Ku and Spears [20]. The LHS design shown in figure 3 assumes that semiconductor lasers will be multimode and will be controlled to operate in the quiet regions of the TDL output spectrum. Control of "excess noise" regions by controlling operating temperature and injection current has been discussed previously [21]. In figure 6 comparative measurements of the RF noise power in dB as a function of frequency from 5 to 500 MHz is shown for a typical tunable semiconductor laser and a CO_2 laser local oscillator for a signal-induced photocurrent of 400 microamperes. For reference, the preamplifier and dark current noise level for the photomixer and if processing electronics is shown. As noted earlier, it has been observed that in "quiet" regions of the laser tuning range, the RF noise power can still be more severe for semiconductor lasers than for CO_2 lasers.

Measurements were performed using a semiconductor laser in which greater than 80 percent power was in the central mode and the remaining power distributed over three other axial modes. Typical values of the degree to which TDL RF power exceeds that from equivalent CO_2 laser-induced RF power are on the order of 2 dB. This residual noise has been used in the ITF in computing the uncertainty in the retrieved concentrations of various trace

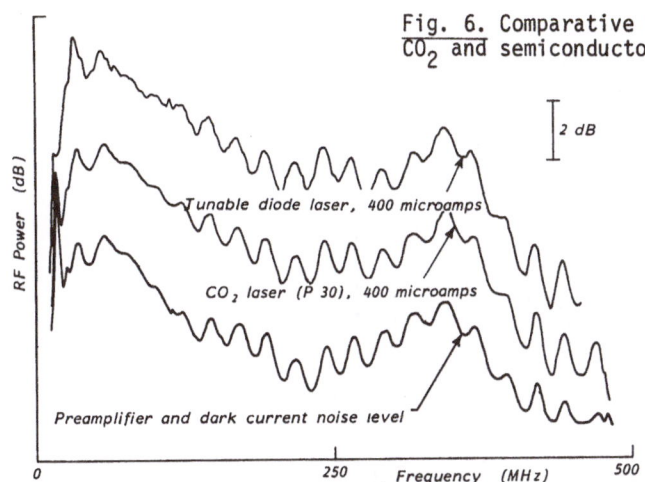

Fig. 6. Comparative RF noise measurements for CO_2 and semiconductor lasers

Tunable diode laser, 400 microamps

CO_2 laser (P 30), 400 microamps

Preamplifier and dark current noise level

2 dB

RF Power (dB)

Frequency (MHz)

molecules in the stratosphere using the LHS experiment. For completeness, the measured dependence of the photomixer efficiency as a function of frequency for two photomixers studied in our laboratory is shown in figure 4. In the ITF, photomixer A has been modeled as the expected performance criteria for the LHS experiment.

LHS Retrieved Concentration Profiles (Simulated)

Figure 7 shows results of the LHS sensitivity studies for measuring stratospheric O_3 from Spacelab altitudes, using the ITF function discussed previously. For this simulation, an lo wavelength at 1129.4420 cm^{-1} was selected to correspond to the peak of a relatively intense O_3 molecular transition lying within an atmosphere window. Five IF channels were selected to detect radiances within the O_3 line in order to optimize the sensitivity of the measurement over the entire altitude range. The position of the if channel centers relative to the lo wavelength are listed as DNU (cm^{-1}); the bandpass of each channel is listed as BETA (MHz). An integration time, TAU (sec), of 100 ms is used for each channel. For each tangent altitude, two channels are used to invert the radiance data. One channel lies within the O_3 absorption line, and the other channel lies outside the absorption line to account for fluctuations in the background radiance and for continuum absorption effects. For upper altitudes where O_3 attenuation is relatively low, channels near line center are used; for lower altitudes where the O_3 attenuation is high, channels in the wing of the O_3 line are used. The channels selected for various altitudes are listed to the right of figure 7.

In figure 8, the two solid profiles represent the initial and mean retrieved O_3 profile. Error bars indicate ±1 standard deviation of 20 sample retrievals generated by perturbing the simulated instrument current profiles with a random Gaussian noise source to simulate random instrument errors. The error bars are an estimate of the uncertainty in the LHS measurement of ozone at various tangent altitudes. Figure 8 shows the fractional error of the mean mixing ratio for the simulated measurements as a function of altitude for the profile in figure 7. Measurement

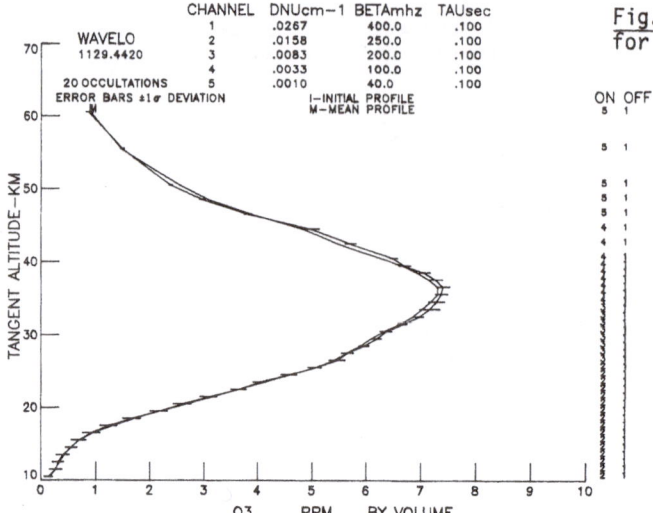

Fig. 7. LHS simulations for stratospheric ozone

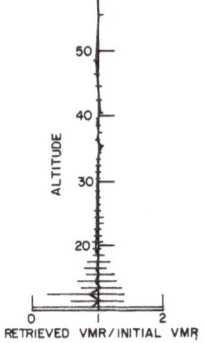

Fig. 8. Fractional error of the mean mixing ratio for ozone; error bars equal ±1 σ deviation

uncertainties $\leq$ 7 percent can be achieved for altitudes $\geq$ 20 km. For altitudes below 20 km, the fractional error increases due to the relatively small O_3 concentrations in this altitude range and the large attenuation by O_3 in the outer shells. To obtain a fractional error of the mean $\leq$ 7 percent below 20 km requires the use of a second local oscillator tuned to a weaker O_3 transition in order to increase the solar radiance incident upon the Hg-Cd-Te photomixer. Use of two lo's for the ozone measurement provide measurement uncertainties $\leq$ 7 percent over the total profile.

Figures 9 and 10 show similar results for ClO for an lo wavelength of 856.499 cm-1. In figure 9, two retrieved profiles are shown for the same initial profile. In figure 9(a), the conventional inversion algorithm was used as described in [1]. In figure 9(b), a data smoothing cubic spline routine was used to reduce the random errors associated with the retrieved profile. The results show a significant improvement in the retrieved profile where error bars representing ±1 standard deviation are significantly reduced (factor of 6). Although use of the cubic spline

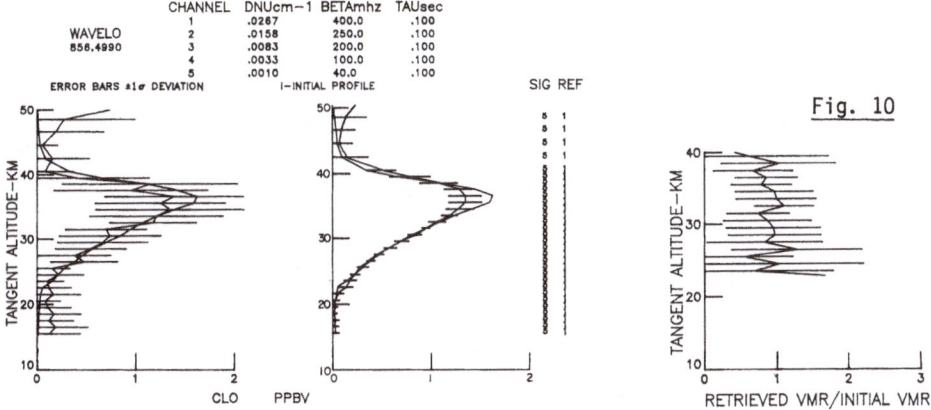

CHANNEL	DNUcm−1	BETAmhz	TAUsec
1	.0267	400.0	.100
2	.0158	250.0	.100
3	.0083	200.0	.100
4	.0033	100.0	.100
5	.0010	40.0	.100

Fig. 9. LHS simulations for stratospheric chlorine monoxide. (a) Without data smoothing. (b) With data smoothing (cubic spline technique)

Fig. 10. Fractional error of the mean mixing ratio for $C_\ell O$. Error bars equal +1 σ deviation

technique significantly reduces the error bounds, use of the technique will reduce the capability of the experiment to measure spatial variability in the vertical profile for a spatial extent < 2 km. Figure 10, which shows the fractional error of the mean mixing ratio for ClO, corresponds to the unsmoothed retrieved profiles. A similar analysis has been performed for H_2O_2.

For these simulations, it has been assumed that atmospheric pressure at the tangent altitude can be measured to an accuracy of ±3 percent, and the temperature profile is determined to an accuracy of ±3 K. Results of the studies performed on O_3, ClO, and H_2O_2 with the current ITF indicate the following: O_3 retrievals over the vertical profile from 20 to 50 km can be obtained with accuracies exceeding 93 percent without data smoothing and with accuracies approaching 96 percent with data smoothing; for ClO, accuracies ≥ 50 percent can be achieved over 20 to 40 km without data smoothing and ≥ 85 percent with data smoothing; for H_2O_2, accuracies ≥ 90 and 80 percent can be achieved over 20 to 40 km with and without use of data smoothing techniques, respectively. These sensitivity analyses have also been performed for measurements from a balloon platform where the integration time can be increased substantially above that of an orbiting space platform. For various orbital conditions that have been analyzed, integration times for the LHS experiment are in the range of 100-400 ms, to obtain a vertical resolution ≤ 2 km. For a balloon platform, the integration time can be extended to 4s to achieve a similar vertical resolution. while 100s may be desirable to obtain high precision measurements of trace gases at selected altitudes. In table 1, a summary of the LHS sensitivity analyses from a balloon platform for several atmospheric molecules is given. Column 2 lists that position of the spectroscopic absorpiton line analyzed. Column 3 lists the assumed concentration of the molecules at the selected upper atmospheric altitudes listed in column 4. The predicted accuracy for an integration time of 4 to 100 s is given in columns 5 and 6, for a balloon altitude of 40 km. In general, an LHS experiment has the capability to detect with fairly high precision those tenuous molecules that are marginal or difficult to detect with lower spectral resolution

Table I. Summary of the LHS Sensitivity Studies for Various Stratospheric Gases (Balloon Platform.

SUMMARY OF LHS SENSITIVITY ANALYSES (BALLOON)

MOLECULE	WAVELENGTH (CM^{-1})	CONCENTRATION (PPB)		ALTITUDE (KM)	PREDICTED ACCURACY (%)	
					τ (4s)	τ (100s)
$C\ell O$	856.499	1.6	(Sunset)	36	≥85	≥95
		1.0	(Sunset)	30	≥80	≥85
		.50	(Sunset)	25	≥80	≥85
$HOC\ell$	1256.940	.350		36	≥80	≥85
		.250		30	≥75	≥80
		.150		25	≥75	≥80
H_2O_2	1244.010	2.0		32	≥90	≥98
		1.8		30	≥85	≥93
		1.0		25	≥85	≥93
HO_2	1097.0	.71		37	SPECTROSCOPIC	
		.19		35	MEASUREMENTS	
		.07		30	IN PROGRESS	
O_3	1129.442	7200		36	≥95	≥98
		6000		30	≥93	≥96
		4500		25	≥93	≥96
HNO_3	895.630	2.6		30	≥90	≥95
		5.3		24 (Peak)	≥85	≥93
		2.7		20	≥80	≥85

techniques. A collaborative experiment with a spectrally scanning interferometer and a multiple gas LHS intrument can provide measurements of source/sink and radical molecular species important to obtain improved understanding of the chemistry of the upper atmosphere. A similar experiment from a space platform would provide global contours of these species to provide science measurements on atmospheric dynamics as well as chemistry.

Ground-Based Solar Measurements of Atmospheric Species

A ground-based, solar viewing, tunable laser heterodyne spectrometer (TLHS) has been assembled and operated from a laboratory environment. This system which has been described in detail [22], utilizes a single wavelength scanning TDL as lo, and has been used to investigate TDL ground-based systems to verify atmospheric transmissivity calculations and molecular line parameter data and LHS system sensitivities. Solar absorption spectra have been obtained over a number of spectral regions at which tunable diode lasers have exhibited spectral modes suitable for lo operation. Figure 11 illustrates an unnormalized spectra of the atmosphere obtained with the TLHS. The upper two curves were obtained from the wavelength identification section of the TLHS with HNO_3 in the reference gas cell. The upper curve is for pure HNO_3, while the lower curve is for HNO_3 broadened by approximately 50 Torr of air. Comparison of the reference and atmospheric spectra indicates that many of the fine features in the atmospheric spectra are due to stratospheric HNO_3. The wavelengths identified on the figure were obtained from tunable diode laser spectra of HNO_3 [4]. The broad feature at 896.507 cm^{-1} is due to absorption by H_2O and CO_2. With the exception of the overall shape of the experimental data due to the TDL power variation,

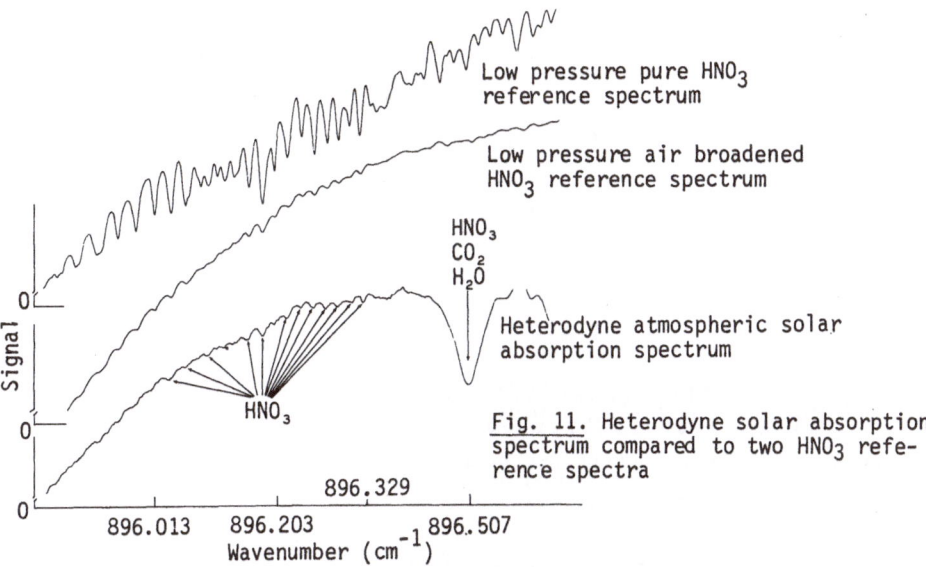

Low pressure pure HNO_3 reference spectrum

Low pressure air broadened HNO_3 reference spectrum

HNO_3
CO_2
H_2O

Heterodyne atmospheric solar absorption spectrum

HNO_3

Signal

896.329

896.013   896.203   896.507
Wavenumber (cm^{-1})

Fig. 11. Heterodyne solar absorption spectrum compared to two HNO_3 reference spectra

agreement between synthetic spectrum in this region and the experimental spectrum is excellent.

The data shown in figure 11 is presented in a format (e.g., relative transmissivity in arbitrary units versus TDL current) which is, in general, useful only for determining the presence or relative position of particular spectral features. Quantitative comparison of different spectral features or for that matter, the same feature obtained at different times can be misleading due to changes in instrument response with respect to wavelength or time. To avoid this dilemma, the more recently obtained spectra shown in figures 12(a) and 12(b) have been power normalized, corrected for the solar induced shot noise and converted to absolute transmissivity using calibration scans with a 1300K and 300K blackbody. In addition, each spectrum has been corrected for changes in the solar intensity during a given wavelength scan. Conversion to absolute transmissivity is based on a solar temperature of 5000K and a linear extraction of the TLHS response curve obtained using the calibration blackbody. In figures 12(a) and 12(b), the left and right ordinates give the relative and absolute transmissivity respectively, while the abscissa gives the absolute wave number. The lower curve in each figure is obtained from the etalon detector in the wavelength identification section and the upper curve is from the reference gas cell containing SO_2. The apparent power variation in the reference spectra is due to the etalon formed by the windows in the reference gas cell. The absolute wavelength is evaluated from known absorption features in the SO_2 reference spectra or atmospheric spectra (e.g., CO_2, figure 12(b)) and the free spectral range of the etalon (e.g., 0.0247 cm^{-1}). The relative accuracy with respect to the reference line is expected to be less than .001 cm^{-1} while the absolute accuracy is dependent upon the accuracy of the reference lines. The date and time at which each spectra was obtained is given at the top of each figure.

Within the spectral region covered in figures 12(a) and 12(b), the identifiable features are due to O_3 and CO_2. The CO_2 spectra exhibits relatively broad, but well defined absorption features characteristic of an atmospheric constituent with a relatively large mixing ratio in both the

65

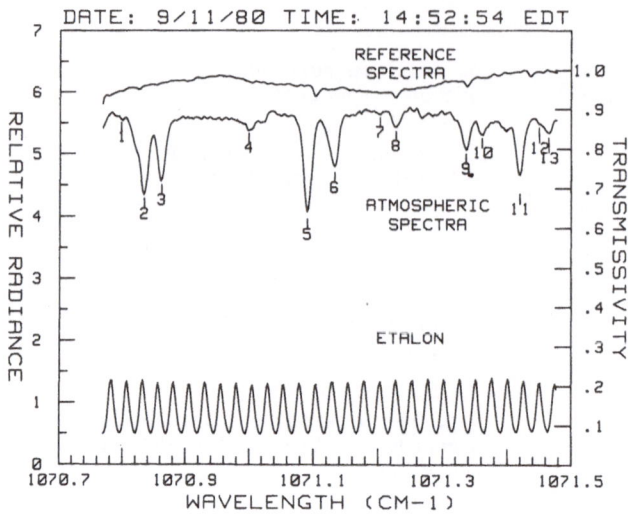

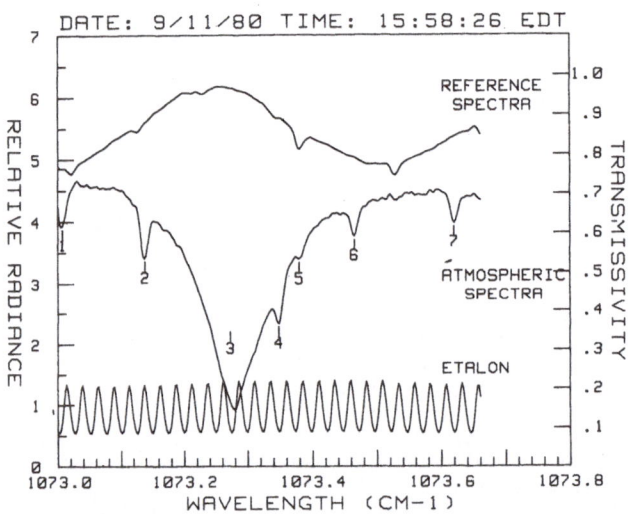

Fig. 12. Heterodyne solar absorption spectrum for (a) 1.3 air masses, (b) 1.6 air masses. In each figure the upper curve is the SO_2 reference spectrum and the lower curve is from the etalon detector. Numbered points indicate absorption features. Number 3 in (b) is due to CO_2; all others are due to O_3

troposphere and stratosphere, while O_3 exhibits much narrower absorption features characteristic of a constituent concentrated primarily in the stratosphere. It should be noted that high-resolution measurements of individual spectral lines similar to those presented here can be used to infer the vertical distribution of a given atmospheric constituent.

Moreover, the spectra presented here provide high-resolution atmospheric
extinction measurements over a wider continuous spectral region than
heretofore available. With coordinated balloon sonde measurements of
temperature, pressure, water vapor, and ozone, these spectra can be used for
quantitative evaluation of high-resolution atmopheric transmissivity models.

To evaluate the accuracy of atmospheric transmissivity evaluation
performed with currently available molecular data, synthetic spectra have
been calculated for the appropriate viewing angle, and included the water
vapor absorption model for the 10μm spectral region, and a line-by-line
evaluation of absorption coefficients. In general, the agreement between
measured and synthetic spectra is good, particularly with respect to the
position of O_3 lines and has been reported [22]. The agreement provides a
high level of confidence to the modeling studies which provide the
sensitivity analyses for the space and airborne simulations of the LHS
experiment capability, as well as our ability to analyze future data from
the LHS flight test.

CONCLUDING REMARKS

In this paper, we have reviewed the tunable laser heterodyne
spectroscopy program at the Langley Research Center. The goal of this
program is to provide an enhanced measurement capability for the NASA Upper
Atmospheric Research Program to complement techniques such as gas filter
correlation radiometry and interferometry. The focus of the current LHS
program is to measure those trace gases which are difficult or impossible to
measure with more conventional techniques. Due to the ultra-high resolution
capability of the instrument, and the high signal-to-noise ratio for solar
temperatures, laser heterodyne spectroscopy can play a major role to improve
current understanding of atmospheric chemistry and dynamics by measuring the
vertical and horizontal distribution of short-lived radical species which
control the chemistry of the three major chemical families in the upper
atmosphere ($C\ell x, Hx, Nx$).

A parameter which can be used to characterize the enhanced measurement
capability of the laser heterodyne technique is the Noise Equivalent -
Equivalent Width (NEEW) [12]. NEEW is defined as the ratio of the
instrumental line width to the achievable signal-to-noise ratio (i.e. NEEW
$= \Delta \nu/S/N$). NEEW as a performance parameter is also important since it
translates directly to a minimum measurable column density of a gas through
the line strength of the vibrational-rotational absorption line.

Source photon statistics is the limiting noise source in the ideal
laser heterodyne signal-to-noise ratio. Achievable resolving power can be
as low as 10^{-3} to 10^{-4} cm^{-1}, which provides a capability for the heterodyne
technique to achieve NEEW's in the range of 10^{-6} to 10^{-7}, for a
signal-to-noise ratio of 10^3, even for relatively short observation times.
This heuristic description highlights the power of heterodyne techniques in
detecting narrow line widths which exist in the upper atmosphere.
Inherently, the heterodyne technique detects absorption of a single
molecular-vibrational transition, distinct from overlapping absorption
lines.

However, NEEW can be rapidly degraded when the observable line width
exceeds the instrumental line width due to collisional broadening or

overlapping of neighboring transitions (i.e. polyatomic molecules). Therefore, for polyatomic molecules at any altitude or "simple" molecules in the lower atmosphere, the unique measurement capability of heterodyne techniques can rapidly degrade, and may become noncompetitive with lower resolution techniques. The requirement for solar tracking also inhibits deployment scenarios for specific applications.

If one utilizes a tunable infrared laser source instead of the sun as radiation source as discussed in several papers at this conference, the heterodyne technique becomes attractive for local or regional detection of polyatomic molecules in the lower atmosphere. In this case, the laser source must have sufficient tunability to generate a minimum number of absorption lines to uniquely "tag" the molecular species. By comparing the experimental spectra generated by the tunable source, with theoretical spectra of various toxic chemicals, correlation techniques performed on-line with high-speed processing and memory storage devices can be used to generate a level of confidence in the identification and detection of toxic molecules. This level of confidence depends upon experimentally measured spectroscopic parameters (i.e. line width, strength, position) under high resolution and at conditions simulating the atmospheric absorption path. To achieve this measurement capability, technological advancements in tunable laser sources in the 10μm region are required, as well as development of the appropiate system level advances to handle the high-speed/high capacity processing requirements for this concept.

REFERENCES

1. Allario, F.; Hoell, J.M.; Katzberg, S.J.; and Larsen, J.C.: "An Experiment Concept to Measure Stratospheric Trace Constituents by Laser Heterodyne Spectroscopy". Applied Physics 23, pp. 47-56, 1980.

2. Linden, K.J.; Butler, J.F.; Nill, K.W.; and Reeder, R.E.: "Development of Lead Salt Semiconductor Lasers for the 9-17 Micron Spectral Region". NASA CR-165682; March 1981.

3. Harward C.N. and Sidney B.D.: "Excess Noise in $Pb_{1-x}Sn_xSe$ Semiconductor Lasers". Proceedings of the Heterodyne Systems and Technology Conference, NASA CP-2138, pp. 129-142, 1980.

4. Brockman, P.; Bair, C.H.; and Allario, F.: "High Resolution Spectral Measurement of the HNO_3 11.3-μm Band Using Tunable Diode Lasers". Applied Optics, Volume 17, Number 1; January 1, 1978.

5. Hoell, J.M.; Harward, C.N.; Bair, C.H.; and Williams, B.S.: "Ozone Air Broadening Coefficients in the 9μm Region". Accepted for Publication in Optical Engineering, May/June 1982.

6. Allario, F.; Katzberg S.J.; and Laren, J.C.: "Sensitivity Studies and Laboratory Measurements for the Laser Heterodyne Spectrometer Experiment". Proceeding of the Heterodyne Systems and Technology Conference, NASA CP-2138, pp. 221-239, 1980.

7. LANGLEY RESEARCH CENTER, INTERNAL DOCUMENT FOR EXPERIMENTAL FLIGHT PROGRAM.

8. Abbas, M.M.; Shapiro, G.L.; and Alvarez, J.M.: "Inversion Technique for IR Heterodyne Sounding of Stratospheric Constituents from Space Platforms". Applied Optics, Volume 20, p. 3755, November 1, 1981.

9. Majumdar, A.K.; Menzies, R.T.; and Jain, S.L.: "Stratospheric Trace Constituent Profile Retrievals Using Laser Heterodyne Radiometers IR Limb Sensing Spectrum". Applied Optics, Volume 20, Number 3, pp. 505-513; February 1, 1981.

10. Blaney, T.G.: "Signal-to-Noise Ratio and Other Characteristics of Heterodyne Radiation Receivers". Space Science Reviews, 17, pp. 691-702. D. Reidel Publishing Company, Dordrecht-Holland. 1975.

11. Brockman, P.; Hess, R.V.; Staton, L.D., and Bair, C.H.: "Dial with Heterodyne Detection Including Speckle Noise: Aircraft/Shuttle Measurements of O_3, H_2O, NH_3 with Pulsed Tunable CO_2 Lasers". Proceedings of the Heterodyne Systems and Technology Conference, NASA CP-2138, pp. 557-568, 1980.

12. Murcray, D.G., University of Denver, and Alvarez, J.M., Langley Research Center, Editors: "High Resolution Infrared Spectroscopy Techniques for Upper Atmospheric Measurements". Proceedings of a Workshop Held at Silverthorne, Colorado; July 31 - August 2, 1979.

13. Rogowski, R.S.; Bair, C.H., Wade, W.R.; Hoell, J.M.; and Copeland, G.E.: "Infrared Vibration-Rotation Spectra of the CIO Radical Using Tunable Diode Laser Spectroscopy". Applied Optics, Volume 17, Number 9; May 1, 1978.

14. Hoell, J.M.; Bair, C.H.; Harward, C., and Williams, B.: "High Resolution Absorption Coefficients for Freon-12". Geophysical Research Letters, Volume 6, Number 11; November 1979.

15. Jaeckel, H., and Guekos, G.: "High Frequency Intensity Noise Spectra of Axial Mode Groups in the Radiation from CW GaAlAs Diode Lasers". Opt. and Quant. Elect., 9, 233; 1977.

16. Harwood, C.N. and Sidney, B.D.: "Excess Noise in Pb1-xSnxSe Semiconductor Lasers". Proceedings of the Heterodyne Systems and Technology Conference, NASA CP 2138, pp. 129-142; 1980.

17. Fales, C.L. and Robinson, D.M.: "Spatial Frequency Response of an Optical Heterodyne Receiver". Proceedings of the Heterodyne Systems and Technology Conference, NASA CP 2138, pp. 495-510; 1980.

18. Melchior, H.; Fisher, M.B., and Arams, F.R.: "Photodetectors for Optical Communication Systems". Proceedings of the IEEE, Volume 58, Number 10; October 1970.

19. Kowitz, H.R.: "Comparative Performance of HgCdTe Photodiodes for Heterodyne Application". Proceedings of the Heterodyne Systems and Technology Conference, NASA CP 2138, pp. 297-308; 1980.

20. Ku, R.T., and Spears, D.L.: "High Sensitivity Heterodyne Radiometer Using a Tunable-Diode-Laser Local Oscillator". Opt. Lett., pp. 84-86, 1977.

21. Rowland, C.W.: "Excess Noise in Tunable Diode Lasers". NASA TP 1935; 1981.

22. Hoell, J.M., Jr.; Harward, C.N.; and Lo, W.: "High-Resolution Atmospheric Spectroscopy Using a Diode Laser Heterodyne Spectrometer". Optical Engineering, Paper 1791; March/April 1982.

2.2 Interferometric Measurements of Atmospheric Species

D.G. Murcray, A. Goldman, F.H. Murcray, and F.J. Murcray
Department of Physics, University of Denver, Denver, CO 80208, USA

INTRODUCTION:

A number of problems of current interest in atmospheric chemistry have drawn attention to the desirability of obtaining data on the atmospheric concentration of many compounds present in the atmosphere at very low levels. Many of these species have moderate to strong infrared absorption bands. The possibility exists of using infrared techniques to obtain data on the presence and concentration of these species in the atmosphere. In an initial attempt to assess possible detection limits for a given species one generally assumes that the absorption will be weak and in the linear region. Under those conditions the following expression holds:

$$\int_{\Delta\nu} A\,(\nu)\,d\nu = Su \qquad (1)$$

In this expression $A\,(\nu)$ is the absorption at frequency ν, the integration is performed over the line. S is the line strength in $cm^{-2}atm^{-1}$ and u is the amount of absorbing gas in atm-cm.

Converting this expression into volume mixing ratio and air mass yields

$$\int_{\Delta\nu} A\,(\nu)\,d\nu = S\beta m \cdot 8 \cdot 10^5 \qquad (2)$$

or

$$\beta \simeq \frac{\int_{\Delta\nu} A\,(\nu)\,d\nu}{Sm} \cdot 10^{-6} \quad . \qquad (3)$$

Several of the other papers at this workshop will discuss various techniques for increasing sensitivity by decreasing $\int_{\Delta\nu} A\,(\nu)\,d\nu$ and also for achieving larger values of m. One of the easiest techniques for obtaining large values of m is to use the sun as the infrared source and to take data at large solar zenith angles. This technique is particularly useful for stratospheric measurements where it is difficult to obtain large values of m by other techniques. Figure 1 shows typical air mass values obtained from balloon altitudes during sunset. Current interferometer systems used to obtain balloon-borne solar spectra achieve $\int_{\Delta\nu} A\,(\nu)\,d\nu$ values on the order of 10^{-4}. These values coupled with the optical paths shown in Figure 1 yield values for β on the order of 10^{-10} for S = 1. This sensitivity is more than adequate for many species of stratospheric interest. It is expected that this figure will be reduced by almost an order of magnitude in the next year. This should increase the altitude range over which data can be obtained and also increase the list of species which can be measured by this technique.

The main advantage of the interferometric technique over many of the other techniques is that spectral data are obtained over a fairly wide spectral region. This is particularly important when data are sought for several constituents simultaneously. The wide spectral coverage is also of interest for survey type measurements where data are sought on the presence of any compounds present in the atmosphere which would have absorption features present in the

71

spectral region covered. Finally, data are needed concerning possible interference between absorption features due to other atmospheric species and the
compound of interest. Examples of these applications are given below.

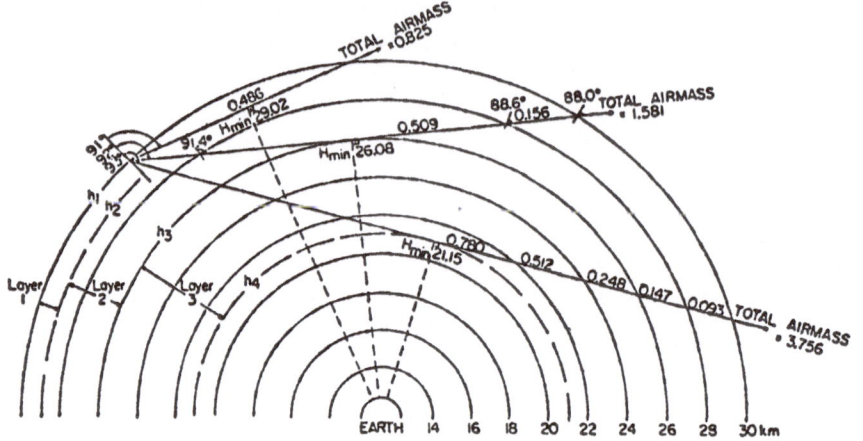

Fig. 1. Ray geometry and airmass values for 2 km shells along 91°, 92° and
93° paths from 30 km. Only one-half of the symmetrical airmass distribution
is shown. The minimum altitudes define the layers for the inversion

INSTRUMENTATION:

 The interferometer used in our solar spectral studies is a moving
mirror Michelson unit employing the basic Bomem mirror movement and servo
alignment system. The unit was constructed for field use and incorporates a
number of changes which are required to get reliable operation over the temperature and pressure range encountered during a balloon flight. The system employs
a copper doped germanium detector. Various filters are used with the detector
to limit the spectral region covered on any given flight. The data are digitally
filtered on board and the filtered data are transmitted via a PCM telemetry
system. The on-board filtering reduces the telemetry data rate.

 Data are also recorded on board with a micro-processor controlled digital
recording system which uses 4 instrumentation cassette recorders. Solar radiation is kept on the interferometer's entrance aperture by means of a servo-
controlled heliostat. All units are designed to operate off 28 vdc which is
supplied by silver zinc battery pack. All flights with this system made to date
have been launched from Holloman Air Force Base. For these flights the Air Force
command and control system is included in the gondola along with the ballast
needed to control the flight. When used with maximum displacement of the moving
mirror, the system is capable of an unapodized resolution of 0.01 cm^{-1}.

RESULTS:

 Figure 2 shows several spectra obtained with the unit during sunset on a
flight performed on the 10th of October 1979. The solar zenith angles at which
the data were obtained are given on the figure. These data were obtained with
the balloon floating at an altitude of 33 km. The spectral region contains a

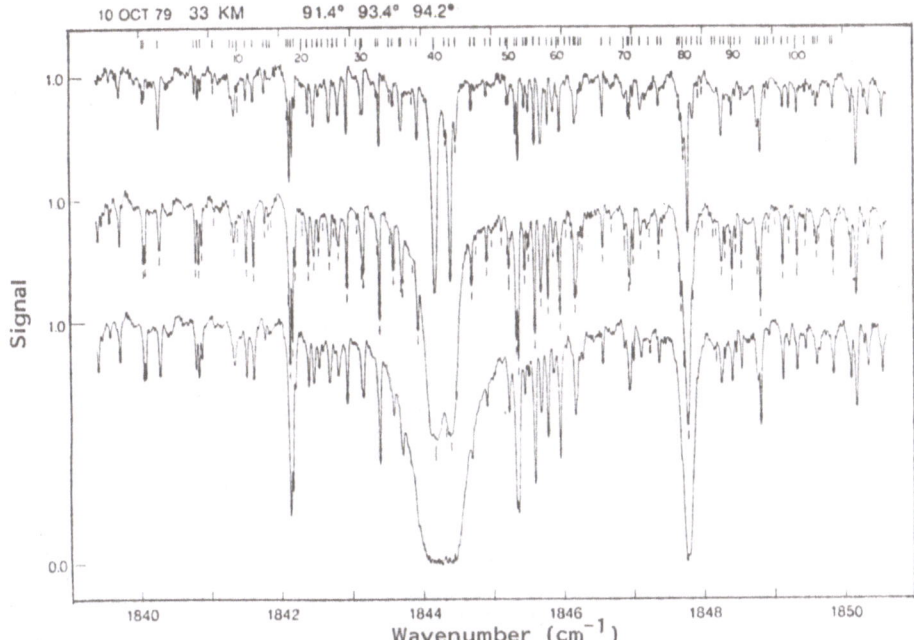

Fig. 2. Solar spectra observed from an altitude of 33 km at various zenith angles near Alamogordo, New Mexico (Oct. 10. 1979)

number of features which are due to NO and illustrates why some of the early attempts to measure NO using instruments of lower resolution gave erroneous results.

Figure 3 shows several spectra in the 1600 cm^{-1} region obtained during the same flight. The major absorption features in this region are due to NO_2 and H_2O. The H_2O lines are easy to identify by the way they grow in the spectra obtained at large zenith angles where the minimum height along the optical path is close to or below the tropopause. The triplet at 1603.8 cm^{-1} is due to an O_2 quadrupole transition.

Figure 4 shows an additional spectral region covered during this flight. This region contains absorption features due mainly to HNO_3 and O_3. Thus, the spectra obtained during this flight contain the information required to obtain altitude profiles for NO, NO_2, HNO_3, O_3 and H_2O simultaneously. These data have been used to compare with a 1D photochemical model developed by Dr. Shaw Liu of NOAA. The results of this comparison will be published later this year. The individual profiles and the identification of the O_2 lines have been published in Geophysics Research Letters.[1] Figure 5 shows the profiles obtained for NO and NO_2 and points up one of the problems with the solar occultation technique. This is the difficulty of interpreting the data for constituents such as NO which exhibit a rapid diurnal variation.

Figure 6 illustrates another aspect of the data obtained with an interferometer. The data were obtained during a balloon flight made to obtain spectral data over the spectral region from 750 cm^{-1} to 1300 cm^{-1}. In examining the data obtained in the 1283 cm^{-1} region it became evident that another species

73

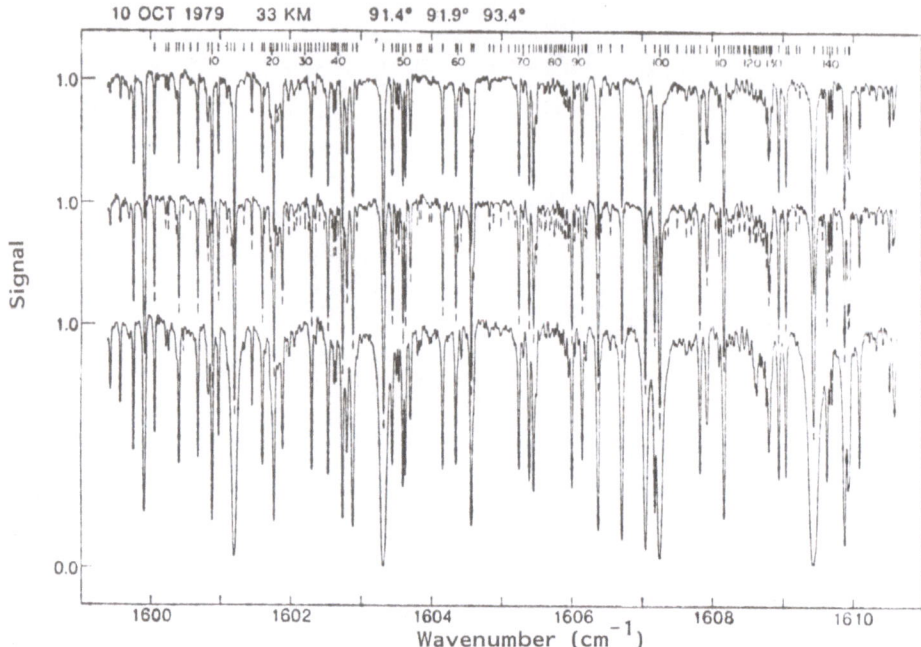

Fig. 3. Solar spectra observed from an altitude of 33 km at various zenith angles near Alamogordo, New Mexico (Oct. 10, 1979)

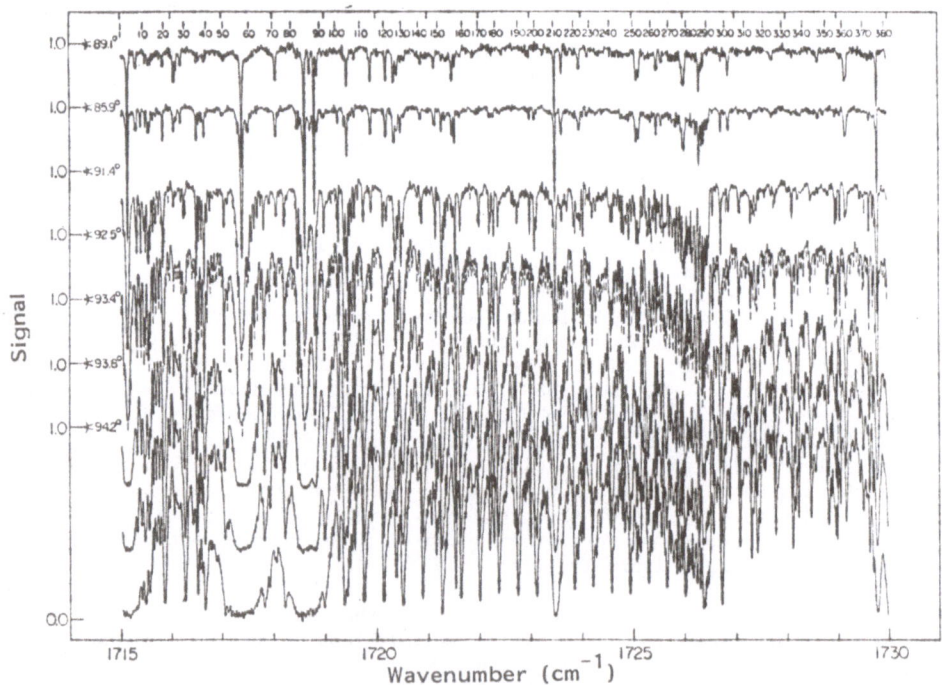

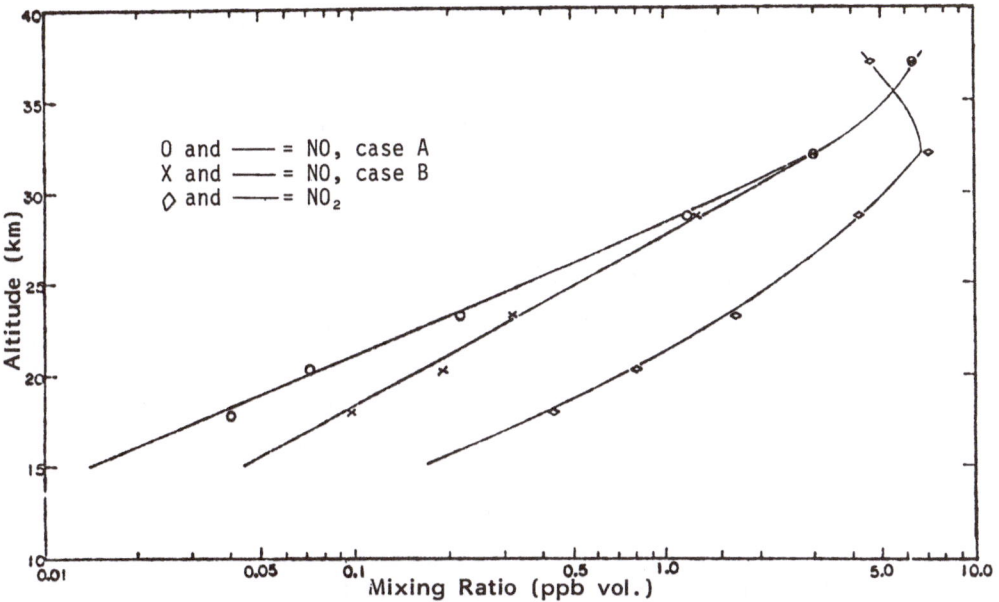

<u>Fig. 5.</u> NO and NO_2 profiles from balloon flight of October 10, 1979

in addition to CH_4 and N_2O was causing absorption in this spectral region. After obtaining suitable laboratory data for a number of candidate species it was determined that the absorption is due to CF_4. This species is very inactive photochemically[2] and appears to be uniformly mixed at 75 pptv. Other species which have been initially detected in the stratosphere using this technique include acetylene[3] and F-22.[4] The search for "new" species entails attempting to identify all of the absorption features present in the solar spectra. This has turned out to be a major project and one that is quite important for other instruments attempting to measure particular species by obtaining data over a narrow spectral interval. Attempts at identification start with the AFGL tape of line parameters and proceeds from there. Unfortunately the tape was not designed for optical paths such as one obtains at sunset, nor for the resolution being achieved in these studies. As a result the initial work resulted in the identification of many lines which were not on the tape or for which the intensity and center frequency were in error. A great deal of additional work has been done by the AFGL personnel to correct the deficiencies, and current versions of the tape are much better for assessing possible interference in solar occultation measurements than the earlier versions. A great deal still remains to be done in this area and the tape must be used with caution.

Infrared solar spectra obtained from ground-based sites can also be used to study the composition not only of the lower atmosphere but also of some stratospheric species. Ground-based measurements are limited by the interference from water vapor and the other minor constituents. Figure 7 shows a portion of a solar spectrum obtained from the South Pole. The South Pole is an excellent site for such studies since the altitude of the site is 2900 m and the cold ambient temperatures reduces the water vapor absorptions. It has the disadvantage that spectra can only be obtained at one solar zenith angle. We are also

<u>Fig. 4.</u> Solar spectra observed from an altitude of 33 km at various zenith angles near Alamogordo, New Mexico (Oct. 10, 1979)

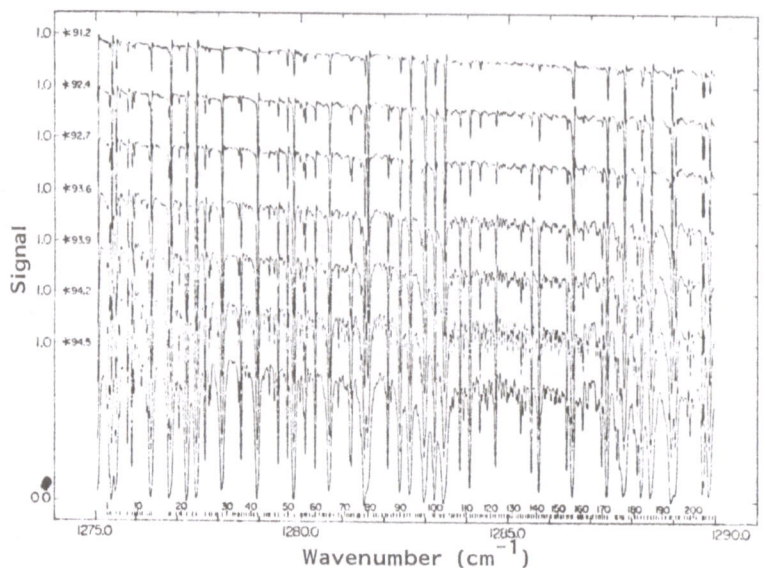

Figure 6. Infrared solar spectra obtained from 38 km at various solar zenith angles.

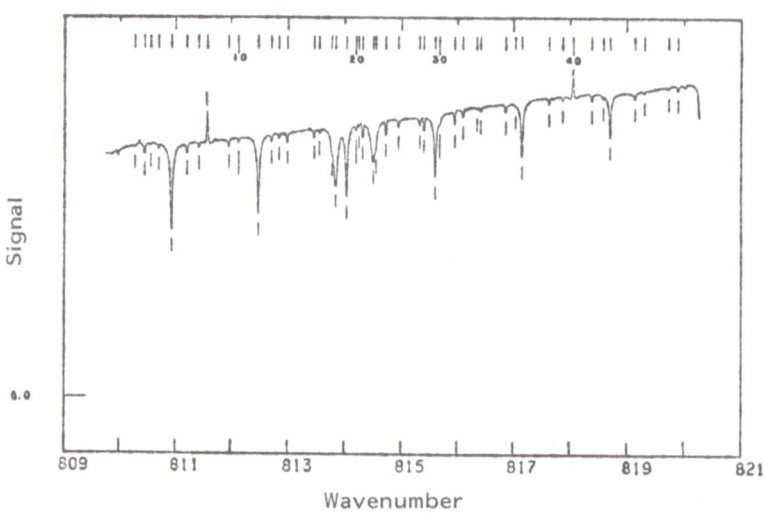

Figure 7. Portion of an infrared solar spectrum obtained from the ground at the South Pole.

attempting to identify all of the features present in these spectra. Table I shows a portion of the listing for the spectrum shown in Figure 6.

In summary, the wide wavelength coverage available with FTIR instruments make them ideal for many applications. These applications will increase as new instruments capable of even higher resolution become available.

Table I.

$810-820 \text{ cm}^{-1}$

Seq. No.	ν(observed) (cm^{-1})	Identification	Seq. No.	ν(observed) (cm^{-1})	Identification
1	810.274	O_3	25	814.725	ΘOH
2	810.443	O_3	26	814.949	O_3
3	810.553	O_3	27	815.325	O_3
4	810.692	O_3	28	815.400	ΘOH
5	810.927	CO_2	29	815.596	CO_2
6	811.202	O_3	30	815.676	O_3
7	811.417	O_3	31	815.949	ΘOH
8	811.575	Θ emission, Si VIII ?	32	816.096	O_3
9	811.960	O_3	33	816.345	O_3
10	812.132	O_3	34	816.414	O_3
11	812.483	CO_2	35	816.865	O_3
12	812.713	O_3	36	817.035	O_3
13	812.846	O_3	37	817.152	CO_2
14	812.989	O_3	38	817.629	O_3
15	813.460	O_3	39	817.869	O_3
16	813.553	O_3	40	818.058	Θ emission, Mg VIII ?
17	813.774	O_3	41	818.390	O_3
18	813.845	H_2O	42	818.595	O_3
19	814.039	CO_2	43	818.711	CO_2
20	814.205	O_3	44	819.144	O_3
21	814.256	O_3	45	819.311	O_3
22	814.321	ΘOH	46	819.733	O_3
23	814.508	H_2O	47	819.896	O_3
24	814.553	O_3			

Note: Emission line near 810.3 cm^{-1} was inadvertantly omitted.

REFERENCES:

1. Goldman, A., (DU), J. Reid (McMaster Univ.), L.S. Rothman (USAFGL), Identification of Electric Quadrupole O_2 and N_2 Lines in the Infrared Atmospheric Absorption Spectrum Due to the Vibration-Rotation Fundamentals, Geophysical Research Letters, **8**, 1, 77-78, 1981.

2. Cicerone, R., Atmospheric Carbon Tetrafluoride: A Nearly Inert Gas, Science, **206**, 59-61, 1979.

3. Goldman, A., F.J. Murcray, R.D. Blatherwick, J.R. Gillis, F.S. Bonomo, F.H. Murcray, and D.G. Murcray (DU), and R.J. Cicerone (NCAR, Boulder), Identification of Acetylene (C_2H_2) in Infrared Atmospheric Absorption Spectra, J. of Geophysical Research, **86**, 12,143-12,146, 1981.

4. Goldman, A., F.J. Murcray, R.D. Blatherwick, F.S. Bonomo, F.H. Murcray, and D.G. Murcray, Spectroscopic Identification of CHCℓF$_2$ (F-22) in the Lower Stratosphere, Geophysical Research Letters, **8**, 9, 1012-1014, 1981.

2.3 Remote Sensing by Infrared Heterodyne Spectroscopy

Michael J. Mumma

Head, Infrared and Radio Astronomy Branch, Code 693,
NASA/Goddard Space Flight Center 20771, USA

Infrared heterodyne spectroscopy is a convenient technique for measuring atomic and molecular spectral lines with high sensitivity and specificity. The instrumental spectral resolving power can be made arbitrarily high although signal-to-noise considerations limit the maximum useful resolving powers $(\lambda/\Delta\lambda)$ to $\sim10^7$ for passive sensing. Nevertheless, this provides the capability to resolve completely individual spectral lines, even when the line shapes are doppler-limited at temperature/molecular mass ratios as low as $\sim2K/amu$. Since the heterodyne process beats the source radiation against a laser local oscillator whose frequency is precisely known (typically to better than $1/10^7$), the methodology provides very precise internal frequency calibration enabling great specificity in line identification and measurement of source motion.

Astonishing advances in the technology of detectors,[1] isotopic CO_2 lasers, and acousto-optic signal processors[2] now permit construction of a brief-case sized heterodyne spectrometer for use in the 9-12μm region, meaning that such instruments are eligible for flight on free-flying spacecraft and may be useful for certain ground-based applications where portability and compactness are important. CO_2 laser heterodyne spectrometers have now been used extensively on the ground for remote sensing[3,4] and laboratory spectroscopy,[5] on aircraft,[6] and on balloon-borne[7] payloads. Work outside the 9-12μm region is less well developed since a different local oscillator is required. Semiconductor diode lasers have been used as local oscillators,[8] but they are not yet completely developed and so are difficult to work with. Other local oscillators have been proposed (e.g. difference frequency generation, sum frequency, generation) but are still in the laboratory development stage.

A number of interesting results have been obtained to date with infrared heterodyne spectroscopy. Some examples are:

o Discovery of a Natural CO_2 Laser on Mars[4]

o Discovery of non-thermal CO_2 emission on Mars and Venus[9]

o Measurement of the Stratospheric O_3[3], HNO_3[10,11], and ClO[11,12] abundances

o Measurement of NH_3 and C_2H_4 in infrared stars[13,14] o Measurement of the temperature-pressure profile in sunspots[15] from 8μm SiO lines

o Measurement of the mixing ratio profile[16] of C_2H_6 near the Jovian south pole from 12μm observations.

o Discovery[17] of intense non-thermal (probably lasing) NH_3 emission from the Jovian auroral zones near 10μm.

o Measurement of stratospheric haze-thickening on Venus.[18]

In addition to the papers cited in references 9-21, the serious worker is referred to some instrumental papers and references cited therein (see ref. 19-22).

References

1. D. L. Spears, "IR Detectors: Heterodyne and Direct," this proceedings.

2. G. Chin, D. Buhl, J. M. Florez, "Acouso-Optic Spectrometer for Radio Astronomy," Heterodyne Systems and Technology, NASA CP-2138 (1980).

3. M. M. Abbas, T. Kostiuk, M. J. Mumma, D. Buhl, V. G. Kunde, and L. W. Brown, "Stratospheric Ozone Measurement with an Infrared Heterodyne Spectrometer," Geophys. Res. Letters 5, 317 (1978).

4. M. J. Mumma, D. Buhl, G. Chin, D. Deming, F. Espenak, T. Kostiuk, and D. Zipoy, "Discovery of Natural Gain Amplification in the 10μm CO_2 Laser Bands on Mars: A Natural Laser", Science, 212, 45 (1981).

5. J. J. Hillman, T. Kostiuk, D. Buhl, J. L. Faris, J. C. Novaco, and M. J. Mumma, "Precision Measurements of NH_3 Spectral Lines near 11μm Using the Infrared Heterodyne Technique," Optics Letters 1, 81 (1981).

6. M. S. Shumate, R. T. Menzies, W. B. Grant, and d. S. McDougal, "Laser Absorption Spectrometer: Remote Measurement of Tropospheric Ozone," Appl. Opt. 20, 545 (1981).

7. R. T. Menzies, C. W. Rutledge, R. A. Zanteson, and D. L. Spears, "Balloon-borne Laser Heterodyne Radiometer for Measurements of Stratospheric Trace Species," Appl. Opt. 20, 536 (1981).

8. M. Mumma, T. Kostiuk, S. Cohen, D. Buhl, and P. C. von Thuna, "Infrared Heterodyne Spectroscopy of Astronomical and Laboratory Sources at 8.5μm," Nature, 253, 514 (1975). D. A. Glenar, T. Kostiuk, D. E. Jennings, D. Buhl, and M. J. Mumma, "A Tuneable Diode Laser Heterodyne Spectrometer for Remote Observations Near 8 Microns," Applied Optics 21, 253 (1982).

9. M. A. Johnson, A. L. Betz, R. H. McLaren, E. C. Sutton, and C. H. Townes, "Non-thermal 10μm CO_2 Emission Lines in the Atmospheres of Mars and Venus," Ap. J. 208, L145 (1976).

10. C. N. Harward and J. M. Hoell, "Atmospheric Solar Absorption Measurements in the 9-11μm Region using a Diode Laser Heterodyne Spectrometer," Heterodyne Systems and Technology, NASA CP-2138 (1980).

11. J. D. Rogers, M. J. Mumma, T. Kostiuk, D. Deming, J. J. Hillman, J. Faris, and D. Zipoy, "Is there any Chlorine Monoxide in the Earth's Stratosphere?", Science (submitted).

12. R. T. Menzies, "Remote Measurement of CℓO in the Stratosphere," Geophys. Res. Lett. 6, 151 (1979).

13. R. A. McLaren and A. L. Betz, "Infrared Observations of Circumstellar Ammonia in OH/IR Supergiants, Ap. J. 240, L159 (1980).

14. A. L. Betz, "Ethylene in IRC 10216," Ap. J. 244, L103, 26 (1981).

15. D. A. Glenar, D. Deming, D. E. Jennings, T. Kostiuk, and M. J. Mumma, "Diode Laser Heterodyne Observations of SiO in Sunspots," Solar Physics (submitted).

16. T. Kostiuk, M. J. Mumma, F. Espenak, D. Deming, D. Jennings, W. Maguire, and D. Zipoy, "Infrared Heterodyne Observations of 12μm Ethane Emission Lines Near the South Pole of Jupiter," Icarus (submitted).

17. T. Kostiuk, M. J. Mumma, D. Buhl, L. Brown, J. Faris, and D. Spears, "NH_3 Spectral Line Measurements on Earth and Jupiter Using a 10μm Superheterodyne Receiver," Infrared Physics 17, 431 (1977).

18. D. Deming, F. Espenak, D. Jennings, T. Kostiuk, and M. J. Mumma, "Evidence for High Altitude Haze-Thickening on the Dark Side of Venus from 10 Micron Heterodyne Spectroscopy of CO_2", Icarus (in press).

19. M. J. Mumma, T. Kostiuk, D. Buhl, D. Deming, and G. Chin and D. Zipoy, "Infrared Heterodyne Spectroscopy", SPIE 280 (Infrared-Astronony) p. 111 (1981) and Optical Engineering (in press). Also see M. J. Mumma, T. Kostiuk, D. Buhl, "A 10μm Laser Heterodyne Spectrometer for Remote Detection of Trace Gases", Optical Engineering 17, 50 (1977).

20. D. A. Glenar, T. Kostiuk, D. E. Jennings, D. Buhl, and M. J. Mumma, "A tuneable Diode Laser Heterodyne Spectrometer for Remote Observations Near 8 Microns," Applied Optics 21, 253 1982). See also M. Mumma, T. Kostiuk, S. Cohen, D. Buhl, and P. C. von Thuna, "Heterodyne Spectroscopy of Astronomical and Laboratory Sources Using Diode Laser Local Oscillators," Space Science Reviews, 17, 661 (1975).

21. M. Abbas, M. J. Mumma, T. Kostiuk, and D. Buhl, "Sensitivity Limits of an Infrared Heterodyne Spectrometer for Astrophysical Applications," Appl. Opt. 15, 427 (1976).

22. Heterodyne Systems and Technology, NASA CP 2138 (1980).

2.4 Detection of Trace Gases Using High-Resolution IR Spectroscopy

Alexander S. Zachor, Atmospheric Radiation Consultants, Acton, MA 01720, USA

Theodore Zehnpfennig, Visidyne, Inc., Burlington, MA 01803, USA

A.T. Stair, Jr., Air Force Geophysics Laboratory, Hanscom Air Force Base, MA 01731, USA

INTRODUCTION

The capability of a Fourier Transform Spectrometer (FTS) system to detect and characterize trace gases in a localized cloud, such as the effluent from a stationary source, is discussed in this paper. An FTS sensor with imaging capability can measure the spectrum of the spatial contrast produced by the cloud (Figure 1). Of course, the cloud temperature must be different from the background brightness temperature (higher or lower), and the resultant spatial contrast must be distinguishable from background clutter. The contrast can be measured by simply subtracting the signals measured in different pixels, or by the Background Optical Suppression Scheme (BOSS) described in Part II.

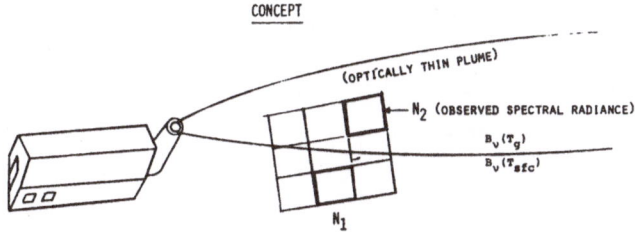

CONCEPT

(OPTICALLY THIN PLUME)

N_2 (OBSERVED SPECTRAL RADIANCE)

$B_\nu(T_g)$

$B_\nu(T_{sfc})$

N_1

CONTRAST $\Delta N_\nu = D\,\tau_\nu \alpha_{g\nu}$ + NOISE

τ_ν = ATMOSPHERIC TRANSMITTANCE

$\alpha_{g\nu}$ = TARGET ABSORPTION COEFFICIENT

DETECTABLE QUANTITY $D \equiv u\overline{\Delta B}_\nu(T_g, T_{sfc})$

u = COLUMN THICKNESS OF TARGET GAS (MOLEC/CM^2)

$\overline{\Delta B}_\nu$ = AVERAGE PLANCK SPECTRAL RADIANCE DIFFERENCE BETWEEN TARGET AND SURFACE

Fig. 1. Get estimate d´of d by least-squares fit of meas, ΔN_ν to theoretical $\tau_\nu \alpha_{g\nu}$

PART I

Previous studies (References 1, 2, and 3) determined the detection limits for a variety of trace gases under ideal conditions. By implication, the gases are optically thin and the background is treated as spatially uniform. With these assumptions, the spectrum of the radiance contrast $N_2 - N_1$ is approximately equal to a "detectable quantity" D times a "reference spectrum" $\tau_\nu \alpha_{g\nu}$, where τ_ν is the atmospheric spectral transmittance, and $\alpha_{g\nu}$ is the target gas spectral absorption coefficient. The detectable quanity D is:

$$D = u\overline{\Delta B}_\nu \; ,$$

the product of the column thickness u of the trace gas along the line-of-sight and the difference of the average Planck spectral radiances corresponding to the gas temperature and background brightness temperature.

More generally, the observed contrast spectrum is a weighted sum of reference spectra for multiple target species and interferents in the cloud, plus a residual background spectrum (Figure 2). Using the known reference spectra and a parameterized characterization of the background residual, one can solve for the unknown weights (detectable quantities) by standard least-squares techniques. Figure 3 shows a typical reference spectrum for the ν_3 band of N_2O, computed using Air Force Geophysics Laboratory's FASCOD1 program.

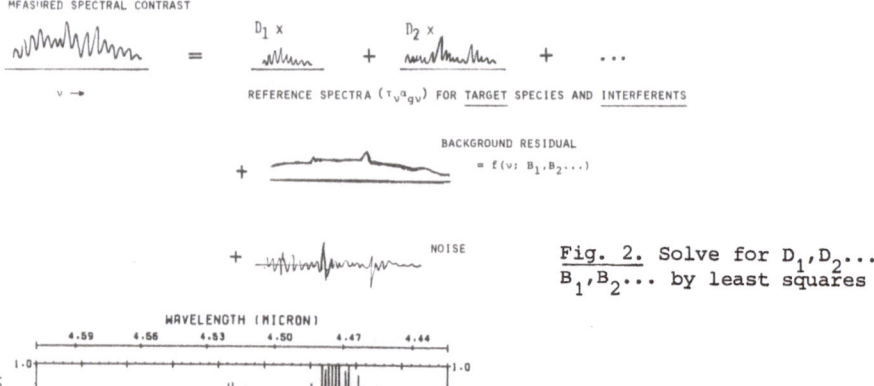

GENERALIZED CONCEPT

MEASURED SPECTRAL CONTRAST

REFERENCE SPECTRA ($\tau_\nu \alpha_{g\nu}$) FOR TARGET SPECIES AND INTERFERENTS

BACKGROUND RESIDUAL

$= f(\nu; B_1, B_2 \ldots)$

NOISE

Fig. 2. Solve for $D_1, D_2 \ldots$; $B_1, B_2 \ldots$ by least squares

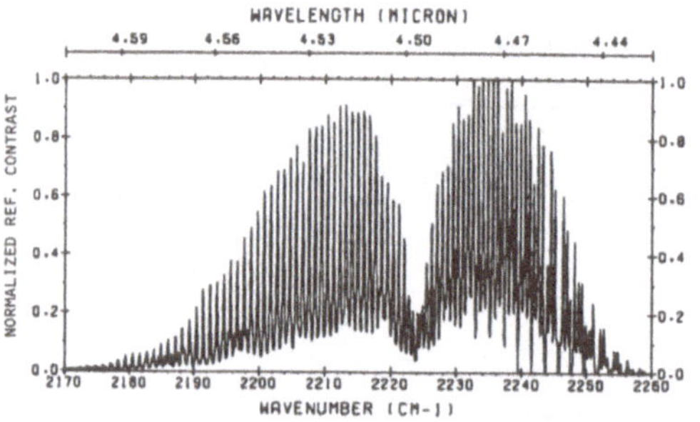

WAVELENGTH (MICRON)

NORMALIZED REF. CONTRAST

WAVENUMBER (CM-1)

Fig. 3. Normalized Reference ($\tau_\nu \alpha_{g\nu} R_\nu$) Spectrum (Sept. Tests) Air Temperature = 21.8° C.

The background residual will have spectral structure due to spatial variations in atmospheric composition and temperature. The random component of atmospheric spatial variations can result in substantially reduced detection probability and increased susceptibility to false detections, corresponding to an increase in the Minimum Detectable Quantity (MDQ). No background suppression method can eliminate entirely the effects of variations that are uncorrelated on a spatial scale equal to the IFOV footprint size. In our approach, the problem is essentially avoided by using a resolution $\Delta\nu \approx 0.1$ cm, and discarding all resolution elements having low predicted atmospheric transmittance τ_ν. If, on the other hand, the atmosphere were spatially uniform, a much lower resolution ($\Delta\nu$ of the order of wave numbers) would be optimum.

The relative uncertainty in the detected quantity for the ideal case of a single trace species (no interferents) and perfect background subtraction is approximately:

$$\frac{\Delta D}{D} \simeq \frac{\sqrt{2}}{\sqrt{M}} \frac{\overline{NESR}}{D\sigma_R} = \frac{1}{\sqrt{M}} \cdot \frac{1}{\text{spectral S/N}}$$

where $\overline{NESR}$ is the FTS average noise-equivalent spectral radiance over the detection band, M is the number of resolved spectral elements, and σ_R is the standard deviation of the reference spectrum. If we define the target gas rms spectral radiance $D\sigma_R$ as the spectral "signal", it is evident that low relative uncertainties can be achieved for sufficiently large M even when S/N << 1.

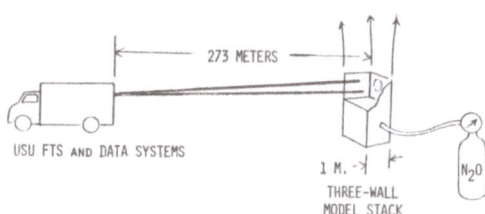

Fig. 4. Field test validation of concept

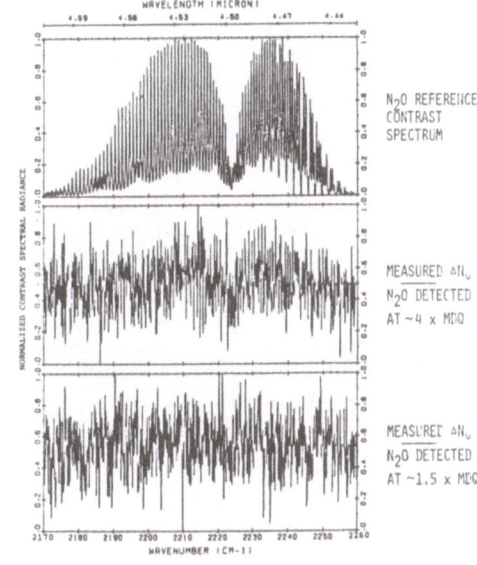

Fig. 5

The detection concept and MDQ prediction formulas were validated in field tests, which used an FTS and data system developed by the Air Force Geophysics Laboratory and Utah State University (Figure 4). The 0.12 cm system observed a model stack emitting a controlled flow of ambient temperature N_2O in front of a heated "background" plate. Figure 5 shows the $N_2O(\nu_3)$ reference spectrum computed for the field test conditions, and two examples of measured contrast spectra that resulted in successful N_2O detection. (The figure shows the "flow-off" case minus the "flow-on" case; hence, N_2O spectral absorption corresponds to enhanced positive contrast). The N_2O absorption is barely discernible in the middle panel of the figure although the N_2O amount is four times the computed MDQ. The bottom panel corresponds to an N_2O amount 1.5 times the MDQ. The spectral S/N at the MDQ level was ~0.15 for these tests.

Some of the spectral data obtained by USU in the field tests was analyzed by Atmospheric Radiation Consultants without knowledge of the N_2O flow rate or concentration. The deduced amounts converted to volume concentrations are compared to the actual concentrations in Figure 6. Also shown are the predicted uncertainty and the pre-computed MDQ and threshold detection levels correspondng to 95 percent detection probability and one percent false detection probability. Note that there were no missed detections or false detections in this series of 12 runs.

Figure 7 gives some values of minimum detectable concentration computed for an airborne FTS sensor and a scenario defined in the table foot-

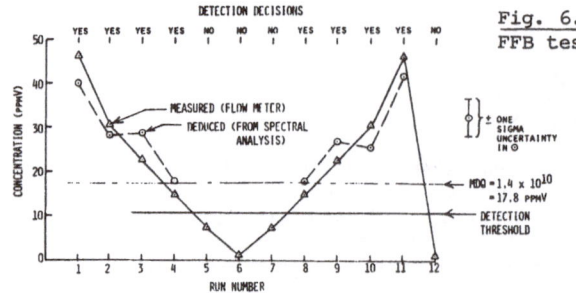

Fig. 6. Analysis results for 19
FFB test

SPECIES (BAND)	MIN. DETECTABLE CONC. (PPMV)
NO_2 (ν_2)	63
NH_3 (ν_2)	3.5
HO (1-0)	1900
N_2O (ν_3)	65

* GAS TEMPERATURE ≃ 300K

GAS BACKGROUND ΔT = 5°C

GAS CLOUD THICKNESS = 3m

SENSOR AIRSPEED = 350 MPH

SENSOR AC COUPLED

** COLLECTOR DIAMETER = 9 in.

SENSOR ALTITUDE = 10 Kft

IFOV FOOTPRINT ≃ 1m x 1m
  (IFOV = 0.35 x 0.35 mr)

RESOLUTION ≃ 0.1 cm^{-1}

MEASUREMENT TIME ≃ 30 secs.

DETECTION PROBABILITY = 95%

FALSE DETECTION PROBABILITY = 1%

Fig. 7. Minimum detectable concentrations* for an airborne fits sensor**

notes. NO is difficult to detect below the parts-per-thousand level because of strong H_2O absorption in the fundamental band near 5.6 μm.

PART II

So far we have considered only an ideal detection scenario. In the real world, we must contend with spatially and spectrally structured backgrounds. Two promising techniques for suppressing these backgrounds will now be described: spectral background suppression using double-beam Fourier Transform Spectroscopy; and spatial filtering using tailored pairs of modulation transfer functions.

The upper half of Figure 8 shows the major components of a Fourier Transform Spectrometer of the Michelson form, and a typical interferogram which might be observed as mirror M_1 is translated. The lower part of this figure shows a variation on the basic form, in which Beamsplitter 1 is initially illuminated from the backside. The corresponding interferogram is inverted (or complementary) with respect to the first interferogram, as shown. This is because the different sequence of reflections and transmissions through Beamsplitter 1 in this case results in a half-wave phase shift between the two arms of the interferometer which is not present in the front-illuminated configuration.

The interferometer can be reconfigured with two entrance apertures, A and B, so that the beamsplitter is illuminated from the front and the rear simultaneously, as shown in the upper half of Figure 9. Then, assuming that both entrance apertures are viewing identical backgrounds, the two interferograms previously shown will add to give a constant, unmodulated signal. Then, when a spectral imbalance such as a target is added to just one input, as shown in the lower part of Figure 9, the resulting modulation in the interferogram is due to that target alone. Thus, the background has been effectively suppressed.

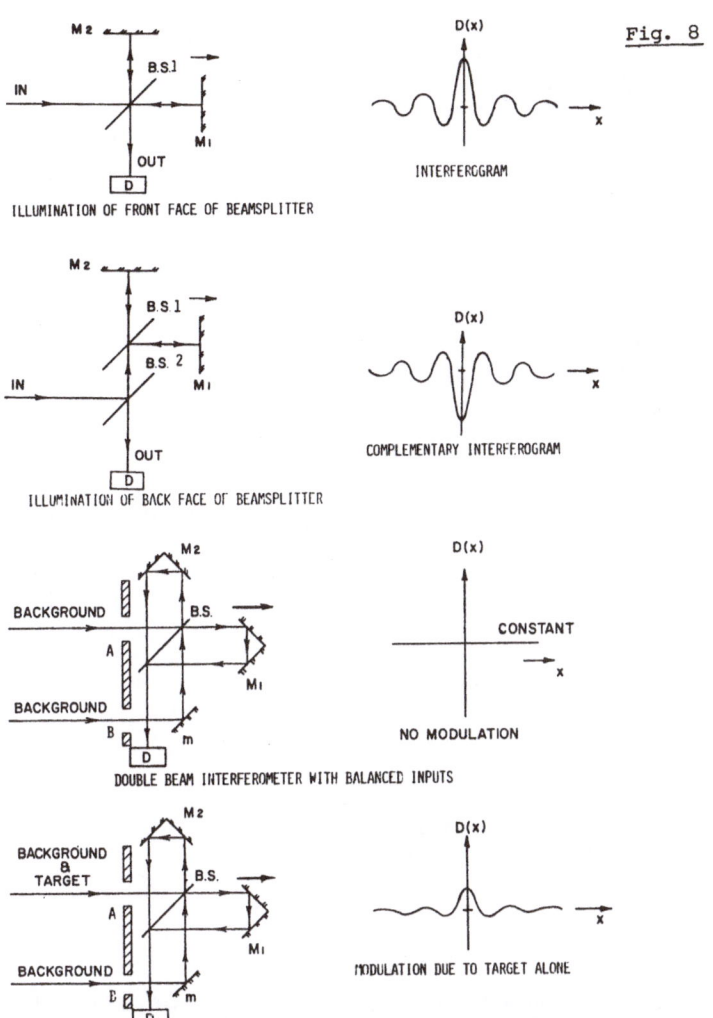

Fig. 8

D(x)

INTERFEROGRAM

ILLUMINATION OF FRONT FACE OF BEAMSPLITTER

D(x)

COMPLEMENTARY INTERFEROGRAM

ILLUMINATION OF BACK FACE OF BEAMSPLITTER

D(x)

CONSTANT

NO MODULATION

DOUBLE BEAM INTERFEROMETER WITH BALANCED INPUTS

D(x)

MODULATION DUE TO TARGET ALONE

INPUTS UNBALANCED BY THE PRESENCE OF A TARGET IN ONE BEAM

Fig. 9

Laboratory demonstration of these characteristics performed at Visi-dyne, Inc., is shown in Figure 10. The left half of this figure shows the single-beam interferogram of the background through aperture A alone, the corresponding single-beam interferogram through B alone, and the combined double-beam interferogram with both apertures open (A + B). The lack of modulation in the double-beam interferogram indicates that the single-beam interferograms were indeed complementary, and thus added to nearly a DC level. On the right side of the figure this sequence is repeated, but with a target in beam A. The modulation in the resulting double-beam interferogram is evidently due to the target alone.

In the spectral domain, these suppression characteristics are shown in Figures 11 and 12. Figure 11 is the single-beam spectrum of a gray-body source, along with absorption features from water vapor in the laboratory air and ammonia contained in a sample cell, which was located between the source

Fig. 10

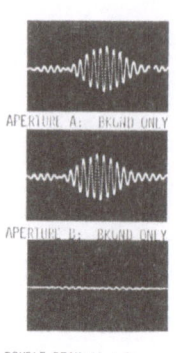

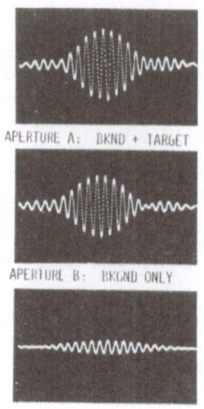

DOUBLE BEAM (A + B)

DOUBLE BEAM (A + B)

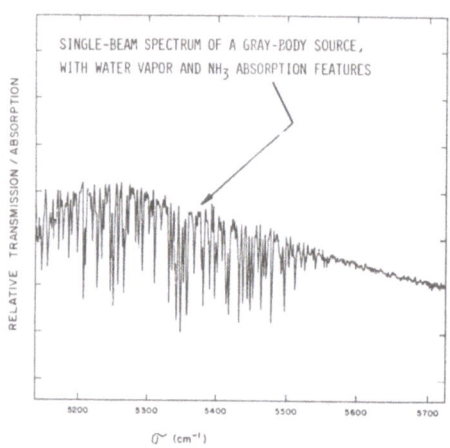

Fig. 11

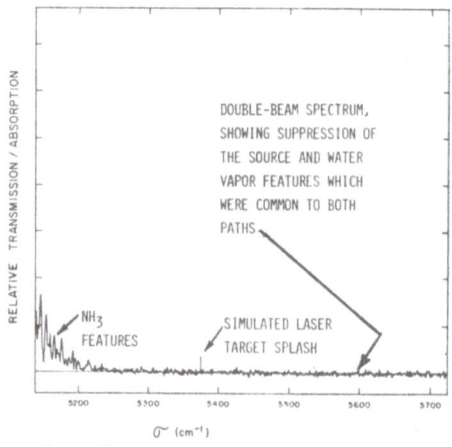

Fig. 12

and the entrance aperture. In Figure 12, with both apertures open, the double-beam spectrum demonstrates very effective suppression of the source and water vapor features. The ammonia features are still evident because the ammonia sample cell was located in one beam only. A single weak emission line has been drawn in to this spectrum to represent a laser target splash in a remote sensing scenario. It is evident that such a feature would be very difficult to detect in the complex single-beam spectrum (Figure 11), but it would stand out clearly with double-beaming.

Figure 13 lists some of the advantages of double-beaming in remote sensing applications. Detection of weak targets, such as the previous laser target splash, in the presence of strong, spectrally complex backgrounds, is facilitated. The dynamic range required of the signal processing system is reduced because the amplitude of the strong zero-retardation peak in the interferogram is greatly reduced. Sensitivities to drive nonlinearities and channel spectra are both proportional to the amplitude of the modulation in the interferogram. Thus, by removing the large amplitude modulation due to strong backgrounds, sensitivity to these sources of error is reduced. Finally, since the background suppression is done in real time on an optical level, we avoid

- IMPROVED DETECTION OF WEAK TARGETS   IN THE PRESENCE SPECTRALLY
     COMPLEX BACKGROUNDS.
- REDUCED DYNAMIC RANGE REQUIREMENTS.
- REDUCED SENSITIVITY TO DRIVE NON-LINEARITIES AND CHANNEL SPECTRA.
- ELIMINATES THE REQUIREMENT FOR PRECISE SPECTRAL OVERLAY NEEDED
     WITH SINGLE-BEAM SUBTRACTION.

the spectral overlay problem which occurs when backgrounds are suppressed by subtraction of two sequential single-beam spectra. In such subtractions, if the spectral scales of the two measurements do not line up exactly, then line shapes are distorted, false features may be introduced, and weak target features can be completely masked by strong background features.

We will now discuss how spatial filtering characteristics can be added to a double-beam FTS in order to suppress spatial structure in the backgound, as well. Figure 14 shows a rudimentary form for such an instrument. The two entranace pupils have been modified so that the two corresponding modulation transfer functions match very closely over a range of low spatial frequencies, as shown, but differ strongly at higher spatial frequencies. Here, this was done by making entrance aperture A an annulus with a nominal diameter of unity and a central obstruction of diameter .54. The diameter of aperture B was stopped-down to .45. With these dimensions, the MTF's will match out to about 25 percent of the cutoff frequency. Since the double-beam FTS responds only to differences in the two inputs, the resulting Net Transfer Function of the instrument is the difference between the two MTF's, as shown. Thus, we have made a high-pass spatial filter which operates on an optical level, suppressing the low spatial frequencies in the background. Response to the higher spatial frequencies is still intact, and it is at these higher frequencies that the target-to-background signal ratios are highest for point-like targets. Thus, the detectability of point targets has been enhanced. The imaging element shown here forms two real images, one for input A and one for input B, of the distant scene of interest on the face of the detector, which is assumed to be a mosaic array. (The imaging element is shown conceptually as a lens, although in fact it would probably consist of reflecting elements.) For effective background suppression, the interferometer must be aligned so that the two images overlay each other precisely.

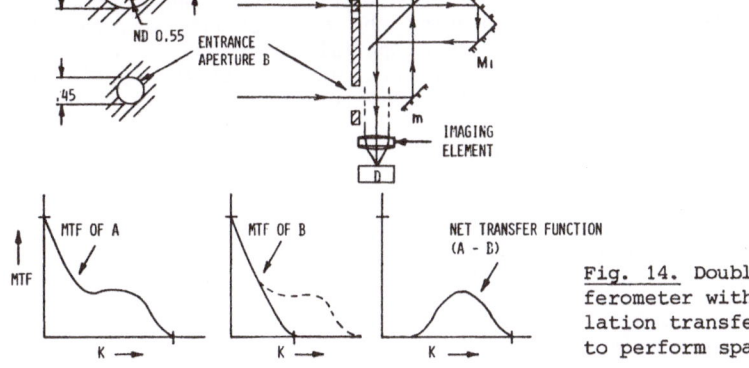

Fig. 14. Double-beam interferometer with matched modulation transfer functions to perform spatial filtering

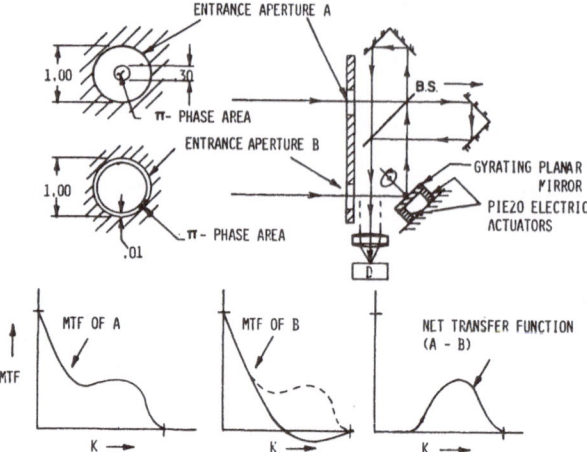

Fig. 15. An improved matched MTF system, using a gyrating planar mirror and π-phase areas on the entrance apertures. [T. Zehnpfennig, S.Rappaport, R. Wattson: Proc. Spie *253*, 8 (1980)].

There is a disadvantage to this rudimentary form of the device. In order to balance the effective collecting areas of the two inputs, a neutral density filter of ND 0.55 must be placed over aperture A, which sacrifices much of the potential collecting area and sensitivity. In Figure 15, this drawback is avoided by tailoring the MTF's in a different manner. Here, certain areas of the two entrance apertures are coated or otherwise modified so as to increase the optical path length through them by 1/2 wave. These are labeled "π-phase areas" in this figure. Also, the image corresponding to aperture B is caused to undergo small amplitude, high frequency circular gyrations on the detector face. The radius of these gyrations is typically on the order of 75 percent of radius of the first dark ring in the stationary (ungyrated) diffraction pattern. The gyration frequency should be high compared to the modulation frequencies expected in the interferograms. As shown here, these image motions are produced by gyrating one of the turning mirrors with a set of piezoelectric actuators. The resulting net transfer function has the desired high frequency response and low frequency suppression, without accompanying sacrifices in the effective collecting area.

Figure 16 lists some of the advantages of adding spatial background suppression, using matched MTF's, to the basic double-beam Fourier Transform Spectrometer. The spatial filter eliminates most of the lower spatial frequency background clutter thereby enhancing the detectability of point-like targets, which are characterized by higher spatial frequencies. Next, this type of spatial filter operates isotropically. That is, unlike spatial filtering schemes which utilize moving grilles and reticles, the background sup-

• ADDITIONAL ADVANTAGES WHEN SPATIAL BACKGROUND SUPPRESSION, USING MATCHED
  MODULATION TRANSFER FUNCTIONS, IS ADDED TO SPECTRAL SUPPRESSION:

   • ELIMINATES MOST OF THE LOW SPATIAL FREQUENCY BACKGROUND CLUTTER,
      THEREBY ENHANCING THE DETECTABILITY OF POINT-LIKE TARGETS.
   • OPERATES ISOTROPICALLY AND IN THE PRESENCE OF DETECTOR PATTERN NOISE.
   • CAN BE USED IN CONJUNCTION WITH ELECTRONIC PROCESSING TO ENHANCE
      OVERALL CLUTTER REJECTION FACTORS.   Fig. 16

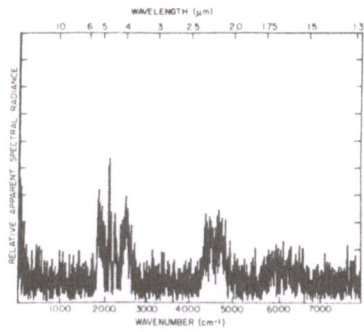

SINGLE-BEAM, PLUME PLUS BACKGROUND

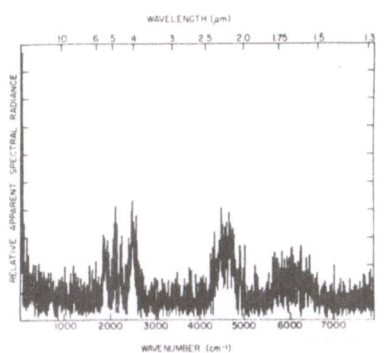

SINGLE-BEAM, BACKGROUND ONLY

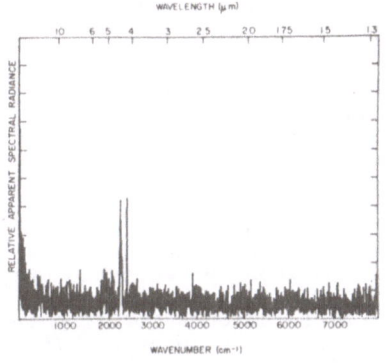

DOUBLE-BEAM, PLUME PLUS BACKGROUND

Fig. 17

SPECTRA OF THE EMISSION FROM A NATURAL GAS
HEATING PLANT, VIEWED AGAINST A SKY BACKGROUND.

REFERENCES: O. SHEPHERD, ET AL., APPL. OPT. 20, 3972
(1981)
T. ZEHNPFENNIG, ET AL., APPL. OPT. 18,
1996 (1979).

pression characteristics are the same for spatial frequencies running in any
direction in the image plane. Also, the background suppression characteristics are not affected by the presence of detector pattern noise. This is
because this type of spatial filtering is performed without the requirement
for intercomparing the outputs of adjacent pixels, which is the source of pattern noise limitations in various other spatial filtering techniques. Finally, pre-processing with an optical spatial filter does not preclude further
electronic spatial filtering in order to make the clutter rejection factors
even larger. Alternatively, optical pre-processing might be used to relieve
part of the workload of the electronic signal processor, thereby freeing it to
perform higher level forms of processing and tracking.
    Figure 17 shows some actual field data taken with a double-beam FTS.
The target was the plume of the AFGL heating plant viewed at about 1/2 kilometer range against a sky background. This heating plant is fired by natural
gas. Comparison of the single-beam spectra with and without the target shows
that the target signature is deeply buried in the emission features of the sky
background. However, in the double-beam mode, the background is effectively
suppressed, and the red and blue spikes which are the target plume signature
stand out prominently.

1. A.S. Zachor, B. Bartschi, and M. Ahmadjian, Proc. SPIE 277, 86 (1981).
2. Air Force Geophysics Laboratory Reports AFGL-TR-81-0134, AFGL-TR-80-0236,
        and AFGL-TR-80-0237.
3. W. Herget, Appl. Opt. 21, 635 (1982).

2.5 Gaseous Correlation Spectrometric Measurements

J.H. Davies, A.R. Barringer[1], and R. Dick
Barringer Research, Toronto, Canada

1 - INTRODUCTION

Certain basic criteria must be considered when designing remote sensors. The gaseous species of interest may have characteristic spectra in various bands, so that some flexibility may exist for instrument selection and measurement concept. Major design parameters are:
- (a) Instrument sensitivity, i.e. adequate signal-to-noise ratio.
- (b) Instrument specificity, i.e. the ability to discriminate against interferents within the spectral bandpass.
- (c) Instrument measurement time to receive and process the signal, e.g. finite spectral scan time or interferometer delay.
- (d) Complementary data requirements e.g. temperature.
- (e) Data inversion requirements.
- (f) Size, weight, power and environmental constraints.

Correlation spectrometric instrumental techniques have been developed which go a long way towards meeting these various and often conflicting design criteria.

This paper describes two electro-optical techniques for true remote sensing of gases via their characteristic absorption spectra. If incoming radiation is collected and analyzed the gases can be detected.

The fundamental vibration-rotation bands of the majority of pollutant gas molecules appear in the thermal IR spectral region and consist of a distribution of separate lines whose locations provide unambiguous fingerprints of the target molecule in question. It is the sum of the strengths of lines detected by a remote sensor that leads to the quantitative determination of gas burden in the sensor field of view.

The line shapes of trace vapors in the lower troposphere are primarily determined by collisions, and for this reason line emissivities can often be reasonably well described using a Lorentz line profile.

2 - INSTRUMENTAL TECHNIQUES FOR REMOTE SENSING

Several instrumental techniques have been used to make passive remote measurements of trace gases in the earth's atmosphere. These have included conventional spectrometers and interferometers as well as the generally more sensitive and specific sensors using correlation techniques. These correlation sensors are more sensitive since they are dedicated to repeatedly measuring only selected spectral samples where the target gas has a strong signature. Correlation sensors include dispersive, interferometric, and nondispersive gas filter devices.

---

[1] Barringer Resources, Golden, CO 80112, USA

Two instrumental techniques developed by Barringer Research for remote detection include the dispersive mask spectrometer (COSPEC) and the nondispersive gas filter correlation spectrometer (GASPEC). For remote sensing measurements special attention has to be given to the background radiation source, the transmission of the atmosphere in the spectral region of interest; the integrated band strengths of the species at their various spectral regions; the performance of detectors and optical elements at these wavelengths together with other systems engineering constraints. For some species detection at several spectral regions are possible and it takes study to determine the most useful region when all measurement and environmental factors are considered and to then select the best instrumental approach.

3 - CORRELATION SPECTROSCOPY (COSPEC)

When the relative strengths and spectral locations of lines of particular species are known, the entrance or exit slits of a dispersive spectrometer can be replaced by physical masks (see for example the SO_2 absorption spectrum shown in Figure 1) to produce an instrument slit function which is a matched filter, i.e. it has a high degree of correlation with the spectrum of a particular gas, and a low degree of correlation with all other overlapping spectral structure. If the incoming radiation contains absorption/emission lines of this gas, scanning the mask past the spectrum produces a signal which is largely determined by the concentration of the particular gas. One successful type (COSPEC) employs a mask spun in the exit plane of a

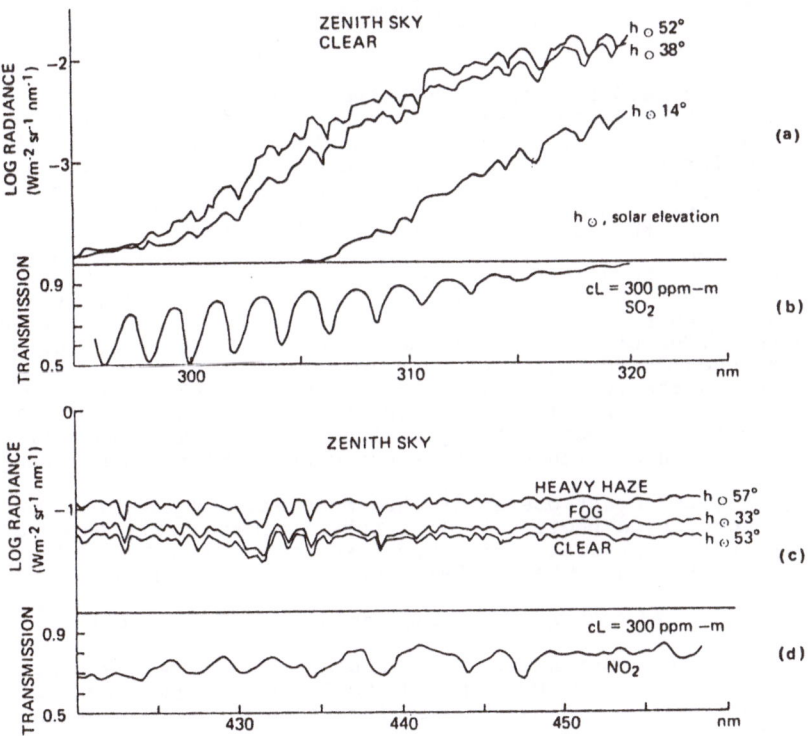

Fig. 1. SO_2 and NO_2 transmission spectra, each with three typical backgrounds. Resolution 0.2 nm

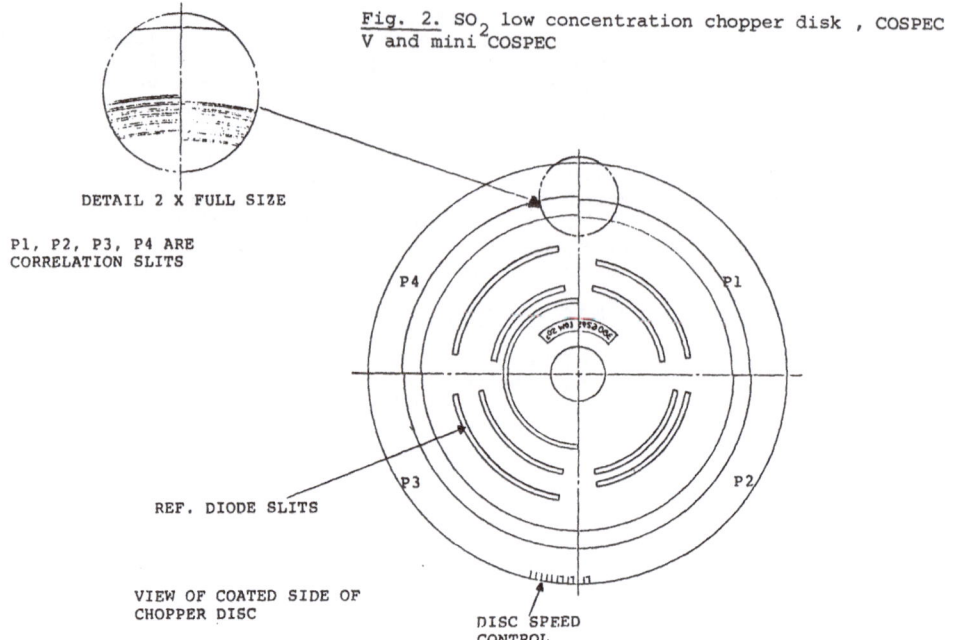

Fig. 2. SO_2 low concentration chopper disk , COSPEC V and mini COSPEC

DETAIL 2 X FULL SIZE

P1, P2, P3, P4 ARE
CORRELATION SLITS

P4

P1

PROCESS-10W-30

P3

P2

REF. DIODE SLITS

VIEW OF COATED SIDE OF
CHOPPER DISC

DISC SPEED
CONTROL

spectrometer and is used for SO_2 and NO_2 remote measurements. While achieving a high degree of specificity, the throughput of the instrument is increased because of the multiple exit slits. When the line separation for the gas of interest is constant or slightly varying, (e.g. SO_2 in the UV), multiple entrance slits can also be used. Thus a correlation spectrometer has both high specificity and high "etendue", with the advantage of real time output.

The mask design must take into account the actual solar radiation as background, complete with Fraunhofer lines and other atmosphere effects. Furthermore the absorption bands saturated at different concentrations so that masks for low and high concentration detection may be required. The solar spectral content of course changes during the day and produces effects, which must be compensated.

Figure 2 shows the arrangement of such a COSPEC mask. The COSPEC comprises a quarter meter folded Ebert spectrometer. In the exit plane of the spectrometer is located the optical correlator mask upon which are enscribed the lines matching the selected absorption maximum and minima.

The output of the spectrometer is thus modulated by the disc, and the output detected by a photomultiplier tube. Timing marks encoded on the periphery of the correlator disc provided synchronous detection signals to diode pickups. The complete scheme shown in Figure 3.

Calibration is achieved by inserting known gas cells into the instrument.

The COSPEC has been successfully commercialized for remote sensing of SO_2 and NO_2 using solar radiation (300 to 330 nm for SO_2 and 400 to 440 nm for NO_2). Thus remote sensing is only possible during daylight hours.

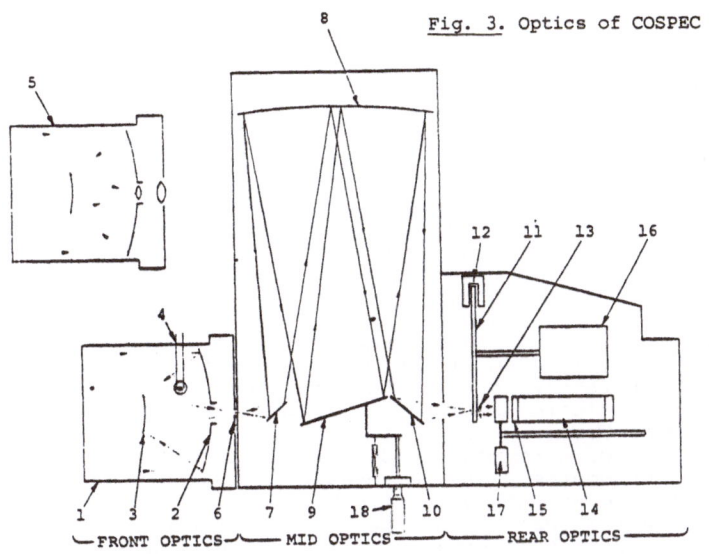

Fig. 3. Optics of COSPEC

FRONT OPTICS — MID OPTICS — REAR OPTICS

1. Cassegrain Telescope
2. Primary Concave Mirror
3. Secondary Convex Mirror
4. Reference Lamp
5. Maxwellian Telescope (Optional Replacement for 1.)
6. Entrance Slit Array
7. & 10. Plane Mirrors
8. Spherical Mirror
9. Plane Diffraction Grating

11. Correlation Disc
12. LEDs and Detectors
13. Condensing Lens
14. Photomultipliers
15. Optical Filter
16. Motor
17. Low and High Calibration Cells
18. Grating Angle Control

4 - GAS FILTER CORRELATION SPECTROMETER (GASPEC)

The nondispersive analyzer, or gas filter correlation spectrometer has a pair of gas cells. One cell, referred to as the sample gas cell, contains a judiciously chosen quantity of the target gas to be sensed, while the second cell, the reference gas cell, contains a spectrally inactive gas. The optical depth of target gas in the sample cell is optimized for a maximum product of the modulation of target gas energy and average transmission. Incoming radiation characteristic of the target gas is selectively filtered by being absorbed in the sample cell, but is readily transmitted through the reference cell. Therefore the radiance transmitted through the sample cell is largely independent of the presence of target gas signature in the received spectrum, whereas the radiance transmitted through the reference cell is strongly dependent upon the presence of the target gas. The difference in spectral transmittance between the two cells is thus a sensitive indicator of the amount of target gas signature in the radiation received by the sensor.

Gas filter correlation spectrometers described previously have used single detectors, and the differential transmittance was derived by time sharing the incident energy through the two gas cells. The time sharing design has two potential problem areas when used for high sensitivity detection from a downward looking fast moving platform, or other viewing geometries where the source radiance changes rapidly.

93

The necessary thermal stability of such designs is difficult to achieve, since they are generally aperture chopped systems with the aperture and chopper imaged onto the detector, so that changes in the chopper correlate with spatial responsivity differences at the detector creating error signals. Second and more fundamentally, the time share systems must either chop very quickly (not easy for large aperture) or use image motion compensation so as to avoid scene radiance changes as the sensor footprint moves.

A unique feature of GASPEC is the use of two detectors that receive amplitude shared source signals at the source chopping frequency. This arrangement makes the two gas cell arms insensitive to thermal radiation originating within the instrument. This is due to the optical configuration where the chopper is located near the sensor field stop. Only radiance originating at the objective lens contributes coherently with radiance originating outside the sensor. Because of the low lens emissivity, only a small amount of radiation is emitted, and the flat spectral character of this radiation does not correlate with the sample gas transmission spectrum. Thermal changes within the instrument are thus overcome.

In addition, the use of a common field stop and chopper results in simultaneous sampling of source radiance through both sensor arms. This enables the sensor to operate effectively even in the presence of rapidly changing source radiance. Radiance from two internal blackbodies at two different temperatures is alternately chopped to generate a constant reference signal. This reference signal is at a different frequency from the source signal, but at a fixed frequency ratio with respect to the source frequency due to the use of a single reflective chopper disc with a double set of annular holes. The reference signal and the source signal are optically combined and then follow a common path through the sensor to the detectors. Coherent detection is employed to separate electronically the two signal frequencies. The reference signals derived from the detectors are fixed to a constant ratio to provide a precise and continuous opto-electronic gain stabilization.

An additional feature of the GASPEC design is the use of a further adjustable reference blackbody. During the half-cycle when the source radiance is blocked by the chopper, the reflective inner side of the chopper views radiance from this blackbody. This blackbody may be adjusted in temperature until its radiance is comparable to

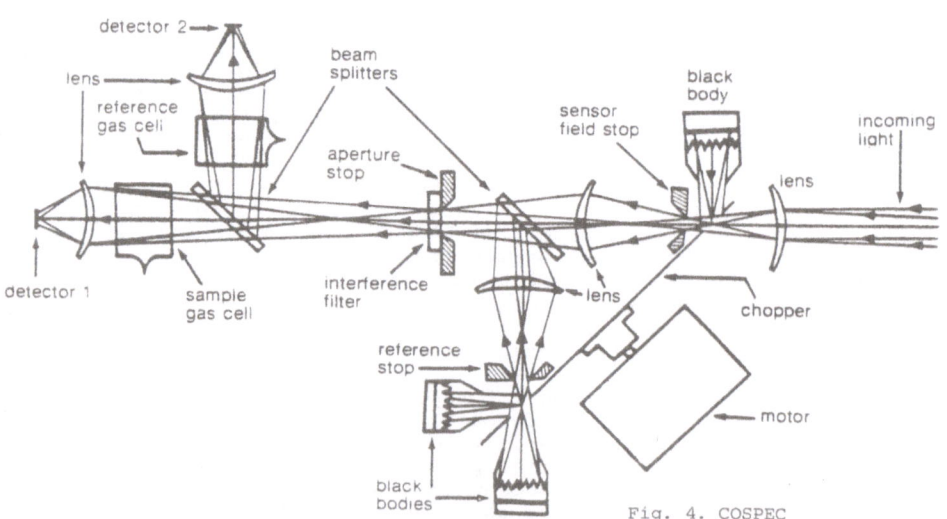

Fig. 4. COSPEC

the source radiance. This greatly facilitates the operation of the electronics signal processing by keeping the chopper radiance to a minimum. The GASPEC optical layout is shown in Fig. 4. Source radiance enters the sensor via objective lens. The size of the objective is determined by the desired sensor field of view. Radiance from a distant source is imaged in the objective focal plane where a field stop common to both detectors is located. A 45 degree reflective chopper is located adjacent to the field stop.

A relay lens images the objective lens at the aperture stop and also images the sensor field stop through the interference filter and the gas cells onto the field lenses via the beam splitter. The field lenses image the aperture stop and the objective onto the detectors thus avoiding imaging any scene hot spots onto the detectors. A second relay lens images via beam splitter the reference stop onto the field lenses coincident with the sensor field stop.

The target gas output signal S from GASPEC may be written as the difference in the source signals S_1 and S_2 from the two gas cell detector-amplifier arms. The GASPEC is balanced and insensitive to noncorrelating radiance changes when $S_1 = S_2$ (in the absence of target gas radiance). The electronic signal processing maintains this balance. The radiometer output S_1 can be interpreted in terms of the background source effective temperature.

The GASPEC concept provides two important advantages. First, it is fully multiplexed, that is, all wavelengths are in phase at the detectors. Thus there is no signal power cancellation due to phase shifts of different wavelengths of radiation, as occurs, for example, in an interferometer, and so GASPEC does not suffer these multiplexed disadvantges. Second, the device has a large etendue advantage, since the gas cell that determines the resoltuion limit has none of the spectrometer or interferometer angular restrictions, and the diameter of the gas cells may be large without great expense or technical difficulty. The etendue limitation is governed by the required optical thickness of the sample cell, by the imaging optics, and the bandpass interference filters. Also, the spectral resolution of GASPEC is high, being governed by the absorption line width of the gas in the sample cell.

The opto-electronic gain balance and stabilization are achieved by synchronously detecting the frequency and phase of the signals derived from the two internal reference blackbodies and holding these reference signals at a fixed ratio for the detector pair. By maintaining a fixed ratio of the signals from the reference blackbodies, by means of a feedback loop, one maintains a zero response in the presence of noncorrelating radiance. This null is maintained for detector and electronic gain changes as well as optical gain changes within the loop. Only correlating radiance (i.e., target gas derived radiance) upsets this balance and provides the desired signal.

The GASPEC has most recently been test flown on the Columbia shuttle flight, being used to monitor the vertical burden of CO. The GASPEC is also the technique selected for other satellite measurement experiments.

GENERAL REFERENCES

I. COSPEC Technology

1. Moffat, A. M., Robbins, J. R., and Barringer, A. R. (1971), Electro-Optical Sensing of Environmental Pollutants. Atmos. Environ. 4, 511-525.

2. Millan, M. M., Gallant, A. J. and Turner, H. E. (1976) The Application of Correlation Spectroscopy to the Study of Dispersion from Tall Stacks. Atmos. Environ. 10, pp 499-511.

3. Millan, M. M. and Hoff, R. M. (1977) Dispersive Correlation Spectroscopy: A Study of Mask Optimization Procedures. Applied Optics 16 pp 1609-1618.

4. Millan, M. M. and Chung, Y.S. (1977) Detection of a Plume at 400 km from the Source. Environ., 11 949-944.

5. Haulet, R., Zetwoog, P. and Sabroux, J.D. (1978) Sulphur Dioxide Discharge From Mount Etna. Nature 268 pp 715-77.

6. Sandroni, S. and Cerutti, C. (1978) Long Path Measurements of Atmospheric Sulphur Dioxide by a Barringer COSPEC III. Atmos. Environ., 11 pp 1235-1232.

7. Hamilton, P. M., Varey, R. H. and Millan, M. M. (1978) Remote Sensing of Sulphur Dioxide. Atmos. Environ. 12 pp 127-133.

8. Millan, M. M. and Hoff, R.M. (1978) Remote Sensing of Air Pollutants by Correlation Spectroscopy - Instrumental Response Characteristics. Atmos. Environ. 12, pp 853-864.

9. Van Egmond, N. D., Tissing, E., Onderlinden, D. and Bartels, C. (1978) Quantitative Evaluation of Mesoscale Air Pollution Transport. Atmos. Environ. 12, 2279-2287.

10. Guillot, P. et al. (1979) First European Community Campaign for Remote Sensing of Atmospheric Pollution Lacq (France), 7-11 July 1975. Atmos. Environ. 13, 895-917.

11. Sandroni, S. and De Groot, M. (1980) Intercomparison of Remote Sensors of Sulphur Dioxide at the 1979 European Community Campaign at Turbigo. Atmos. Environ. 14, 1331-1333.

II. GASPEC Technology

1. Infrared, Correlation and Fourier Transform Spectroscopy 1977. In "Computers in Chemistry and Instrumentation Series" Vol. 7. Marcel Dekker Inc. New York.

2. Ward T. V. and Zwick H. H. , Gas Cell Correlation Spectrometer: GASPEC, Appl. Opt. 14, 2896, 1975.

3. Jones, E. P., Ward, T. V. and Zwick, H. H., A Fast Response Atmospheric CO_2 Sensor for Eddy Current Correlation Flux Measurements, Atm. Env. 12, 845, 1978.

4. Russell, J. M., Park, J. K., and Drayson, S. R., Global Monitoring of Stratospheric Halogen Compounds from a Satellite using Gas Filter Correlation Spectroscopy in the Solar Occultation Mode. Appl. Opt. 16, 607, 1977.

5. Wallio, H. A., Reichle, H. G. Jr., Casas, J. A. and Gormsen, B. M., A New Method for Inferring CO Concentrations from Gas Filter Radiometer Data, 4th Conf. on Atm. Radiation, Toronto 1981. America Meterological Society, Boston MA.

2.6 Measurements of Atmospheric Trace Gases by Long Path Differential UV/Visible Absorption Spectroscopy

U. Platt* and D. Perner

Kernforschungsanlage Jülich, Institut für Chemie 3, Atmosphärische Chemie, D-5170 Jülich, Fed. Rep. of Germany

Introduction

Absorption spectroscopic measurement methods for trace gas analysis have a number of clear advantages over sampling methods, including absence of wall losses, greater specificity and the potential for real time measurements of several different species with a single type of instrument. Nonetheless, only a relatively small number of spectroscopic measurements of atmospheric trace gases have been reported in the past, using infrared or UV/visible absorption spectroscopy (Hanst 1971, Tuazon et al. 1981, Connell et al. 1980, Noxon et al. 1980, 1981, Bonafé et al. 1976, Millán et al. 1978).

The present paper describes a relatively simple instrument capable of detecting and measuring a number of important trace gases at tropospheric concentration levels by observing their structured UV/visible absorption features. The absorption is monitored over a long, open air path stretching several kilometers between an artificial light source and the receiving system containing the spectrometer. This setup ensures the coverage of a large air volume by the instrument.

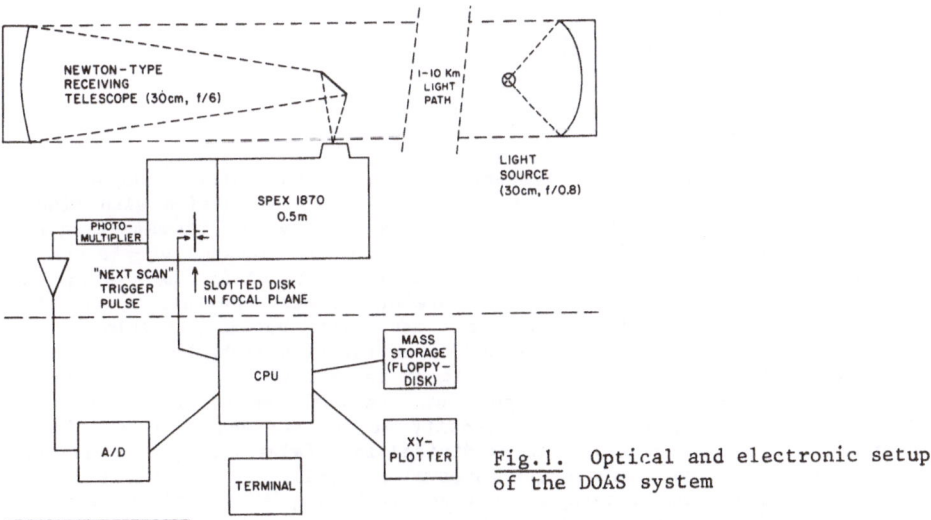

Fig.1. Optical and electronic setup of the DOAS system

*Present Address: Statewide Air Pollution Research Center
                    University of California, Riverside, CA 92521

The Differential Optical Absorption (DOAS) Spectrometer

Optical Setup. The basic DOAS system optical setup monitors the absorption of atmospheric trace gases contained in a long (~1-10 km) optical path, it therefore consists of two separate units (Figure 1):

(1) The light source, designed to emit a parallel beam of light, free of spectral structure in the wavelength ranges of interest (i.e., emitting "white" light). This is accomplished by either using Xe-high pressure lamps (Osram XBO 450, Hanovia 959C1980) or incandescent lamps (150-240 W quartz iodine) in combination with a "search light" type focussing mirror (0.3 m diameter, 0.25 m focal length) or, in special cases a frequency doubled CW-dye-laser.

(2) In the receiving unit at the other end of optical path, the light is collected by a Newton-type telescope and focussed into the entrance slit of a spectrograph. Instruments with focal lengths from 0.2 to 0.85 m equipped with 600-2160 groove/mm gratings are being used, depending on the required resolution. Also a minicomputer with appropriate peripherals (PDP 11/2 or 11/23) is an integral part of the instrument.

The Rapid Scanning Device. A 6 to 40 nm segment of the dispersed spectrum produced in the exit focal plane of the spectrograph is scanned by a series of moving exit slits etched radially in a thin metal disk (slotted disk) rotating in the focal plane. At a given time, one particular slit is used as an exit slit. The light passing through the exit slit is received by a photomultiplier tube, the output signal of which is digitized by a high-speed analog-to-digital converter and read by a minicomputer. During one scan (i.e., one sweep of an exit slit over the spectral interval of interest), several hundred digitized signal samples are taken. Consecutive scans are performed at a rate of approximately 100 scans per second and are signal averaged by the software.

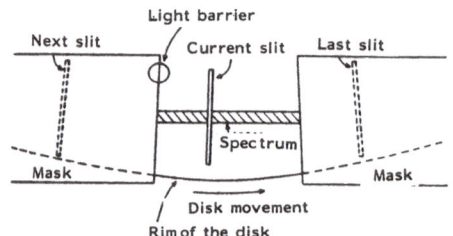

Fig.2. The "slotted disk" rapid scanning device located in the focal plane of the spectrometer

The process of taking one spectral scan and the interaction of hardware and software thereby is best illustrated in conjunction with Figure 2. The central wavelength of the scan is selected by the spectrograph setting and the width of the scan region by a mask located very close to the slotted disk. The distance between the slits along the rim of the disk is slightly larger than the aperture of the mask, so that at any time no more than one slit is irradiated. As a slit becomes visible at the left edge of the mask, it is detected by an infrared light barrier located there (see Figure 2), and a trigger signal is sent to the computer. While the slit then sweeps over the spectrum, the computer continuously takes digitized samples of the light intensity at the current position of the slit. Thus, one sweep of a slit is divided into several hundred channels, each associated with a wavelength interval several times narrower than the resolution of the spectrograph. During each scan, the digitized samples are added to the corresponding channels in the computer memory, thereby all consecutive scans are superimposed in the computer memory (signal averaged).

After finishing one scan, the computer waits for the next trigger pulse to indicate the next slit (shown in dashed lines in Figure 2) approaching the left edge of the mask. In order to preserve the spectral resolution while superimposing a large number of individual scans, the rotational speed of the slotted disk is kept constant to within ± 0.1%. The signal variations seen by the computer during a single spectral scan can be caused by several effects:

(1) The light absorption by trace gases varies with wavelength.
(2) The light losses due to mirror reflectivities, etc., may vary with wavelength.
(3) The output of the light source may vary with wavelength and time.
(4) Atmospheric refraction may change with time (due to turbulence).
(5) Random noise is added to the signal by the photomultiplier, preamplifier and A/D converter.

Since a single scan takes less than 10 msec, the effect of atmospheric scintillations is very small, because the frequency spectrum of atmospheric turbulence close to the ground peaks around 0.1-1 Hz and contains very little energy at frequencies above 10 Hz (Haugen 1973). In addition, typical spectra obtained during several minutes of integration time represent an average over 10-40 thousand individual scans. Thus, effects of noise and temporal signal variations are very effectively suppressed. In fact, even momentarily blocking the light beam entirely (e.g., due to a vehicle driving through the beam) has no noticeable effect on the spectrum.

Due to the very narrow field of view of the instrument (approximately $0.3 \ 10^{-4}$ by $1.1 \ 10^{-4}$ steradians), solar stray light levels have been found to be extremely low, with the exception of very hazy conditions. In the latter cases, however, the stray light can easily be cancelled by frequently interrupting the measurement and taking a "straylight spectrum" with the telescope pointed beside the light source. Since only an offset of a fraction of a degree is required, this does not change the amount of stray light entering the system, which then can then be subtracted from the total spectrum.

Remaining undesired influences on the shape of the spectrum are eliminated by the mathematical treatment of the spectrum and the way the information on trace gas concentration is extracted, as discussed below.

Processing of Absorption Spectra

Usually the raw spectra will show an overall slant caused by slight variations of lamp output or atmospheric scattering over the observed wavelength interval. In order to remove this overall feature, a polynomial (first to fifth order) is fitted to the spectrum and subsequently the spectrum is divided by this polynomial. The "narrow" features (extending only over a narrow wavelength range) caused by trace gas absorption are not much affected by the process.

The trace gas concentration C is, as usual with optical methods, derived from the spectrum by applying Beer's law:

$$C = (\log I_o/I)/(\varepsilon \ L) \tag{1}$$

where I_o = light intensity without absorption by the trace gas
I = light intensity, reduced due to absorption by the trace gas
ε = absorption coefficient of the trace gas
L = length of the light path .

INTENSITY

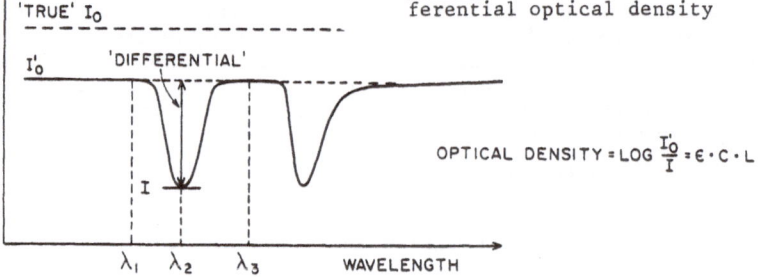

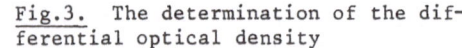
Fig.3. The determination of the differential optical density

Since the true light intensity I_o, without any absorption, cannot be obtained with this method (see Figure 3), the "differential" optical density is used to evaluate the trace gas concentration. The differential optical density is defined by $\log I_o'/I$ where I_o' is the intensity (at wavelength λ_2) in the absence of the particular absorption structure, rather than in the absence of any absorption at all (as indicated by the dashed line in Figure 3). According to this definition, I_o' can be calculated from $I(\lambda_1)$ and $I(\lambda_3)$:

$$I_o' = I(\lambda_1) + [I(\lambda_3)-I(\lambda_1)] \cdot \frac{\lambda_2-\lambda_1}{\lambda_3-\lambda_1} \qquad (2)$$

Of course, the absorption coefficient ε has to reflect the above definition of I_o', accordingly the "differential absorption coefficient" of a trace gas will generally be lower than the total absorption coefficient at the same wavelength.

In the practical application of the instrument, the precision of the measurements is improved by least squares fitting a reference spectrum of the substance under consideration to the observed absorption spectrum, thus all absorption bands of a given substance in the scanned spectral region, as well as the information contained in their relative strengths and particular shapes (the "fingerprint" of the substance), are used. This method even allows the deconvolution of overlapping spectra of different species to obtain the concentration of the contributing components with good accuracy. Examples of this process will be given below.

Software

The software for the DOAS spectrometer performs three major tasks (Figure 4):

(1) The individual spectral scans are signal averaged in the computer memory and displayed on an oscilloscope screen.

(2) System housekeeping (e.g., monitoring signal intensity, counting of scans, etc.) is carried out as a "background" task.

(3) A large number of mathematical functions can be applied to the acquired absorption spectra in order to extract the trace gas concentrations from the spectra. Also, spectra can be stored and recalled on mass storage devices (e.g., disks).

The program is written in MACRO 11 and FORTRAN IV for use with Digital Equipment PDP 11 computers.

The mathematical functions available for manipulation of the absorption spectra include multiplication and division by a constant factor,

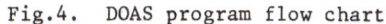

Fig.4. DOAS program flow chart

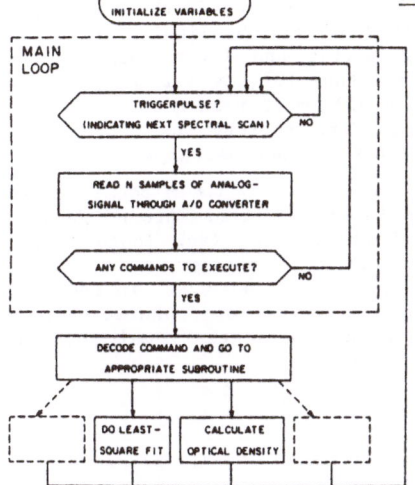

DUVVS—DATA—AQUISITION PROGRAM FLOWCHART

addition and subtraction of two spectra, division by a least square fitted polynomial, shifting the spectra in wavelength, change of the wavelength scale ("expansion" and "compression" of a spectrum), calculation of the differential optical density of an indicated absorption band, least square fitting of up to four reference spectra to a given spectrum, and taking the logarithm of a spectrum.

In addition to that, several hardware functions like changing the wavelength setting of the spectrometer or moving a filter into the light path are controlled by the software. All software functions can be executed in automatic sequences, thus allowing unattended operation of the instrument.

Detection Limits of the DOAS System

The detection limit for a particular substance can be calculated according to equation (1), if the differential absorption coefficient, the minimum detectable optical density (OD) and the length of the light path are known. While for a given minimum detectable OD the detection limit improves proportionally to the length of the light path, this does not mean that the actual detection limit always improves with a longer light path. This is because with a longer light path the received light intensity $I(\lambda)$ tends to be lower, increasing the noise associated with the measurements of $I(\lambda)$ and thus increasing the minimum detectable OD. A few simple considerations may clarify this relationship and also allow the derivation of detection limits.

The amount of light emitted by a search-light type light source (see Figure 1) varies enormously with power and type of lamp, as well as with focal length and size of the mirror. However, the size of the light emitting area of the lamp and the focal length of the mirror only affect the divergence of the beam. Thus the intensity of the beam (in W/sr) only depends on the luminous intensity of the lamp (W/mm^2) and the area of the mirror.

Furthermore, the size of the light source image projected onto the entrance slit of the spectrometer by the receiving system is proportional to the size of the light source mirror, the focal length of the telescope, and inversely proportional to the distance L between light source and receiving system. Nothing is gained by making the light source image larger than the height of the entrance slit, thus for given light path and slit dimensions, the maximum usable light source mirror size and receiving telescope focal length are also determined (and thereby the mirror diameter, since it has to match the aperture of the spectrometer).

For a typical DOAS system with an 0.3 m diameter, 1.8 m focal length receiving telescope and 0.2 mm entrance slit height, there are no geometrical light losses up to path lengths of about 3 km (for longer light paths, the size of the light source mirror could easily be increased, since an inexpensive metal mirror can be used here). However, in addition to geometrical light losses, light attenuations due to atmospheric absorption as well as to scattering affect the magnitude of the received light signal:

$$I_{received} = I_{source} \cdot \exp(-L/L_0) \qquad (3)$$

where the absorption length L_o reflects the combined effects of broad band atmospheric absorption and Mie- and Rayleigh scattering.

Since the noise level of a photomultiplier signal is essentially dependent on the number of photons received and thus is proportional to the square root of the light intensity. The signal-to-noise ratio of a DOAS system as a function of the light path length L can be expressed as:

$$S/N \sim L \exp(-L/2L_0) \qquad (4)$$

This relationship holds for the case of "no geometrical light losses" and yields an optimum S/N ratio for a light path length $L = 2 \, L_o$. While there are no lower limits for L_o in the atmosphere (fog), the upper limits are given by Rayleigth scattering, and the scattering by atmospheric background aerosol which indicate optimum light path lengths ($2 \, L_o$) in excess of 10 km for wavelengths above 300 nm.

With a typical DOAS system (using an Xe light source), for wavelengths > 300 nm minimum detectable optical densities of $(5-10) \, 10^{-5}$ (base 10) can be achieved in a few minutes of averaging time. The minimum detectable OD can be higher, if it is necessary to subtract overlapping absorption features of other species (see below); however, in most practical cases, the detection limits are only slightly degraded, since overlapping bands can usually be removed to 99-99.9%. Thus, to noticeably increase the detection limit, the overlapping absorption must be ~100 times stronger than the absorption under consideration. Table 1 gives detection limits for a number of atmospheric trace constituents, for a 10 km ground based light path and a minimum detectable OD of 10^{-4}. The detection limits are usually in the part per trillion (ppt) range, and for studies in polluted air much shorter light paths (e.g., 1-3.5 km) have been used. All substances in Table 1 have already been observed in the ambient atmosphere (Platt et al. 1979, Perner and Platt 1979, Platt et al. 1980a,b,c, Platt et al. 1981, Harris et al. 1982).

<u>Absorption Spectra of Substances Important to Tropospheric Chemistry</u>

Figures 5 through 7 show examples of absorption spectra of trace gases which are important in the polluted as well as in the unpolluted troposphere. Nitrous acid (HONO) and the nitrate radical (NO_3) have been detected in the troposphere for the first time using this method.

Table 1. Atmospheric Trace Components Observed by Differential UV-VIS Spectroscopy

Substance	Wavelength Range [nm]	Differential Absorption Coefficient[a] [cm²/molec]	[at nm]	Detection for 10 km Light Path [ppt]
SO_2	200–230,290–310	5.7×10^{-19}	300	17
CS_2	200–220,320–340	4×10^{-20}		240
NO	215,226	2.3×10^{-18}	226	400[b]
NO_2	330–500	1.0×10^{-19}	363	100
NO_3	623,662	1.8×10^{-17}	662	0.5
HNO_2	330–380	4.2×10^{-19}	354	20
O_3	220–330	4.5×10^{-21}	328	2100
HCHO	250–360	7.8×10^{-20}	340	120
OH	308	2×10^{-16}[c]	308	0.05

[a]0.3 nm, spectral resolution; [b]1 km light path, minimum detectable O.D. $= 10^{-3}$; [c]0.003 nm spectral resolution

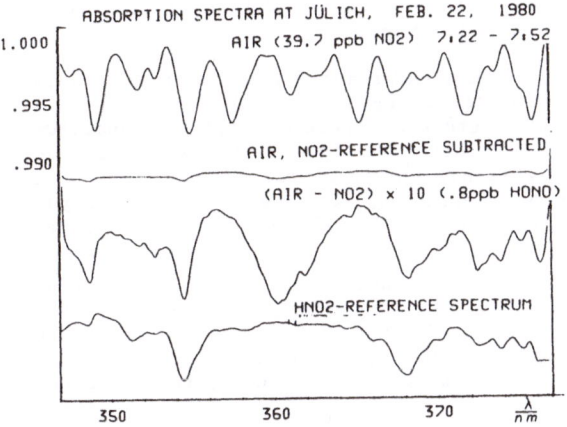

Fig.5. Deconvolution process of an air spectrum (taken at Jülich, Germany, 3.5 km light path) showing NO_2 and HONO absorption features

Figure 5 shows the spectral region from 348 through 378 nm in which NO_2 and HONO are the dominant absorbing species. The uppermost trace in Figure 5 shows the "air" spectrum which is almost entirely due to NO_2. After the subtraction of a suitably weighted NO_2 reference spectrum, the residual spectrum is an almost featureless trace. The next trace (third from top, Figure 5) is identical to the second, with 10 times expanded vertical scale. Here the absorption bands of nitrous acid (354 nm and 368 nm) become clearly visible, agreeing well in position and shape with the pure HONO reference shown in the lowest trace. In addition, the strong, wide absorption feature at 360 nm is due to the light absorption by collision pairs of oxygen $(O_2)_2$ (Perner and Platt 1980).

The example in Figure 6 covers the wavelength region from 325 nm to 350 nm, where the most dominant absorption usually is due to NO_2, and additional absorption features of ozone and formaldehyde are observed.

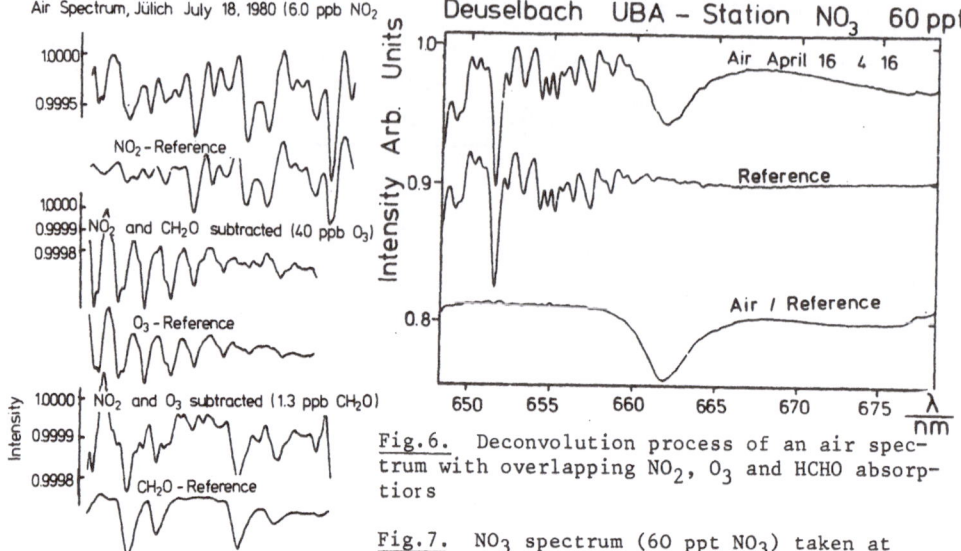

Fig.6. Deconvolution process of an air spectrum with overlapping NO_2, O_3 and HCHO absorptions

Fig.7. NO_3 spectrum (60 ppt NO_3) taken at Deuselbach, Germany, 4.8 km light path. The narrow features between 650 and 660 nm are due to water vapor

The spectrum in Figure 6 was taken under moderately polluted conditions in Jülich, Germany and shows (from top to bottom) three pairs of traces: (1) air spectrum and NO_2 reference spectrum, (2) air spectrum with fitted amounts of NO_2 and HCHO subtracted, leaving the ozone absorption feature, and a pure ozone reference spectrum, (3) air spectrum with fitted amounts of NO_2 and ozone subtracted, leaving the formaldehyde absorption features as compared with a pure formaldehyde reference spectrum.

In practice, of course, the concentrations of all three contributing species (NO_2, O_3, HCHO) are obtained in one step by simultaneously fitting the reference spectra to the air spectrum. The procedure shown in Figure 6 of partial subtraction of features, however, provides a good test of whether the fit works properly and also shows, in this case, that no other absorptions are present in the spectrum.

The last example (Figure 7) shows one absorption line of the nitrate radical (NO_3) at 662 nm. Here the feature is already clearly visible in the air spectrum (taken at Deuselbach, Germany). However, some overlap occurs with strong water bands in the 650 to 660 nm range. Since the nitrate radical concentration is extremely low at daytime due to its rapid photolysis (Magnotta and Johnston 1981), a daytime spectrum can be taken as a reference to ratio out the water lines, as shown in the lowest trace of Figure 7.

The application of the DOAS system to studies of trace pollutants and atmospheric constituents in the Los Angeles Basin is described in a companion paper (Harris et al. 1982).

Conclusion

In several years of practical use for the measurement of ambient trace gas concentrations, DOAS systems have been proven to be reliable and adaptable instruments. Continuing efforts to improve the system are aimed at lowering the detection limits, the construction of a compact system using a folded path (White optical system) and the expansion of the number of detectable trace constituents. Atmospheric components likely to be detectable by DOAS instruments, in addition to those already observed, include benzaldehyde, C_2H_2, NH_3 and the ClO radical.

References

Bonafé, G. Cesari, G. Giovanelli, T. Tirabassi and O. Vittori, Atmos. Environ., 10, 469–474 (1976).

Connell, P. S., R. A. Perry and C. J. Howard, Geophys. Res. Lett., 7, 1093–1096 (1980).

Hanst, P. L., Adv. Environ. Sci. Technol., 2, 91 (1971).

Harris, G. W., W. P. L. Carter, A. M. Winer, J. N. Pitts, Jr., U. Platt and D. Perner, Env. Sci. Technol., in press (1982).

Harris, G. W., A. M. Winer and J. N. Pitts, Jr., Presented at Workshop on Optical and Laser Remote Sensing, Monterey, CA, February 9–11, 1982 (following paper).

Haugen, D. A. (editor), Workshop on micrometeorology, Science Press, Ephrata, PA (1973).

Millán, M. M. and R. M. Hoff, Atmos. Environ., 12, 853–864 (1978).

Noxon, J. F., R. B. Norton and W. R. Henderson, Geophys. Res. Lett., 7, 125–128 (1980).

Noxon, J. F., Geophys. Res. Lett., 8, 1223–1226 (1981).

Perner, D. and U. Platt, Geophys. Res. Lett., 6, 917–920 (1979).

Perner, D. and U. Platt, Geophys. Res. Lett., 7, 1053–1056 (1980)

Platt, U., D. Perner and H. W. Paetz, J. Geophys. Res., 84, 6329–6335 (1979).

Platt, U., D. Perner, A. M. Winer, G. W. Harris and J. N. Pitts, Jr., Geophys. Res. Lett., 7, 89–92 (1980a).

Platt, U., D. Perner, A. M. Winer, G. W. Harris and J. N. Pitts, Jr., Nature, 285, 312–314 (1980b).

Platt, U. and D. Perner, J. Geophys. Res., 85, 7453– (1980c).

Platt, U., D. Perner, J. Schroeder, C. Kessler and A. Toennissen, J. Geophys. Res., 86, 11965–11970 (1981).

Tuazon, E. C., A. M. Winer and J. N. Pitts, Jr., Environ. Sci. Technol., 15, 1232–1237 (1981).

2.7 Measurements of HONO, NO₃, and NO₂ by Long-Path Differential Optical Absorption Spectroscopy in the Los Angeles Basin

2.7 Measurements of HONO, NO_3, and NO_2 by Long-Path Differential Optical Absorption Spectroscopy in the Los Angeles Basin

G.W. Harris, A.M. Winer, and J.N. Pitts, Jr.

Statewide Air Pollution Research Center and Department of Chemistry, University of California, Riverside, CA 92521, USA

U. Platt and D. Perner

Kernforschungsanlage Jülich, Institut für Chemie 3, Atmosphärische Chemie, D-5170 Jülich, Fed. Rep. of Germany

Introduction

In recent years, atmospheric chemists and air pollution researchers have increasingly looked beyond the "criteria" pollutants (i.e., those for which air quality standards have been established) such as ozone (O_3), nitrogen dioxide (NO_2), sulfur dioxide (SO_2) and carbon monoxide (CO). Thus, interest has grown in the unambiguous identification and measurement of such species as nitric (HNO_3) and nitrous ($HONO$) acids, formaldehyde ($HCHO$), the nitrate radical (NO_3), the hydroxyl radical, etc. Of particular concern have been the trace nitrogenous species and the need to assess their roles in the chemical cycles of the clean and polluted troposphere, as well as their impacts on biological systems, including human health and vegetation. Recently, the emerging issue of nitrogenous "acid rain" has accelerated interest in these compounds.

Detailed in situ study of many of the trace compounds in the atmosphere has awaited the development of instruments with the requisite specificity and sensitivity. The measurement of stable compounds and radical intermediates, which may be present at part per trillion (ppt) to part per billion (ppb) concentrations in the complex atmospheric system, presents a challenging analytical problem. Beginning in the mid-1970s, researchers at the Statewide Air Pollution Research Center (SAPRC) undertook a program designed to establish the temporal and geographical distribution of trace pollutants in the Los Angeles air basin. This investigation was originally based on the application of kilometer pathlength Fourier transform infrared (FT-IR) spectroscopy using a Michelson interferometer interfaced to an eight-mirror multiple reflection cell with a base path of 22.5 m (1,2). That system yielded the first spectroscopic detection of HNO_3 and HCHO in the polluted troposphere, as well as detailed simultaneous measurements of ambient concentrations of formic acid (HCOOH), peroxyacetyl nitrate (PAN), ammonia and O_3 (1-4).

Since 1979, we have utilized a second long path technique to complement our FT-IR studies. This technique, differential optical (UV/visible) absorption spectrometry (DOAS), was initially developed by researchers at the Institut für Chemie, Jülich, West Germany (5,6). A detailed description of the DOAS system is given in a companion paper (7). Here we describe the application of this spectrometer to the detection and measurement of trace nitrogenous pollutants in the Los Angeles airshed. Emphasis will be given to our measurements of HONO and NO_3, two species which have been included in models of the chemistry of the atmosphere for many years but which we have only recently characterized for the first time in the polluted troposphere using the DOAS technique (8-10).

DOAS Field-Study Methodologies

In any program of atmospheric monitoring, the technology employed must be practical in a field environment. Thus, for example, wet chemical methods for analyzing air pollutants may be reliable in the laboratory but generally suffer from a number of drawbacks when applied to atmospheric measurements. If a specific procedure is required for each of several atmospheric species, it is unlikely that rapid, simultaneous, multi-component data acquisition will be possible. Moreover, the time required for analyses may be long compared to the characteristic times for changes in the concentration of the species under study and hence only "averaged" rather than "dynamic" information will be available. Finally, the problem of interferences due to gas phase or particulate co-pollutants may be serious for both instrumental and wet chemical analyses.

Broad band (or tuneable narrow band) optical absorption techniques such as DOAS (and FT-IR spectroscopy) offer the advantages, by comparison with nonoptical techniques, of very rapid data acquisition and data reduction as well as having the potential for making multi-component measurements. The rapid signal-averaging rate (7) of the DOAS technique in particular, ensures that atmospheric scintillations are averaged out and there is therefore no need to enclose an air mass before analysis. Thus, labile species, or those in rapid photo-equilibrium, can be effectively studied with the DOAS system.

DOAS field studies are applicable to a wide range of atmospheric conditions. These include very clean background air measurements of, for example, HCHO and NO_2 (6) where optical pathlengths of $\geqslant$ 10 km may be employed, as well as measurements in extremely polluted air for which pathlengths of a few hundred meters or less may be adequate. Thus, the criteria for pathlength selection are primarily the sensitivity required for a specific measurement and the extent of broad band atmospheric extinction.

High intensity, short arc xenon light sources provide useful photon fluxes in the spectral region from ~200 nm to > 700 nm. However, at the short wavelength end of this region, molecular scattering limits the maximum pathlength useable even in very clean air to $\leqslant$ 1 km. At longer wavelengths (e.g., ~350 nm), scattering from particles in smoggy air also restricts the maximum light path. Fortunately, poor visiblity due to scattering from particles is usually associated with elevated levels of the nitrogenous pollutants and other species of interest and sufficient sensitivity is therefore still available for monitoring studies in photochemical air pollution episodes. Figure 1 illustrates the use of twin light sources at a monitoring site employed for HONO studies described below. Twin light sources not only provide some degree of spatial resolution but allow measurements to be conducted over a wider range of visibility conditions at the site.

Figure 2 illustrates the mobile field laboratory presently in use for DOAS studies at SAPRC. A monitoring site can be quickly established by one or two workers and the spectrometer system and associated instrumentation are as far as possible under automatic control, requiring only intermittent attention (once every 24-48 hours) by the operator. The collecting mirror shown in Figure 2 is under servo control of the DEC 11/23 minicomputer and dynamically maintains correct alignment for accurate viewing of the remote light source. In principle, reference spectra could be automatically acquired by rotating short (10 cm) absorption cells into the light beam at appropriate intervals, although at present this task must be performed by the operator.

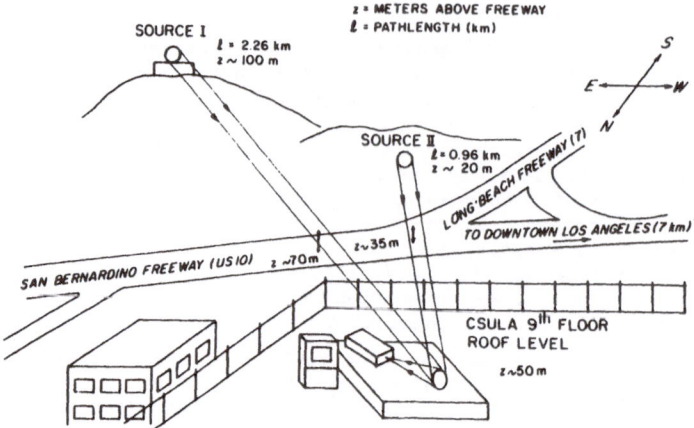

Fig. 1. DOAS field site near downtown Los Angeles with two light sources at different pathlengths

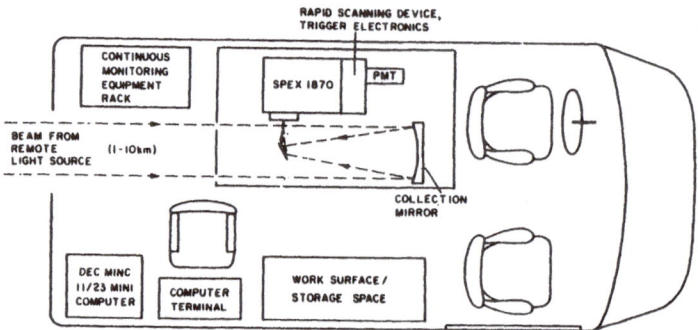

Fig.2. Mobile van containing DOAS system and associated instrumentation

Subtraction of suitably weighted reference spectra of the strongest absorber may be performed automatically as described in the companion paper (7) and time-resolved optical density/concentration data output to mass storage and/or an X-Y recorder. Since not all the species of interest in a particular study have absorption structure in the same 40 nm wavelength region which the DOAS may monitor at any one time, the spectrometer is equipped with a stepper motor wavelength drive which may be directed by the computer to traverse the central wavelength pointer between regions of interest in a programmable sequence.

The following sections describe selected monitoring studies carried out with our DOAS instrumentation over the last two years in southern California and illustrate the chemical insight which may be obtained by such rapid, simultaneous, multi-component monitoring of clean and polluted air masses.

Measurements of the Nitrate Radical

Due to their free radical structure, NO and NO_2, usually referred to collectively as "NO_x," are active in many atmospheric trace gas cycles. However, in order to understand the chemical cycles of NO and NO_2, the chemistry of a number of other nitrogen compounds, which play a role in the production, storage, conversion and sink mechanisms of NO_x have to be investigated as well. These species include the nitrate radical (NO_3), the nitrogen oxyacids (HNO_2 and HNO_3) and the anhydride of nitric acid (N_2O_5). The chemical cycles and reaction sequences involving the chemistry of these nitrogenous species (both during the day and at night) are complex. Since O_3 is usually present at night, NO is rapidly converted to NO_2 which in turn can react slowly with O_3 to form NO_3. The nitrate radical is itself in equilibrium with N_2O_5. While the homogeneous reaction of N_2O_5 with water is known to be very slow (11), the heterogeneous hydrolysis of N_2O_5 may be a major loss mechanism for NO_x at night and hence a potentially important nighttime source of HNO_3 (and acid rain).

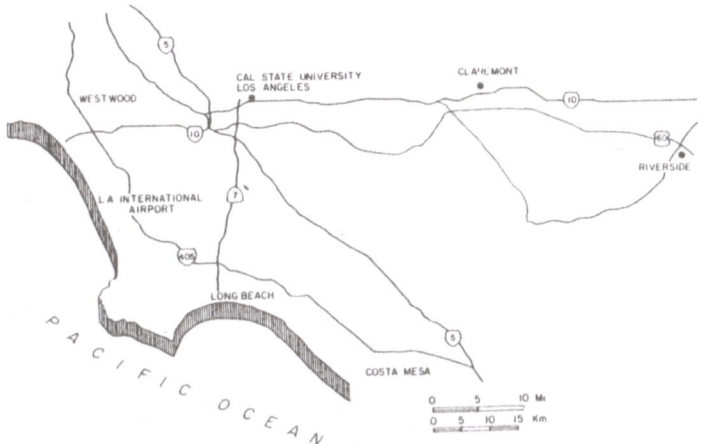

Fig.3. A portion of the Los Angeles airshed including the Riverside, Claremont and downtown Los Angeles sites used in DOAS studies. Not shown is a desert site (Whitewater Hill) ~175 km due east of Los Angeles

Evidence for this mechanism is provided by our measurements of NO_3 time-concentration profiles at four sites in the Los Angeles airshed since August 1979. Studies were conducted at Riverside, Claremont and downtown Los Angeles (see Figure 3), sites usually experiencing air of comparatively high humidity, as well as at a desert site (Whitewater Hill north of Palm Springs) which generally experiences low relative humidities. Our first measurement program was carried out on a total of 15 days in August and September 1979 at Riverside and Claremont, using 970 and 750 m optical paths, respectively. Nitrate radical concentrations as high as 355 ppt were observed in nighttime measurements made at Riverside during an air pollution episode which persisted from September 11 to 19.

Figure 4 shows the development of the 623 and 662 nm NO_3 bands (monitored alternately in 10-minute intervals) on the evening of September 12, 1979. In both wavelength regions, water vapor absorption lines have been cancelled by subtracting early-morning reference spectra taken at times with no ozone (and therefore no NO_3) present. Both the shape and position of the observed bands agree very well with NO_3 reference spectra obtained by placing a 4 cm cell filled with a mixture of O_3 and NO_2 in the light beam, and with the spectrum reported by Graham and Johnston (12).

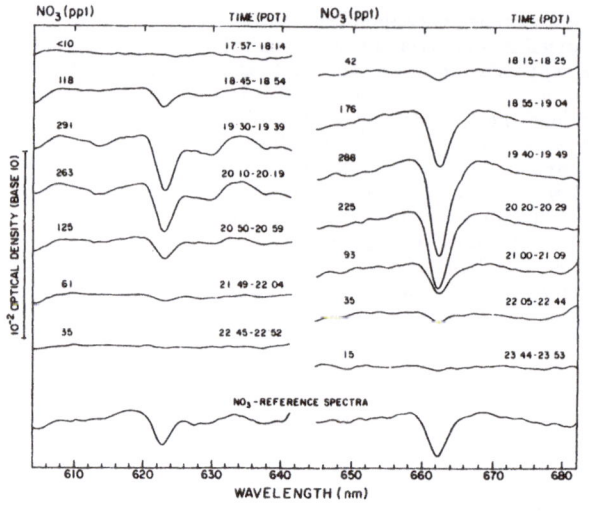

Fig.4. Observation of NO₃ absorption bands by differential optical absorption spectroscopy (8)

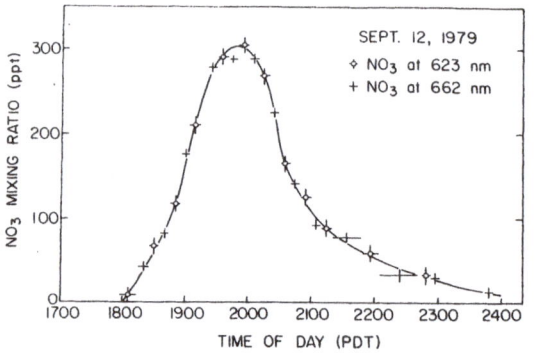

Fig.5. NO₃ time-concentration profile (Riverside, 970 m pathlength) (8)

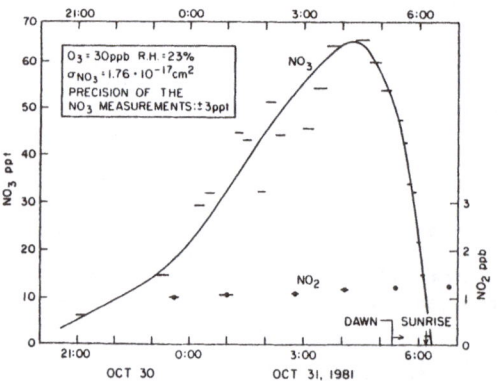

Fig.6. NO₃ night profile obtained with a 2.8 km lightpath at Whitewater Hill, desert site (14)

The concentration-time profile derived from the spectra given in Figure 4 is shown in Figure 5 along with the corresponding NO_2 and O_3 concentrations. Prior to sunset the NO_3 concentration was below the detection limit of ~5 ppt due to the short photolytic lifetime of the nitrate radical (13). After sunset (~1800 PDT), the NO_3 concentration rose rapidly, peaking at ~2000, and then declined to levels below the detection limit after midnight. Similar behavior was observed on many other nights during this study at Riverside and Claremont. The detailed data from this investigation have been reported elsewhere (8).

Another extended set of measurements of ambient NO_3 concentrations was carried out at a desert site (Whitewater Hill) ~170 km east of downtown Los Angeles. These measurements yielded similar peak concentrations (up to 214 ppt) to those observed in the Riverside/Claremont study. However, the concentration-time profiles typically observed at Whitewater Hill (see Figure 6) were very different from those previously measured. In particular appreciable NO_3 concentrations were present throughout the night until dawn when the NO_3 concentration rapidly decreased to levels below our detection limit (14).

Taken together the results from these studies indicate that under conditions of high humidity (> 70%) the lifetime of the nitrate radical is very short (< 1 min) presumably due to the reaction of N_2O_5 with wet aerosol surfaces. On the other hand, for relative humidities below ~50% RH lifetimes as long as ~1 hour occur. Although lifetimes of NO_3 as short as ~1 hr could result from low concentrations of NO at the site, we are able to rule out this possibility using the multi-component monitoring capability of the DOAS technique. Specifically, if sufficient NO were present at the site to account for the observed NO_3 lifetime, then a large increase in the NO_2 concentration (due to the parallel reaction of NO with O_3) would have been observed with the DOAS system. Since this was not the case, we must conclude that a one-hour lifetime cannot be explained on the basis of our present understanding of NO_3 chemistry, suggesting that under dry conditions unknown sinks for NO_3 are operative (see Platt et al. (15). Similar conclusions have also been reached by Noxon et al. (16).

Measurements of Nitrous Acid

The importance of nitrous acid arises not only from its role as precursor for hydroxyl radicals, a key radical intermediate in photochemical cycles in both the clean and polluted troposphere, but also due to its facile reaction with secondary amines to form carcinogenic nitrosamines (17).

HONO is thought to be formed by two major pathways, the recombination of OH radicals with NO

$$OH + NO \overset{M}{\rightarrow} HONO \tag{1}$$

and the direct reaction of nitrogen oxides with water

$$NO + NO_2 + H_2O \rightarrow 2 \; HONO \tag{2}$$

$$NO_2 + NO_2 + H_2O \rightarrow HONO + HNO_3 \tag{3}$$

The latter reactions may proceed either homogeneously or heterogeneously. Reaction (1) proceeds only when significant concentrations of OH radicals are present (i.e., essentially only during daylight). However, the HONO concentrations during daytime are very low, due to its rapid photolysis, and correspond to a lifetime of ~10 min (8). At night in the absence of photolysis, higher concentrations of HONO can accumulate.

In a series of DOAS measurements beginning in the summer of 1979 at Riverside, Claremont and downtown Los Angeles, HONO was identified and measured at all three locations (9,10). The overnight buildup of HONO could be followed, an example of which is given in Figure 7 (see references 9 and 10 for details). Peak HONO concentrations up to 8 ppb (in downtown Los Angeles) were observed in the early morning hours, always followed by a rapid decay to below our detection limit within approximately two hours.

This rapid photolysis of HONO at sunrise produces OH radicals, resulting in an ~2 hr "pulse" of radicals at a time when other sources of OH (e.g., O_3 and HCHO photolysis) are weak (9). Thus, HONO photolysis at sunrise can considerably accelerate the formation of photochemical air pollution and, in order to correctly model photochemical smog formation in urban airsheds, it is clearly necessary to have reliable data concerning HONO concentrations in the early morning hours.

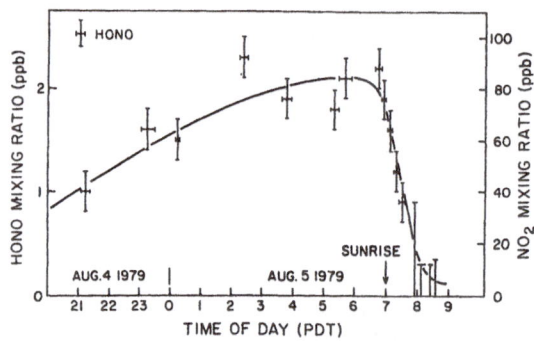

Fig.7. Concentration-time profiles of HONO and NO_2, 4-5 August 1979, Riverside, California (9)

Conclusion

During the past two years, the DOAS technique employed by researchers from the Institut für Chemie (Jülich) and the University of California Statewide Air Pollution Research Center (Riverside) has proven to be a versatile and powerful tool for practical studies of ambient concentrations of such critical species as HONO, NO_3 and HCHO under a wide variety of atmospheric conditions. The DOAS system has also been used to measure the concentrations of the criteria pollutants O_3, SO_2 and NO_2 in clean and polluted atmospheres over long pathlengths, in some cases in a more informative and cost effective manner than possible with point monitors.

In principle many other pollutants and atmospheric species which have structured absorption spectra in the UV, visible and near-IR regions can be identified and measured with the DOAS system. Efforts are underway in our respective laboratories to extend the applications of this instrument to other studies of the chemistry of the clean and polluted troposphere.

Acknowledgement

The authors gratefully acknowledge support of this work by the National Science Foundation (Grants No. PFR 78-01004-A01 and ATM-8001634).

1. E. C. Tuazon, R. A. Graham, A. M. Winer, R. R. Easton, J. N. Pitts, Jr. and P. L. Hanst, Atmos. Environ., 12, 865 (1978).

2. E. C. Tuazon, A. M. Winer, R. A. Graham and J. N. Pitts, Jr., Adv. Environ. Sci. Technol., 10, 259 (1980).

3. E. C. Tuazon, A. M. Winer, R. A. Graham and J. N. Pitts, Jr., Atmospheric Measurements of Trace Pollutants by Long Path FT-IR Spectroscopy," Final Report to U. S. Environmental Protection Agency (Grant No. R-804546), Research Triangle Park, NC, 1981.

4. E. C. Tuazon, A. M. Winer and J. N. Pitts, Jr., _Environ. Sci. Technol._, 15, 1232 (1981).

5. D. Perner and U. Platt, _Geophys. Res. Lett._, 6, 917 (1979).

6. U. Platt, D. Perner and H. W. Pätz, _J. Geophys. Res._, 84, 6329 (1979).

7. U. Platt and D. Perner, "Measurement of Atmospheric Trace Gases by Long Path Differential UV/Visible Absorption Spectroscopy," Presented at Workshop on Optical and Laser Remote Sensing, Monterey, CA, February 9-11, 1982 (preceeding paper).

8. U. Platt, D. Perner, A. M. Winer, G. W. Harris and J. N. Pitts, Jr., _Geophys. Res. Lett._, 7, 89 (1980).

9. U. Platt, D. Perner, G. W. Harris, A. M. Winer and J. N. Pitts, Jr., _Nature,_ 285, 312 (1980).

10. G. W. Harris, W. P. L. Carter, A. M. Winer, J. N. Pitts, Jr., U. Platt and D. Perner, "Observation of Nitrous Acid in the Los Angeles Atmosphere and Implications for Predictions of Ozone-Precursor Relationships," _Environ. Sci. Technol._, in press (1982).

11. E. D. Morris, Jr. and H. Niki, _J. Phys. Chem._, 77, 1929 (1973).

12. R. A. Graham and H. S. Johnston, _J. Phys. Chem._, 82, 254 (1978).

13. F. Magnotta and H. S. Johnston, _Geophys. Res. Lett._, 7, 769 (1980).

14. U. Platt, G. W. Harris, A. M. Winer and J. N. Pitts, Jr., in preparation (1982).

15. U. Platt, D. Perner, J. Schöder, C. Kessler and A. Toennissen, _J. Geophys. Res._, 86, 11965 (1981).

16. J. F. Noxon, R. B. Norton and E. Marovich, _Geophys. Res. Lett._, 7, 125 (1980).

17. J. N. Pitts, Jr., D. Grosjean, K. Van Cauwenberghe, J. P. Schmid and D. R. Fitz, _Environ. Sci. Technol._, 12, 946 (1978).

2.8 Remote Detection of Gases by Gas Correlation Spectroradiometry

J.S. Margolis, D.J. McCleese, and J.V. Martonchik
Jet Propulsion Laboratory, 4800 Oak Grove Drive, Pasadena, CA 91109, USA

We will discuss here the application of a pressure modulated radiometer (PMR) to the remote sensing of trace amounts of gases in the atmosphere as well as to the direct measurement of upper atmosphere winds. The range of altitudes which may be covered with this device vary from ground level up to the lower mesosphere. Gas concentrations corresponding to a mixing ratio of 1 ppm in a 10-meter column can be sensed from geosynchronous orbit. Even higher sensitivity can be achieved at shorter distances. The PMR has an extensive history of flight testing and has been used in earth meteorological satellites, numerous high altitude balloon flights and planetary exploration spacecraft.

The PMR operates as a gas correlation spectrometer. It has some advantages over conventional gas correlation spectrometers particularly in the simplification of the method required for maintaining electrical/optical balance, and in the versatility of the measurements which are carried out. The principle of operation is illustrated in Figure 1. The PMR modulates the light passing through it, but only in the vicinity of absorption lines. The signal at the detector is amplified by a lock-in amplifier which is referenced to the PMR so that only the signals corresponding to wavelengths near absorption lines are amplified. The modulation index is controlled by the amplitude of the mechanical piston which is operated at resonance and which is easily controlled and stabilized electronically. A sketch of the PMR is shown in Figure 2. It exhibits a number of important characteristics.

It possesses high sensitivity owing to its essentially very high effective spectral resolution, and large energy grasp (50-100 lines are typically observed).

It is a passive system and emits no telltale radiations.

It is unlimited in range: the sensitivity is unaffected by increasing distance to the target; however, the spatial resolution does degrade as the range is increased.

It has a high degree of species discrimination depending on the number of interfering absorption lines in the spectral interval chosen to monitor. Spurious signals from interfering species may be reduced if the interfering species is monitored.

It is relatively unaffected by aerosols in the field of view. The PMR signal remains substantially unchanged as the atmospheric aerosol content varies from clear up to an optical density of 0.5.

114

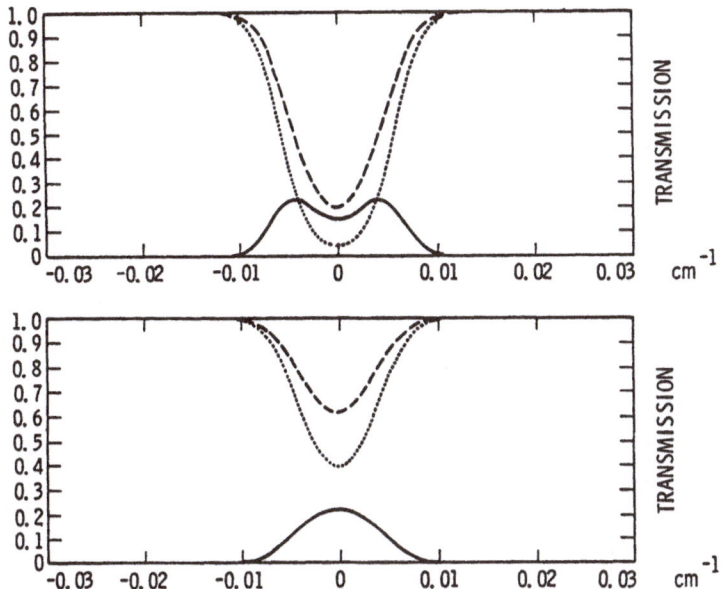

Figure 1   SPECTRAL TRANSFER FUNCTION FOR THE PRESSURE MODULATOR.
        The two cases shown here correspond to different amounts of
gas in the pressure modulated cell. In both cases the dashed lines
demonstrate the transmission through the cell at the pressure extremes;
the solid line is their difference and measures the amount of modulation.
The upper curves, which correspond to a large amount of gas in the cell,
show how the transfer function can be tuned symmetrically somewhat beyond
the core of the absorption lines. This has some advantages in probing
into deep layers of the atmosphere.

        It is light in weight. The pressure modulated cell weighs only 300
gms. The entire package, including electronics can be made to weigh only
a few kgm.

        It is compact and can easily be made to be hand-held.

        The power requirement is less than 1 watt for operation of the pressure
modulator, and only a few watts is required to operate the entire system
including electronics. However, there is an additional power requirement
of a few watts which is required to drive a molecular sieve which is placed
in the cell. This is used to control the pressure of the gas in the cell
and thereby control the altitude peak of the weighting function as mentioned
above.

        A PMR may be used to detect almost any stable gas which can be contained
in the cell. It has been used on the Nimbus 7 satellite to measure the vertical
distribution and abundance of CO, NO, NO_2, N_2O, H_2O, and CH_4; on the Tiros N
satellite to measure the vertical temperature distribution in the upper atmosphere using CO_2 as the sensed gas and in many balloon flights for these gases
as well as HCℓ.

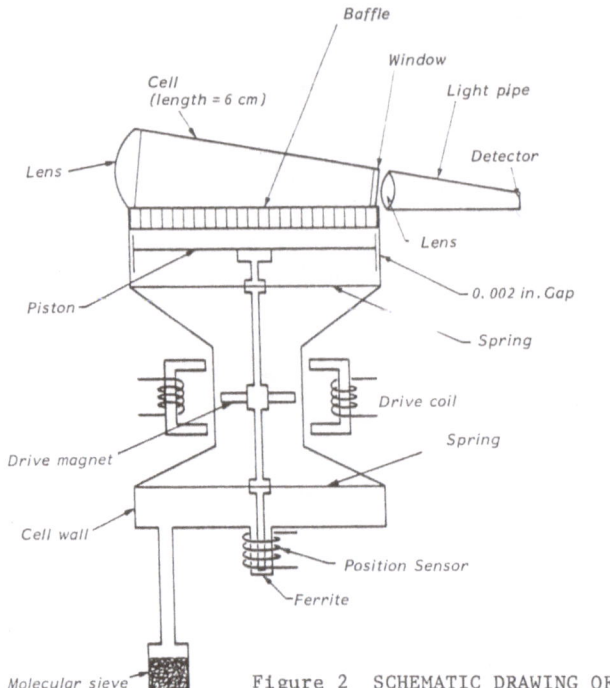

Figure 2   SCHEMATIC DRAWING OF THE PMR.

The type of PMR shown is similar to one flown on Nimbus 6. The overall height of the device is about 5" and it weighs approximately 300 gms. The molecular sieve is used to control the gas pressure in the cell which, in turn, controls the spectral transfer function. The Nimbus 6 PMR was launched in 1975 and is still operational.

At present at JPL we are working to develop new electro/optic correlation methods, as well as new algorithms for analysis of PMR data, and new atmospheric sensing applications for the device. Also, we have just successfully demonstrated in our laboratory that gas correlation spectrometer methods can be used to measure winds with a precision of 3 m/sec in the stratosphere to lower mesosphere. The method requires the measurement of the change in correlation arising from the Doppler shift and requires the use of the new electro/optic techniques we are developing.

The sensitivity of the PMR to minor constituent gas is a function of the thermal control between the gas and the background, the molecular absorption strength of the gas and the noise characteristics of the detector system. The fundamental equations controlling the detection are

$$R_{w.b.} = \int_{\nu} T_{\nu} \, B_{\nu\theta} \, \exp\left\{ - \int \alpha \, dh \right\} d\nu$$

$$R_{pmr} = \int_{\nu} F_{\nu} \, B_{\nu\theta} \, \exp\left\{ - \int \alpha \, dh \right\} d\nu$$

where T = transfer function for the filter
B_ν = Planck function for temperature θ at frequency ν
α = absorption coefficient of gas
h = altitude
F_ν = transfer function for PMR

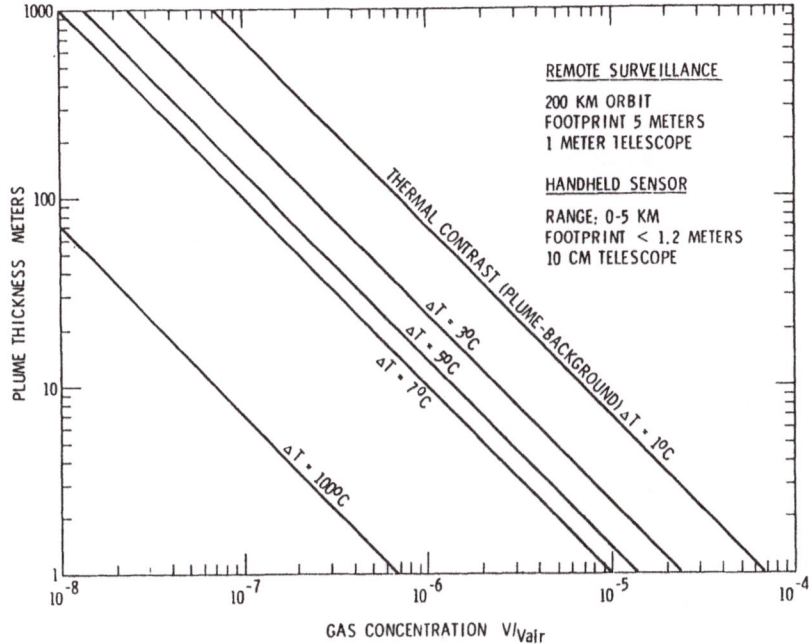

Figure 3 MINIMUM DETECTABLE CONCENTRATION

In Figure 3 we show a family of curves illustrating the sensitivity of the PMR to different concentrations of gas. These curves show the dependence of the sensitivity on thermal contrast (the difference in temperature between the gas being sensed and the background radiation). The sensitivity also depends on the optical parameters of the sensor since these control the energy grasp; this dependence is indicated in the two sets of curves corresponding to observation from a low orbit satellite and from a ground-based compact short range sensor.

Summary

We have briefly described a passive sensor for the remote detection of foreign gas species and minor constituents in the atmosphere. The technique shows considerable promise for long term monitoring of the earth from geosynchronous and polar orbiting satellites as well as having potential for use as a hand held field sensor. While more than a dozen such sensors have flown on satellites to date, field and laboratory tests are still required for verification of sensor performance in detecting the gaseous agents of particular interest to the military community. Supporting spectroscopic data on these agents is also required. Although some work has already been completed, further study is needed of the sensors proficiency in rejecting simulants and obscurants.

Part 3

UV-Visible DIAL Techniques

3.1 Atmospheric Pressure and Temperature Profiling Using Near IR Differential Absorption Lidar

C. Laurence Korb, Geary K. Schwemmer, and Mark Dombrowski

NASA Goddard Space Flight Center, Greenbelt, MD 20771, USA

Chi Y. Weng

Science Systems and Applications, Inc., Seabrook, MD 20706, USA

Introduction

This paper describes differential absorption lidar techniques for remotely measuring the atmospheric temperature and pressure profile, surface pressure, and cloud top pressure-height. The approach to the pressure measurements[1] utilizes a high resolution measurement of absorption in the wings of lines in the oxygen A band where the absorption is highly pressure sensitive through the mechanism of collisional line broadening. The approach for temperature[2] uses a measurement of the absorption at the center of a selected line in the oxygen A band which originates from a quantum state with high ground state energy. The population of the state depends strongly on temperature through the Boltzmann term which produces a highly sensitive temperature determination. Oxygen is used for these measurements since it is uniformly mixed in the atmosphere, which greatly simplifies the measurement approach, and has lines with appropriate strength and energy levels. Also, it is located in a spectral region (760 nm) easily accessible using tunable solid state and dye lasers and efficient detectors.

The theory of temperature measurements and simulation results are given which show accuracies of better than 1K with 2 km vertical resolution from a Space Shuttle platform. The basic theory of the pressure measurement is also given and Shuttle simulations are presented which show accuracies of better than 0.3% for 1 km vertical resolution. The results of horizontal path measurements of both temperature and pressure made with high resolution continuous wave (CW) lasers are presented. Also, a description is given of a lidar system which has recently been built for measuring vertical profiles of temperature and pressure from ground and aircraft platforms.

Temperature

The absorption coefficient of a gas species in the atmosphere at frequency v is given as

$$K(v - v_o) = qnS(T) f(v - v_o) \qquad (1)$$

where q is the molecular mixing ratio of the species, n is the density at pressure p and temperature T, S(T) is the line strength, and $f(v - v_o)$ is the line shape at frequency v. Oxygen is one of the major atmospheric species and is uniformly mixed in the troposphere and lower stratosphere except for small variations introduced by varying water vapor mixing ratio which can be accounted for as $q = q_o (1 - q')$ where q_o is the mixing ratio of oxygen in the dry atmosphere and q' is the mixing ratio of water vapor. The temperature dependence of the absorption coefficient arises mainly through the temperature dependence of the line strength. For oxygen, it is given as

$$S(T) = S(T_o) \frac{T_o}{T} \exp\left[-\frac{E}{k} \left(\frac{1}{T} - \frac{1}{T_o} \right) \right] \qquad (2)$$

120

where E is the rotational energy in the ground state of the transition, k is Boltzmann's constant, and T_o is standard temperature.

In general, the line shape associated with gaseous absorption in the atmosphere can be described by the Voigt line profile which is a convolution of independent collision broadened and Doppler line profiles. At line center, which is the spectral region used for the temperature measurements, the Voigt profile reduces to a relatively simple form[3]

$$f(o) = 2 \frac{f_o \exp(a^2)}{\sqrt{\pi}} \int_a^\infty \exp(-y^2)\, dy \tag{3}$$

where

$$f_o = \frac{1}{b_d} \sqrt{\frac{\ell n2}{\pi}}$$

$$a = (b_c/b_d) \sqrt{\ell n2}$$

and b_c is the collision broadened line half-width which is directly proportional to pressure, and b_d is the Doppler broadened line half-width. From kinetic theory, b_d and b_c have temperature dependencies which are directly and inversely proportional to the square root of the temperature, respectively. The temperature dependence of the absorption coefficient may then be determined from its derivative using Eqs. (1-3) and is given as

$$\frac{dK(o)}{K(o)} = \left(\frac{E}{kT} - \frac{5}{2} - 2a^2 + \frac{a}{\exp(a^2) \int_a^\infty \exp(-y^2)\, dy} \right) dT/T \tag{4}$$

The expression in parentheses in Eq. (4) has values between E/kT − 5/2 and E/kT − 3/2 in the limiting cases of very low and very high pressures. Thus, absorption lines for which E/kT is much greater than 3/2 will exhibit a strong dependence of absorption coefficient on temperature. For the P_{P29} line of oxygen at 12,999.95 cm^{-1}, E/kT is 7.5 at a temperature of 250°K, and the line behaves as if it had a net temperature dependence of $T^{5.6}$.

The absorption coefficient is found experimentally from the ratio of the on-line, E_ℓ (R), and reference, E_r (R), signals backscattered from adjacent ranges R and R + ΔR as[4]

$$\overline{K(R+\Delta R/2)} = \left(\frac{1}{2\Delta R} \right) \ell n \left[\frac{E_\ell (R) / E_r (R)}{E_\ell (R + \Delta R)/E_r (R + \Delta R)} \right] \tag{5}$$

where ΔR is the lidar vertical resolution. For the case of a monochromatic laser measurement, the temperature corresponding to the measured absorption coefficient can be found from Eqs. (1-3) using an iterative technique as

$$T_{i+1} = \frac{E/k}{\ell n \left[\dfrac{T_o S(T_o) q f (o) p \exp (E/kT_o)}{\overline{K} k T_i^2} \right]} \tag{6}$$

where T_{i+1} is the temperature determined from the i + 1 iteration from the prior estimate of the temperature T_i, and where the line shape is an implicit function of temperature. The convergence properties of Eq. (6) are rapid. For the more general case of measurements made with finite laser spectral bandpass about the line center, the problem of temperature determination cannot be evaluated analytically and a numerical approach is required. The results of a calculation of this type[5] are given in Figure 1 for a simulated aircraft or Space Shuttle experi-

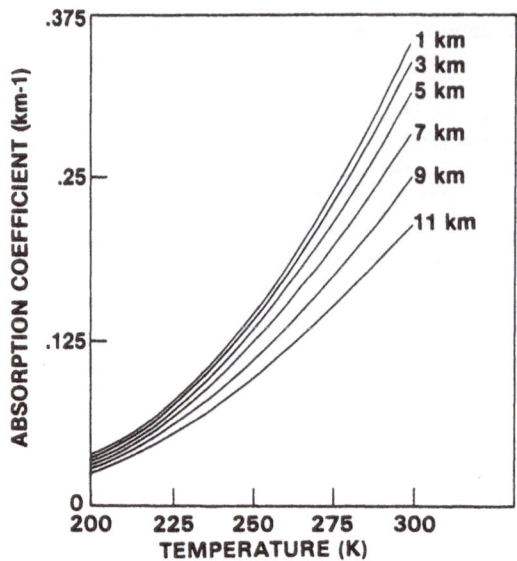

Figure 1. Dependence of the absorption coefficient of the P_{P29} line of oxygen on temperature for various altitudes.

ment for the P_{P29} line of oxygen at 12,999.95 cm^{-1} (769 nm). In this case, the absorption coefficient, K, has been calculated as a function of temperature using the Voigt line profile for various atmospheric layers spanning the altitude range from 0 to 14 km. A spectral bandwidth of 0.02 cm^{-1} was used for these calculations in order to simulate the spectral resolution achievable with current lidar systems.

As an example of the method of temperature determination using Figure 1, the measured absorption coefficient is first located on the vertical axis. The temperature corresponding to the measured value is then found using the appropriate curve corresponding to the altitude of the measurement. These curves were calculated using estimated values of the atmospheric pressure. The absorption is only weakly pressure dependent for measurements centered on a line. Analysis shows[5] that for a 1 percent pressure uncertainty, expected errors in temperature vary from 0.05 K for measurements made near ground level to 0.25 K for measurements at a height of 14 km. Corrections for estimated water vapor amount may also be applied with resulting residual errors of the order of 0.1 K.

Pressure

A measurement in the far wing of the line profile is used for the determination of pressure in order to obtain the maximum variation of absorption with pressure. For the line wing, the Voigt line profile is well represented by a power series expansion of the line width in terms of distance from the line center[6]

$$f(v - v_0) = \frac{b_c}{\pi(v - v_0)^2}\left[1 - \left(\frac{b_c}{v - v_0}\right)^2 + \frac{3}{2\ln 2}\left(\frac{b_d}{v - v_0}\right)^2\right] \quad (7)$$

As may be seen from Eq. (7), at distances greater than several line half-widths from line center, the line profile is dominated by the first term which has the form of a collision broadened line profile in the far wing of a line.

Although the pressure measurements could be made using individual atmospheric layers as was done for temperature, it is much more accurate and efficient to use an integrated path measurement. In this case, the measured absorption over the entire range from the laser to the atmospheric layer of interest is used. It is possible to use this approach since the pressures at various heights in the atmosphere are related by the hydrostatic equation. This same technique may also be used with backscatter from clouds and the earth's surface to obtain either cloud top pressure-height or surface pressure.

The temperature dependence of the pressure measurement can be made very small by choosing absorption lines with energy levels near $E = (3/2)k\overline{T}$, where $\overline{T}$ is an appropriate average temperature in the troposphere. For these lines, it can be shown that the integrated path absorption measured from a satellite to altitude Z is given by

$$\int_{\infty}^{Z} K \, dz = \frac{ap^2(Z)}{(v - v_0)^2} \left[1 - \frac{b_c^2(\overline{T})}{2(v - v_0)^2} \frac{p^2(Z)}{p_0^2} + \frac{3}{2\ln 2} \frac{b_d^2(\overline{T})}{(v - v_0)^2} \right] \quad (8)$$

where the parameter a exhibits a very weak temperature dependence and includes the molecular parameters of the absorption line. Thus, the integrated path absorption coefficient is directly proportional to the square of the pressure to first order, with higher order correction terms contributing only a few percent. Eq. (8) may be solved for pressure versus height using an algebraic technique.

The pressure determination using a wing measurement is not only extremely sensitive to pressure variations, but is also extremely sensitive to small laser frequency variations. The frequency stability required for an accurate determination of pressure can be reduced to 0.005-0.01 cm^{-1}, an improvement of up to a factor of 100, using a trough absorption region between two strong oxygen lines. The resulting absorption coefficient is flat over a significant frequency range, but still allows a measurement in the line wing to be made.

In Figure 2, the integrated absorption coefficient of oxygen in the region near 13,153 cm^{-1} (760 nm) is shown for two-way atmospheric paths from space to ground level, and also to an

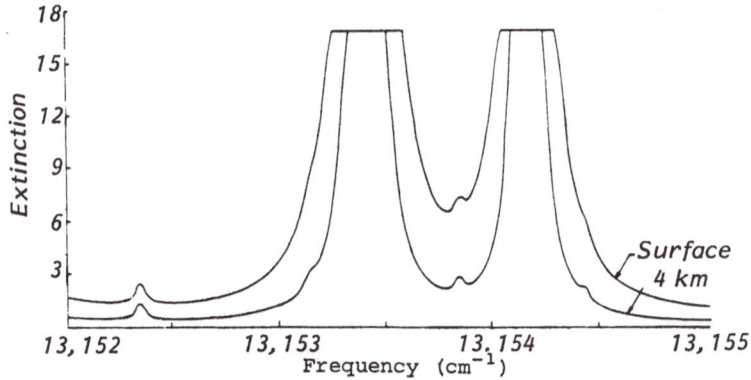

Figure 2. Integrated absorption coefficient (extinction) for oxygen for a two-way vertical path from space to ground level and to an altitude of 4 km.

altitude of 4 km. As shown, trough positions occur at a number of frequencies, including the regions near 13,153.8 cm^{-1}, and 13,152.5 cm^{-1}. For these frequencies, a 2 K shift in the entire atmospheric temperature structure produces errors in the deduced pressure profile of the order of only 0.05 percent.

Results

Simulations of the performance of these lidar techniques have been conducted using the Voigt profile and numerical analysis methods for ground-based, aircraft and Shuttle-based systems. This brief discussion is limited to the effects of noise on Shuttle-based measurements. Figure 3 shows the temperature accuracy for the P_{P29} line of oxygen for a Shuttle altitude of 200 km for the following conditions: a 1 m^2 collector, an overall collection and detection efficiency of 12 percent, a laser energy of 0.1 J per shot, a laser bandwidth of 0.02 cm^{-1}, and a 2 km vertical resolution. A 350 shot average is employed which yields a horizontal resolu-

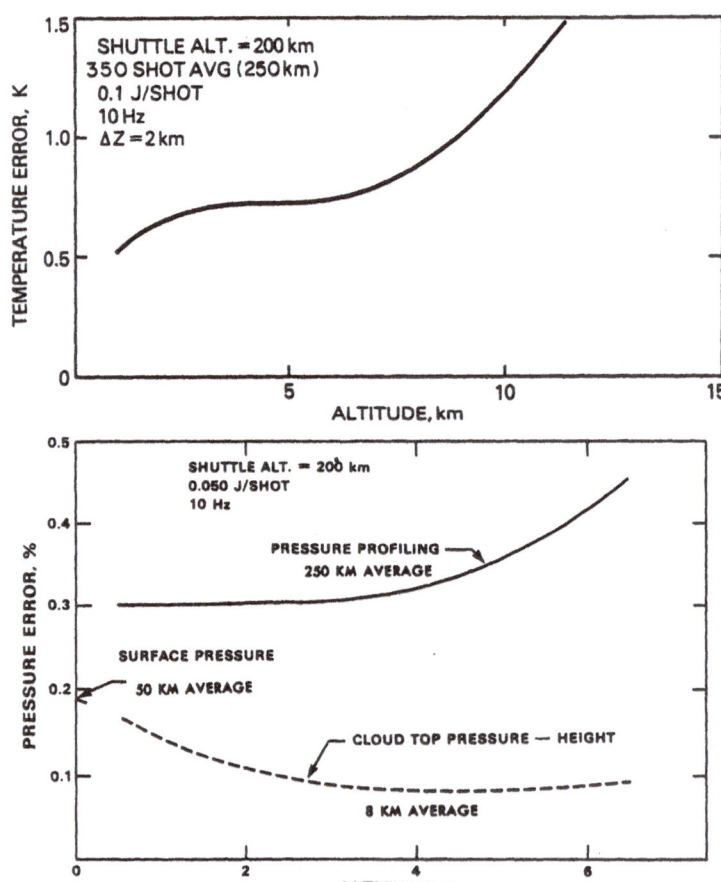

Figure 3. Accuracy of simulated temperature profiling experiment from shuttle for the P_{P29} line of oxygen at 12,999.949 cm^{-1} for $\Delta v = 0.02$ cm^{-1}.

Figure 4. Accuracy of simulated pressure experiments from shuttle for indicated horizontal resolutions.

tion of 250 km at 10 Hz laser operation. These specifications correspond to those of lasers and detectors installed in our lidar system at Goddard, while the telescope corresponds to that proposed for a Space Shuttle lidar facility.[7] As shown in the figure, the temperature errors are near 0.5 K for measurements in the boundary layer and gradually rise to 1 K near 10 km. For the 10-20 km height range, use of the P_{P27} line yields similar accuracies with a slightly poorer vertical resolution.

The results of simulations of the pressure experiments are shown in Figure 4 using the trough absorption at 13,153.8 cm^{-1}, with a laser energy of 0.05 J per shot. As shown, surface pressure can be determined with errors less than 0.2 percent for a 50 km horizontal resolution while measurements from cloud tops give errors in the 0.1 percent range for an 8 km horizontal resolution. The errors in determining the pressure profile are of the order of 0.3 percent in the lower atmosphere for a vertical resolution of 1 km with a horizontal resolution of 250 km.

In order to demonstrate the viability of the temperature and pressure techniques, we have conducted horizontal path atmospheric experiments.[8,9] These experiments used continuous wave (CW) dye lasers to measure the absorption over a 2 km round trip path. A laser with a resolution better than 10^{-3} cm^{-1} was used for the on-line measurements, while a lower resolution, 10^{-2} cm^{-1}, laser was used simultaneously to make the off-line, reference, measurements. Figure 5 shows a comparison of laser and ground truth measurements of temperature taken over a 1 month period. The standard deviation of this data is 1.5 K. It is believed that the single point measurement of ground truth near the transmitter did not properly represent the average temperature of the 2 km path and contributes most of the observed error. The noise level of the laser data was less than 1 percent, which corresponds to better than 0.5 K measurement accuracy. The standard deviation of all laser measurements taken on a particular day was also found to be 0.5 K, which is indicative of the accuracy of this experiment. Measurements of surface pressure have also been made with the CW laser system using a trough position near 13,156 cm^{-1} in conjunction with a nearby reference. Comparisons of the laser measurements with mercury barometer readings give accuracies better than 2.0 mb (0.20%) with noise levels of 1 mb.

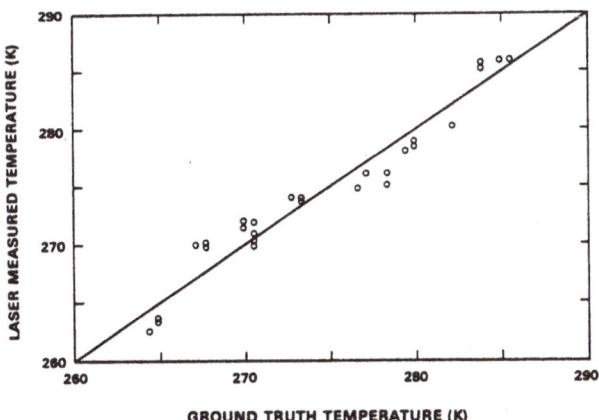

Figure 5. CW laser measurements of the average temperature of a 2 km atmospheric path compared to single point thermistor measurement.

A high resolution lidar system for measuring pressure and temperature profiles from the ground and from aircraft has been recently developed. Figure 6 is a block diagram of the system which utilizes two narrowband lasers, one high resolution (0.02 cm^{-1}) and one medium resolution (0.05 cm^{-1}) for the on-line and reference measurements, respectively. The lasers are fired with a separation of approximately 100 μsec to temporally separate the two return signals which are closely spaced spectrally. A spectral monitor uses the fringes from two Fizeau etalons to measure the spectral distribution and frequency of the transmitted on-line laser signal. This information is used to correct for the effects of laser finite bandwidth and frequency shifts. A narrowband, tunable, solid state Alexandrite laser[10,11] is now being integrated into the system. It can be used to replace either of two ruby pumped dye lasers presently installed. The Alexandrite laser has a spectral bandwidth of better than 0.02 cm^{-1}, an output of 100 mJ per pulse at a 10 Hz repetition rate, a pulse length of 130 nsec, and is tunable from 700 to 800 nm. Compared with the ruby/dye laser system, the Alexandrite laser is smaller and more efficient. At present, the lidar system repetition rate is limited to 1 Hz by the ruby laser, but is capable of being increased to 10 Hz when a second Alexandrite laser is installed. The system will measure profiles up to 4 km altitude when used for ground-based measurements and will provide profiles for most of the troposphere when flown in a medium altitude aircraft.

In conclusion, the theory and methodology of remote measurements of temperature and pressure in the atmosphere has been described using differential absorption lidar techniques. Simulations of temperature and pressure profile experiments from the space Shuttle indicate accuracies in the range 0.5 – 1.0 K, and 0.1 – 0.3 percent, respectively. The results of CW laser experiments for horizontal atmospheric paths have demonstrated the basic feasibility of these experiments. The development of a lidar system for conducting measurements from ground-based and aircraft platforms is currently underway along with the development of efficient solid state tunable lasers for pulsed operations in the 760 nm spectral region.

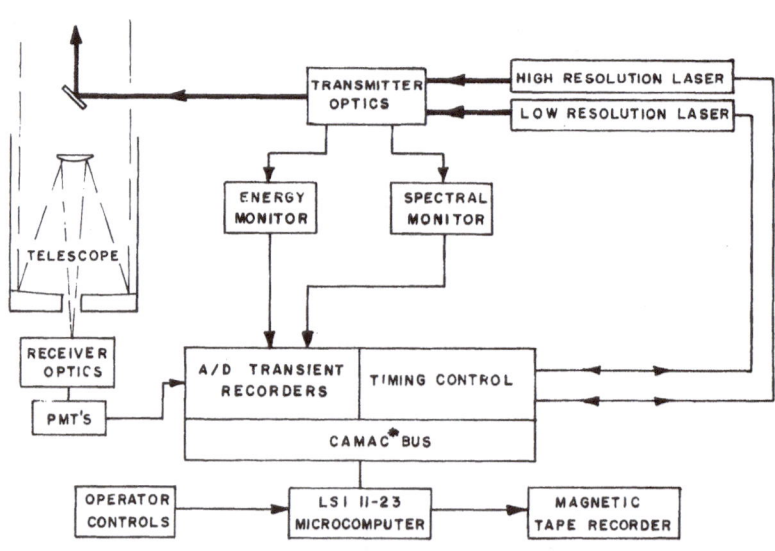

Figure 6. Lidar system layout. Main features include two narrowband tunable pulsed lasers, 45 cm diam. telescope, microprocessor controller and CAMAC data system.

REFERENCES

1. C. L. Korb, "A laser technique for the remote measurement of pressure in the tropo-sphere", 8th International Laser Radar Conference, Conf. Abs., Drexel University, Phila-delphia, PA (June 1977).

2. C. L. Korb and C. Y. Weng, "A two wavelength lidar technique for the measurement of atmospheric temperature profiles", 9th International Laser Radar Conference, Conf. Abs., Munich, Federal Republic of Germany (July 1979).

3. S. S. Penner, Quantitative Molecular Spectroscopy and Gas Emissivities, Chap. 3, Addison-Wesley Publ. Co., Inc., Reading, MA (1959).

4. R. M. Schotland, "Errors in the LIDAR measurement of atmospheric gases by differential absorption", Journal of Applied Meteorology, 13, 71 (1974).

5. C. L. Korb and C. Y. Weng, "A theoretical study of a two wavelength lidar technique for the measurement of atmospheric temperature profiles", Journal of Applied Meteoro-logy, (1982).

6. G. N. Plass and D. J. Fivel, "Influence of Doppler effect and damping on line-absorption coefficient and atmospheric radiation transfer", Astrophys. J., 117, 225 (1953).

7. NASA SP-433, "Shuttle Atmospheric Lidar Research Program", 220 pages (1979).

8. J. E. Kalshoven, Jr., C. L. Korb, G. K. Schwemmer, and M. Dombrowski, "Laser Remote Sensing of Atmospheric Temperature by Observing Resonant Absorption of Oxygen", Applied Optics, 20, 1967 (1981).

9. C. L. Korb, J. E. Kalshoven and C. Y. Weng, "A lidar technique for the measurement of atmospheric pressure profiles", Transactions of the American Geophysical Union, Spring Meeting, Washington, DC (May 1979).

10. J. C. Walling, O. G. Peterson, H. P. Jenssen, R. C. Morris, and E. W. O'Dell, "Tunable Alexandrite Lasers", IEEE J. of Quantum Electronics, QE-16, 1302 (1980).

11. C. L. Sam, J. C. Walling, H. P. Jenssen, R. C. Morris, E. W. O'Dell, "Characteristics of Alexandrite Lasers in Q-switched and Tuned Operations", Proceedings, Soc. of Photo-Optical and Instr. Eng., 247, 130 (1980).

3.2 Ground-Based Ultraviolet Differential Absorption Lidar (DIAL) System and Measurements

J.G. Hawley, L.D. Fletcher, and G.F. Wallace

SRI International Menlo Park, CA 94025, USA

Introduction

A ground-based Differential Absorption Lidar (DIAL) system was developed by SRI International to measure the atmospheric distribution of several pollutants: SO_2, O_3, and NO_2, gases commonly associated with fossil fuel combustion. Although earlier DIAL research was funded by the National Science Foundation (Grant 1974, 1975), the system described in this paper was developed for the Electric Power Research Institute (EPRI), primarily for use in their Plume Model Validation (PMV) Project (Hilst, 1978). The DIAL system is also being used by utility companies to characterize power plant plume diffusion and is available through SRI to other institutions for similar studies.

The DIAL system determines gas concentrations as a function of range by measuring the range-resolved differences in absorption of backscattered laser light at ultraviolet and visible wavelengths. The measurements depend upon laser-based data on the absorption coefficients of the gases that are now being provided by many investigators. The gas concentration as a function of range is related to the logarithm of the backscattered power, as follows:

$$C(R) = \frac{1}{2\Delta k} \cdot \frac{d}{dR} \left[\ln \left(\frac{I_v(R)}{I_p(R)} \right) \right] \quad ,$$

where $C(R)$ is the concentration as a function of range; I_v and I_p are received powers at range R, at the valley (off-line) and peak (on-line) wavelengths with background light levels subtracted; and Δk is the difference in absorption coefficient for the two wavelengths.

Because the technique differentiates measurements of received powers, it can be quite sensitive to noise. Temporal and spatial averaging is used to increase the concentration's signal-to-noise ratio (SNR). Tem-

128

poral averages are obtained by separately averaging I_v and I_p terms, then forming the ratio in the equation. Spatial averaging is performed by a discrete convolution of the logarithm of the ratio with a differential Gaussian function of given half-width. This is equivalent to differentiating and smoothing with a discrete Gaussian function in one step. The half-width of the function defines the range resolution, which usually runs between 15 and 200 m, depending upon atmospheric conditions. For example, high gas concentrations near the power plant stack can be measured at minimum range resolution. Typical averaging times vary between 2.4 s (24 pulse pairs) and 5 min (720 pulse pairs). The range of the DIAL instrument varies between 2.5 to 3 km, depending upon concentration and aerosol levels. SRI's DIAL system has a 500-m inner range, due to unequal convergence of both beams of the transmitter with the receiver field of view.

DIAL System Description

SRI's mobile DIAL system consists of a dual laser system, a telescopic receiver system, a minicomputer for data acquisition, a microcomputer for scanning, and an electronic system for timing and control. This system is housed in a self-contained van with appropriate power and environmental control (Figure 1). Figure 2 is a block diagram of the system, and Table 1 lists the DIAL system specifications.

The two wavelengths are provided by a pair of laser-pumped dye lasers (Quanta-Ray, Inc., Mountain View, CA). The Nd: YAG pump lasers emit about 700 mJ of energy at 1.064 microns at a 10-pulse/s rate. The energy from each pump laser is frequency-doubled in a deuterated type II potassium dihydrogen crystal (KDP) to 0.532 microns in order to pump the dye lasers with about 270 mJ of energy.

For SO_2 or O_3 remote measurements in the ultraviolet spectrum, the dye lasers are tuned to approximately 0.6 microns using the appropriate dye and angle-of-diffraction grating in the dye laser cavity. Each dye laser beam is then frequency-doubled in another crystal to form the appropriate ultraviolet wavelength with about 10 mJ of energy. The exact wavelengths are shown in Table 1.

High-resolution absorption coefficient data as a function of wavelength for SO_2 are given by Thompson (1975), Brassington (1981), and Woods et al. (1980). The absorption coefficient data for O_3 are given by Griggs (1968). By carefully selecting wavelengths, both SO_2 and O_3 may be measured with minimum interference by each other or other gases.

Figure 1　Differential Absorption Lidar (DIAL) for remote measurements of SO_2, O_3, and NO_2

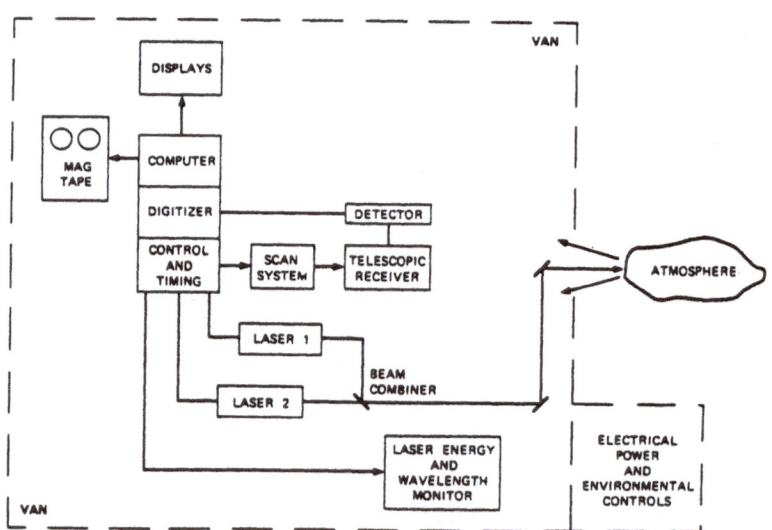

Figure 2　Block diagram of DIAL system for remote measurement of SO_2, O_3, and NO_2

Table 1 DIAL SYSTEM SPECIFICATIONS

Parameter	Specified Value
Transmitter	
Wavelengths and differential absorption coefficients for highest sensitivity	0.29938 and 0.30005 μm (SO_2) 25.40 cm^{-1} – atm^{-1} 0.2914 and 0.30055 μm (O_3) 20.89 cm^{-1} – atm^{-1} 0.4465 and 0.4481 μm (NO_2) 7.2 cm^{-1} – atm^{-1}
Pulse energy	10 mJ
Pulse length	7 ns
Two-wavelength pulse spacing	100 μs
Pulse repetition frequency	10 pps
Beam divergence	1 mrad or 0.3 mrad
Receiver	
Telescope	Newtonian, 50.8 cm diameter by 127 cm focal length primary and a 20 cm diameter secondary with aluminized/MgF surfaces, coaxial configuration
Field of view	2 mrad
Optical filtering	15% transmission, 10-nm wide (SO_2) 10% transmission, 12.5-nm wide (O_3)
Detector	Amperex 2254-B photomultiplier Max. gain = 3×10^7, QE = 17% at 0.3 μ Step gain switched
Data System	
Backscatter digitization	Sequential two-wavelength sampling at 10 MHz with 100-μs delay between wavelength pairs. 512 10-bit words per pulse pair
Processing	HP-1000 system and 8-bit microcomputer
Program storage	Cassette tape and hard disk
Recording	Dual 9-track magnetic tape
Display	Real-time log-range squared corrected backscatter for two-wavelengths versus range (TV)
	Real-time path integrated concentration versus range (TV)
Visual scene	Silicon diode array television with optional recording
Data	
Gaseous concentration	Parts per billion and standard deviation as a function of range from 0.5 to 3 km
Van location	UTM coordinates within 50 m
Date and time	
Range resolution	15 to 200 m
Van heading	±1°
Azimuth angle	±120°
Elevation angle	0 – 90°
Angle increments	0.02° min
Number backscatter returns integrated per angle	1 – 1000

For NO_2 remote measurements in the visible (blue) spectrum, the dye lasers can be tuned to approximately 0.45 microns (exact wavelengths are shown in Table 1). In this case, each pump laser is frequency-tripled to 0.355 microns using two KDP crystals. The dye laser radiation of about 10 mJ is then used without further processing. The absorption coefficient data for NO_2 are given by Wilkerson (1974). NO_2 is not subject to interference by other gases. Although configured for NO_2 we have not yet made measurements with this system. We have made measurements with a previous system (Baumgartner 1979).

The dual wavelengths in all cases are combined in a polarization beamsplitter and transmitted with a 100-μs delay, which minimizes the effect of change in concentration of the plume with time and also reduces the noise associated with scintillation effects in the atmosphere (Killinger 1981). The backscattered returns are then processed sequentially.

The receiver is a 0.5-m diameter scannable Newtonian telescope. The dual laser beams are transmitted coaxially with the receiver. A mirror system brings the beams to the front of the telescope. The detector, a twelve-stage gain-switched photomultiplier tube, is mounted about 10 cm aft of the focal plane of the primary mirror. A field-stop aperture at the focal plane limits the field of view to 2 mrad, and a recollimating 50-mm focal length quartz lens transfers the receiver radiation through an interference filter to the PMT photocathode. A television camera mounted on the receiver provides scene viewing and safety control.

The signal from the anode of the photomultiplier tube is digitized and stored by a 10-bit, 10-MHz analog-to-digital converter. The lidar returns are sequentially stored in the first 256 words of memory for the first laser and, 100 μs later, are stored for the second laser in the second 256 words. The photomultiplier tube is gain-switched 2.5 μs prior to each laser firing and for 35 μs to increase the gain momentarily during the measurement. The average background level measured just prior to the laser firing is subtracted from the lidar return data.

The minicomputer (an HP-1000 system), which operates under the basic control system to allow for maximum speed, writes the lidar returns on tape and computes the concentration-path length integral in ppm-m to be displayed on a television monitor as a function of range. The display is refreshed every 5 s, giving the operators real-time information on plume behavior. The displays have been useful for directing other instruments into the plume area.

During DIAL system design, SRI housed the laser system in a high-quality van, rather than making the complex laser/receiver system sufficiently rugged to endure a less sophisticated conveyance. SRI used a 40-ft-long highway van, which has special air-ride suspension and can be towed by a diesel tractor. In the van's rear half, the laser system is mounted on a 12- by 4-ft optical table, which is mounted on separate suspension. The 0.5-m articulating telescopic receiver base is also mated to the table; the telescope protrudes through the lowered roof of the van (see Figure 1). Collapsible walls and a sliding roof are used to protect the telescope during transport.

The front half of the van carries the signal acquisition computer and displays. Power is provided by a 40-kW diesel generator mounted on the tractor. For deployment, the tractor is pulled forward to decouple generator vibrations from the van, and legs are lowered to level and stabilize the van. Total set-up time is under 1 h.

Representative Results (SO_2)

The DIAL system was used in EPRI's PMV project to measure both plume height on flat terrain 1 to 3 km downwind of a coal-burning power plant and plume ground-level pattern during convective instability. An example of the main data product of such measurements (SO_2 concentration as a function of range) is given in Figure 3, which shows varying plume

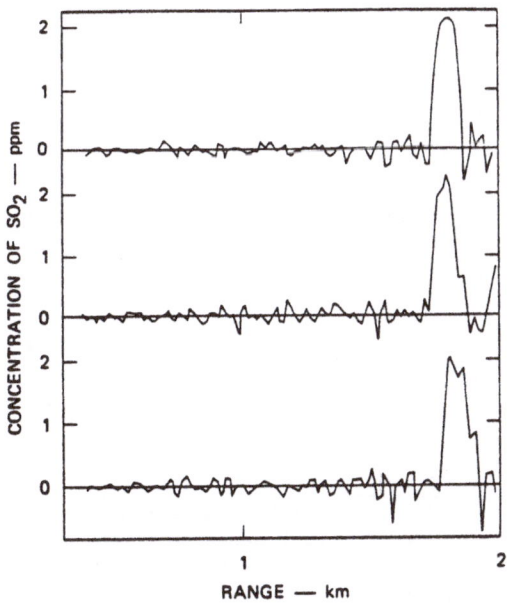

Figure 3 Three measurements of SO_2 profiles in a plume downwind of the stack. (Range to the plume is 1.8 km). 15-m range resolution

133

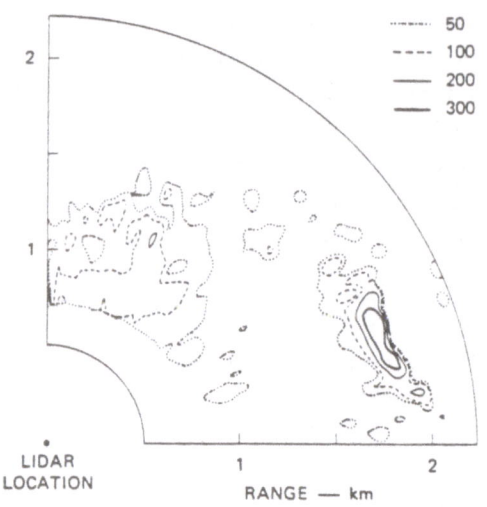

Figure 4 Ground-level map of equal concentrations of SO$_2$ downwind of a coal-fired power plant using the horizontally scanning DIAL system. Plant is located 3 km to the right of the system. (Units are parts per billion)

profiles averaged for 2 min and spaced about 5 min apart. By factoring in the elevation angle, the plume height can be calculated.

Plume touchdown pattern results were similary achieved by horizontally scanning the lidar at a 1° elevation above the horizon in a 90° arc in the plume touchdown zone. A single scan is shown in Figure 4. In this case, 45 profiles were made spaced 2° in azimuth. Each profile consisted of 2.4 s of data (24 lidar return pairs). In this study, six to seven profiles were accomplished each hour to form hourly averages of plume behavior. Signal-to-noise requirements imposed by scanning required some spatial integration by smoothing the concentration versus range data with a 200-m half-width Gaussian filter.

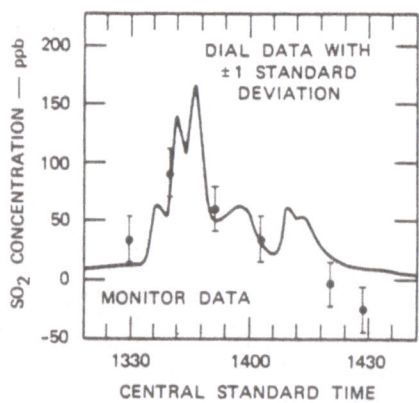

Figure 5

SO$_2$ concentration as a function of time for DIAL and a point monitor. Kincaid, IL, August 17, 1980. (Range = 1.94 km.)

134

Additional data in Figure 5 compare point-monitoring station data with DIAL system data during a horizontal-scanning measurement. The point of closest approach between the lidar beam and the point monitor was 35 m vertically at 1.94-km range. Agreement between the two techniques was very good, with a correlation coefficient of 0.93 and agreement between hourly-averaged values of 6%.

Other tests have further validated DIAL. One test involved a remote sample chamber placed 500 m from the van and into which known amounts of SO_2 were placed. Lidar measurements of differential absorption of these amounts have shown agreement within 10%. Another test involved equal wavelength setting of the tuneable lasers so as to register a zero differential absorption coefficient. In this way, no net concentration of gas is measured, thereby providing a useful means of diagnosing any system errors.

Representative Results (O_3)

Figure 6 shows a sample of correlated ozone data taken by an aircraft-borne monitor (Dasibi 1003) flown in an ascending spiral around a vertical lidar beam. The ozone source was the urban boundary layer over San Jose, California, during a smoggy day. The DIAL profile was made with 360 pulse pairs over a 2-min period, while the aircraft took 12 min to make its measurement. Two-hundred-meter smoothing was applied to these data. SRI is continuing to build this data set by taking hourly averages between the measurement systems. This measurement method could be useful in forecasting smog episodes since ozone is not easily

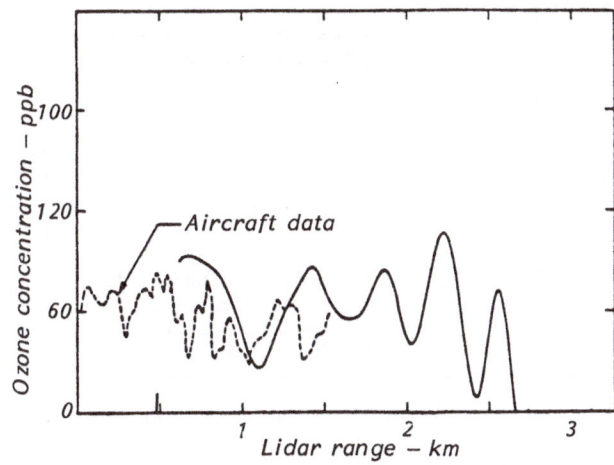

Figure 6
DIAL ozone vertical profile

measured aloft, where it can be stored forming a very polluted, stagnant air mass over cities, especially in the Western United States.

Summary

The ultraviolet ground-based DIAL system measured the spatial distribution of several important pollutants with expected accuracy. Our current DIAL programs are outlined in Table 2. We plan to continue with the EPRI program in moderately-complex and complex terrain. In addition, we may expand the capability of the system to measure range-resolved water vapor, temperature, or hydrocarbons (path-averaged). With the addition of infrared equipment, the DIAL system could measure a large number of additional materials such as ammonia, which is of current interest to utilities because of its combination with SO_2 and NO_2 to form sulfates.

Table 2 SRI DIAL PROGRAMS (AS OF FEBRUARY 1982)

Study	No. of Hours Collected	Comments
Power plant plume model validation project (SO_2)--Midwest	285	Flat terrain, coal-fired plant. Reliability[*] is 89%.
Refinery (SO_2)--Midwest	22	Boiler and process emissions, waste gas flare emissions
Urban ozone--San Francisco Bay Area	8	With an aircraft monitor
Urban--Los Angeles basin	5	Mobile measurements over 26 km
Power plant plume (SO_2)--Oxnard, Calif.	22	Oil-fired plant, sub-visible plume. In conjunction with an SO_2 correlation spectrometer

[*] Defined as the percent hours operational during a 66-day field program. Does not include weather-related delays.

BIBLIOGRAPHY

Baumgartner, R. A., L. D. Fletcher, and J. G. Hawley, "A Comparison of Lidar and Air Quality Station NO_2 Measurements," _Journal of the Air Pollution Control Association_, Vol. 29, No. 11 (November 1979).

Brassington, D. J., "Tuneable Dye Laser Measurements of the SO_2 Absorption Spectrum Between 290 and 317 nm," Central Electricity Research Laboratory Report RD/L/2055N81, Leatherhead, Surrey, U.K. (June 1981).

Grant, W. B., and R. D. Hake, "Calibrated Remote Measurements of SO_2 and O_3 Using Atmospheric Backscatter," _J. Appl Physics_, Vol. 46, No. 7 (July 1975).

Grant, W. B., R. D. Hake, E. M. Liston, R. C. Robbins, and E. K. Proctor, "Calibrated Remote Measurement of NO_2 Using the Differential Absorption Backscatter Technique," _Appl. Physics Letters_, Vol. 24, No. 11 (July 1974).

Griggs, M., "Absorption Coefficients of Ozone in the Ultraviolet and Visible Regions," _J. Chemical Physics_, Vol. 49, No. 2, pp. 857-859 (July 1968).

Hawley, J. G., "Characterization of the EPRI Differential Absorption Lidar (DIAL) System," EA-1267, Final Report, Project 862-14, Electric Power Research Institute, Palo Alto, California (December 1979).

Hawley, J. G., L. D. Fletcher, and G. F. Wallace, "A Mobile Differential Absorption Lidar (DIAL) for Range-Resolved Measurements of SO_2, O_3, and NO_2," Proceedings of the 10th International Laser Radar Conference, Silver Springs, Maryland (October 6-9, 1980).

Hilst, G. R., "Plume Model Validation," Workshop Summary Report, WS-78-99, Electric Power Research Institute, Palo Alto, California (October 19, 1978).

Hinkley, E. D., ed., "Laser Monitoring of the Atmosphere," _Topics in Applied Physics_, Vol. 14, Springer Verlag, New York (1976).

Kildal, H., and R. L. Byer, "Comparison of Laser Methods for Remote Detection of Atmospheric Pollutants, Proc. IEEE 19, 1644 (1971).

Killinger, D. K., and N. Meyuk, "Temporal Correlation Measurements of Pulsed Dual CO_2 Lidar Returns," _Optics Letters_, Vol. 6, No. 6, pp. 301-303 (June 1981).

Schotland, R. M., _Proceedings of Third Symposium on Remote Sensing of the Environment_, University of Michigan, Ann Arbor, Michigan, pp. 214-224 (1964).

Thompson, R. T., J. M. Hoell, and W. R. Wade, "Measurements of SO_2 Absorption Coefficients Using a Tuneable Dye Laser," _J. Appl. Physics_, Vo. 46, No. 7 (July 1975).

Wilkerson, T. D., B. Ercole, and F. S. Tompkins, "Absorption Spectra of Atmospheric Gases," Technical Note BN-748, Institute of Maryland, College Park, Maryland (February 1974).

Woods, P. T., B. W. Joliffe, and B. R. Marx, "High Resolution Spectroscopy of SO_2 Using a Frequency-Doubled Pulsed Dye Laser With Application to Remote Sensing of Atmospheric Pollutants," _Optics Communications_, No. 33, pp. 281-286 (1980).

3.3 Remote Sensing of Tropospheric Gases and Aerosols With an Airborne DIAL System

Edward V. Browell

NASA Langley Research Center, Hampton, VA 23665, USA

INTRODUCTION

A multipurpose airborne differential absorption lidar (DIAL) system has been recently developed at the NASA Langley Research Center to remotely measure the profiles of various gases and aerosols in diverse atmospheric investigations. The capability to rapidly determine the spatial distribution of gases such as ozone, water vapor, sulfur dioxide, and nitrogen dioxide and simultaneously measure the backscattering distribution of aerosols at several laser wavelengths provides the opportunity for developing an extensive data base for examining the complex interaction of atmospheric dynamics and chemistry.

Initial Langley development of the DIAL technique was aimed at ground-based investigations of water vapor[1] and sulfur dioxide.[2] These experiments were the first to demonstrate the DIAL technique in actual atmospheric measurements using the flexibility and efficiency of laser-pumped dye lasers. The knowledge gained from these ground-based DIAL experiments was used in the development of the airborne multipurpose DIAL system. The airborne lidar includes the flexibility to operate in the UV for measurements of ozone or sulfur dioxide, in the visible for nitrogen dioxide, and in the near-IR for water vapor, atmospheric temperature (using water vapor or oxygen absorption lines), and pressure (using oxygen lines). Aerosol backscatter investigations in the UV, visible, and near-IR can be conducted simultaneously with the DIAL measurements. The features of the NASA Langley multipurpose airborne DIAL system are functionally the same as those proposed for an early phase of the NASA Shuttle Lidar Program.[3,4,5] Thus, the experience gained with this system will be useful in a preliminary evaluation of several of the potential Shuttle Lidar investigations[4] and in the eventual design of a Shuttle Lidar system.

The recent emphasis with the airborne lidar has been in measurements of ozone, water vapor, and aerosol profiles. An understanding of the tropospheric ozone budget is essential to establishing a firm knowledge of tropospheric photochemistry and the potential impact of pollutants upon the photochemical system. High resolution remote measurements of water vapor profiles are important for applications such as the initialization of numerical weather forecast models in regions not covered by radiosonde data and in studies of latent heat flux and troposphere/stratosphere exchange, to name only a few. Aerosols can be readily used as a tracer of atmospheric dynamics. Aerosol distributions can provide information on the boundary layer mixing depth, condensation level, cloud-top boundary layer, and plume dispersion parameters. In addition to describing the multipurpose airborne DIAL, this paper discusses the first O_3 profile measurements made with an airborne DIAL system in the lower troposphere[6] and lower stratosphere.[7] The first airborne DIAL measurements of water vapor are presented,[8] and the measurement of aerosol profiles during a regional tropospheric flight experiment[6] are discussed. Potential airborne DIAL measurements of sulfur dioxide and nitrogen dioxide[2] and simultaneous water vapor and temperature[9,10] measurements are also briefly reviewed.

AIRBORNE DIAL SYSTEM

The multipurpose airborne lidar system uses the differential absorption lidar (DIAL) technique in the remote measurement of gas profiles. This technique has been discussed in detail by numerous authors[11-15]; thus only a brief review is presented here. The DIAL technique determines the average gas concentration over some selected range interval by analyzing the difference in lidar backscatter signals for laser wavelengths tuned "on" and "off" a molecular absorption peak of the gas under investigation. The value of the average gas concentration N between range R_1 and R_2 can be determined by taking the ratio of the lidar equations for the on and off wavelengths. This relationship is given by

$$ N = \frac{1}{2(R_2-R_1)(\sigma_{on}-\sigma_{off})} \quad \ell n \quad \frac{P_{off}(R_2)\ P_{on}(R_1)}{P_{off}(R_1)\ P_{on}(R_2)} $$

where $\sigma_{on}-\sigma_{off}$ is the difference between the absorption cross sections at the on and off wavelengths, and $P_{off}(R)$ and $P_{on}(R)$ are the powers received from range R for the on and off wavelengths, respectively. This analysis assumes that the aerosol and molecular scattering parameters are equal at the on and off DIAL wavelengths. If there is an interfering gas which does not have the same absorption coefficient at these wavelengths, the concentration of this gas must be known or determined by a separate experiment. For airborne DIAL measurements of ozone, the laser wavelengths used are in the Hartley continuum of ozone. The absorption cross sections for the on and off wavelengths at 286 and 300 nm are 21.4×10^{-19} and 4.0×10^{-19} cm[2], respectively.[16] The water vapor absorption lines used in the airborne DIAL measurements are in the 720 nm wavelength region.[17] Absorption cross sections in this wavelength region range from 5.0×10^{-24} to 90.1×10^{-24} cm[2].[18]

The airborne DIAL system uses two frequency-doubled Nd:YAG lasers to pump two high conversion efficiency dye lasers. Figure 1 shows a schematic of this system mounted in the NASA Wallops Electra aircraft. The on and off wavelengths which are used in the DIAL measurement are produced in sequential laser pulses with less than 100 μs separation. The backscattered lidar returns at the two wavelengths are sequentially detected by a photomultiplier tube, digitized, and stored on high-speed magnetic tape. Gas concentrations and aerosol backscattering profiles are calculated for each measurement in real time by a minicomputer.

Performance parameters of the system for operation in the UV and near-IR as listed in Table 1. The grating chosen for the dye laser was optimized for H_2O dial measurements near 724 nm where the laser linewidth requirements are more stringent than for O_3 and SO_2 measurements in the UV. The dye lasers are tunable over the wavelength range of the particular dye being used. Dielectric coated steering optics are used to direct the dye laser output through a 40-cm diameter quartz window (used for high UV transmittance) in the bottom or top of the aircraft. The receiver system is composed of a 36-cm diameter Cassegrain telescope and gateable photomultiplier tubes. Gains of each tube are switched

139

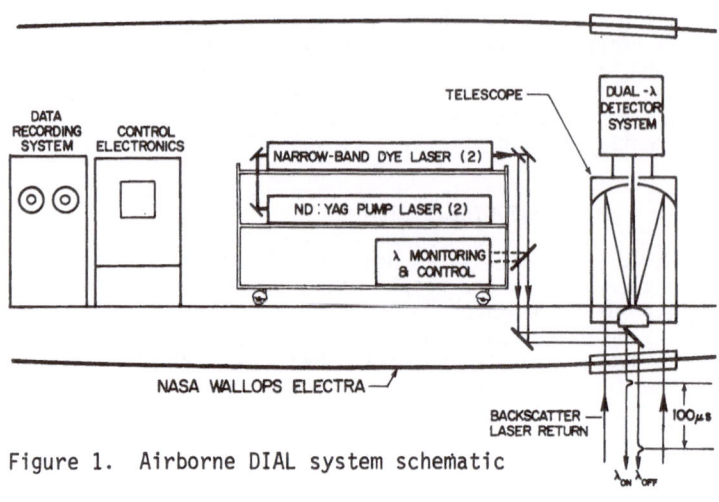

Figure 1. Airborne DIAL system schematic

Table I. Airborne DIAL system characteristics.

Transmitter:

    Two Pump Lasers -- Quantel Model 482

        Pulse Separation -- 100 μs

        Pulse Energy -- 350 mJ at 532 nm

        Repetition Rate -- 10 Hz

        Pulse Length -- 15 nm

    Two Dye Lasers -- Jobin Yvon Model HP-HR

	UV (near 300 nm)	Near-IR (near 720 nm)
Dye Output Energy	157 mJ/pulse near 600 nm	63 mJ/pulse
Doubled Dye Output Energy	47 mJ/pulse	--
Transmitted Laser Energy	40 mJ/pulse near 300 nm and 80 mJ/pulse near 600 nm	50 mJ/pulse
Laser linewidth	<4 pm	<2 pm

Receiver:

Area of Receiver	0.086 m^2	0.086 m^2
Receiver Efficiency to PM7	28%	29%
PMT Quantum Efficiency	29%	4.8%
Total Receiver Efficiency	8.1%	1.4%
Receiver Field of View	2 mrad	2 mrad

at selected intervals to optimize the digitization fo the lidar signal which varies over a large dynamic range. A separate photomultiplier tube is used for aerosol measurements at the dye fundamental wavelength near 600 nm when the frequency-doubled wavelengths near 300 nm are used for a DIAL measurement of O_3 or SO_2. The data system uses three 10-bit transient digitizers operating at a 10 MHz conversion rate to sequentially digitize the on- and off-line DIAL and aerosol return signals. The data are then stored on a 1600 bpi high-speed magnetic tape unit. A PDP 11/34 minicomputer is used for real-time calculation of gas concentration and aerosol profiles. Reduced data are displayed for operator experiment control.

AIRBORNE DIAL MEASUREMENTS

Lower Tropospheric Ozone

The first remote measurement of tropospheric ozone profiles from an aircraft was obtained with the NASA Langley airborne DIAL system in May 1980.[6] Measurements of ozone profiles were obtained on four flights in the vicinity of Wallops Island, Virginia, and the Chesapeake Bay between May 22 and June 5, 1980. The system was operated in a nadir directed mode from a nominal aircraft altitude of 3200 m. An instrumented Cessna 402 aircraft provided the principal means of obtaining in situ correlative ozone measurements. The O_3 concentration increased from about 82 ppb above the mixed layer top at 800 m to 100 ppb in the mixed layer. The airborne O_3 DIAL measurements were within 10 ppb of the in situ measurements.

A major field experiment with the Environmental Protection Agency was conducted in July-August 1980 to study large-scale pollution events in the

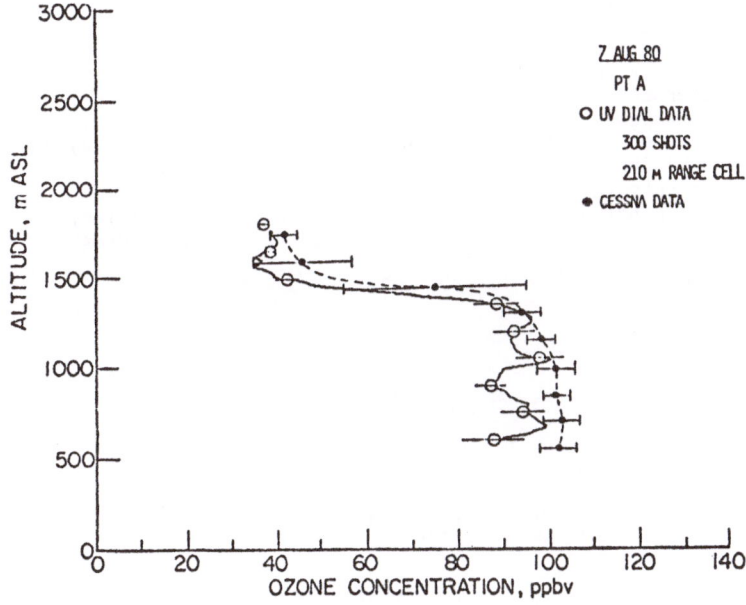

Figure 2.   Comparison of DIAL and in'situ O_3 measurements above and in the boundary layer

Northeast. The airborne DIAL system was used in the study of ozone and aerosols during this program.[19] The objective of this experiment included the characterization of persistent elevated pollution episodes (PEPE) and the evaluation of a four layer regional oxidant model. A comparison of O_3 measurements made with the airborne DIAL system and in situ instruments on a Cessna aircraft is given in Figure 2. The variation in O_3 concentration from 42 ppb above the mixed layer to 100 ppb within the mixed layer was measured remotely by the DIAL system from an altitude of 3200 m. The horizontal bars on the DIAL data represent the standard deviation for the average O_3 profile obtained from 300 laser shots. The Cessna data were obtained during an aircraft spiral performed in the vicinity of Snowhill, Maryland. The in situ data are shown as averages over 200 m vertical increments, and the standard deviation of all measurements in that altitude interval is given. The precision of the in situ measurements was estimated to be about 2 percent while the absolute accuracy was estimated at 10 percent. The DIAL and in situ O_3 measurements agreed very well in the free troposphere and in the boundary layer. Ozone DIAL measurements obtained during this field experiment will be employed to investigate the regional distribution, production, and transport of O_3 in elevated pollution episodes.

Aerosols

Aerosol backscatter profiles at 600 nm are also obtained simultaneously with the UV DIAL data.[6] The aerosol data are processed by subtracting the background signal from the lidar-plus-background signal and then multiplying by the range-squared, which removes the geometrical dependence of the lidar return signal with range. The resulting lidar backscatter profile is indicative of the distribution of aerosols below the aircraft. The backscatter signal level is converted into a 16-level gray scale display line, where stronger scattering is indicated by higher brightness. Sequential gray scale lines are then used to construct a picture of the aerosol vertical distribution along the Electra flight path. An example of this display technique is shown in Figure 3. This picture was produced from 600 laser shots along a 90-km flight track of the Electra. The ground reflection appears as a bright line at the bottom of the picture. The left side of the picture was obtained over land (Virginia) and the right side over the Chesapeake Bay. As shown on the left side of the picture, the lidar detects the presence of clouds and provides a direct measurement of cloud top height. When these clouds are optically thick, signals are obtained to limited ranges in the cloud, and ground returns

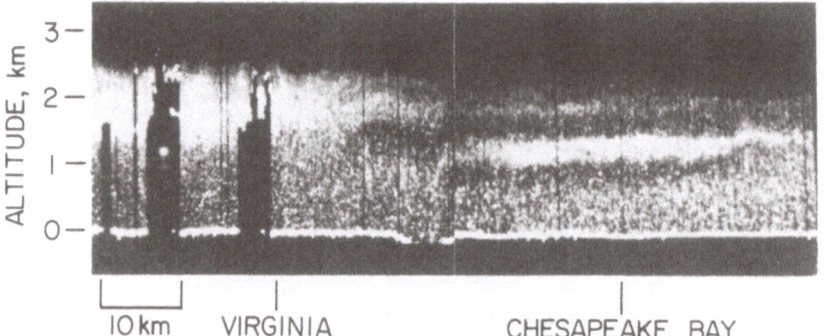

Figure 3. Intensity modulated display of aerosol data taken on July 24, 1980, at about 1500 EDT over Virginia and the Chesapeake Bay

cannot be seen. The abrupt increase in aerosol scattering above the cloud tops indicates a mixed layer height of about 2.5 km over land. It can also be seen from this picture that there is substantially less vertical mixing over the Chesapeake Bay. Since the wind was transporting material from left to right in this picture, the aerosols above the strong stable layer over water may have been advected from the higher boundary layer over land. Boundary layer and tropospheric dynamics can be readily studied using the aerosol distribution information available from the airborne DIAL system.

Ozone Profiles Near Tropopause

The first airborne DIAL measurements of O_3 and aerosol profiles in the tropopause region of the atmosphere were made on August 6, 1981, over Maine.[7] This was the first flight experiment conducted with the DIAL system operating in a zenith-viewing mode from the Electra. The Loring Air Force Base, Maine, site was selected for the tropopause experiment because of its location on the northern side of the jet stream. A National Weather Service station at Caribou, Maine, 17 miles from Loring AFB, was selected for an ozonesonde launch. An ozonesonde launch team from the Wallops Flight Center was flown to Loring AFB along with the DIAL system on the afternoon of August 6.

An ozonesonde was launched from Caribou at 2209 EDT. The ozonesonde data showed a tropopause located at a geopotential altitude of 11.1 km. Two prominent O_3 enhanced layers were unexpectedly found at altitudes of 12.5 and 13.5 km. The peak O_3 concentrations in these layers were found to be about twice the average concentration outside the layers. The depth of the lower layer was about 500 m, and the upper layer had a depth of only 350 m. The airborne DIAL system was flown at an altitude of 8230 m to make O_3 and aerosol profile measurements in and above the tropopause region. The flight over Maine was completed by flying over the ozonesonde launch site at 2348 EDT. A comparison of the DIAL O_3 profile and the ozonesonde data is shown in Figure 4. Also shown in the figure are the potential temperature data from the ozonesonde. The DIAL data represent the average O_3 profile obtained from 300 DIAL measurements along a 6 km horizontal path. The horizontal bars designate the standard deviation of the average O_3 profile. The magnitude and relative position of the O_3 layers are in excellent agreement between the DIAL and ozonesonde measurements. The altitude agreement for these layers is within 150 m, which is very good considering that the DIAL measurements were only in the vicinity of the ozonesonde ascent path. Features in the minimum between the two layers are also accurately determined by the DIAL system. Simultaneous lidar measurements of aerosol scattering were made at a wavelength of 600 nm. The O_3 layers seen by both the DIAL system and the ozonesonde were found to also have an increase in aerosol scattering associated with them. Aerosol scattering ratios of 0.15 and 0.2 were found for the lower and upper O_3 layers, respectively. A cross sectional map of these layers is being constructed along the Electra flight path, which went from the northwestern corner of Maine southeast to the coast. Airborne DIAL data of O_3 and aerosol transport in the upper troposphere will be used to provide new insight into stratosphere/troposphere exchange mechanisms.

Water Vapor Profiles

The first high spatial resolution DIAL water vapor measurements were made in November 1981 with the airborne DIAL system operating in the near-IR.[8] DIAL flights were made from the NASA Wallops Flight Center over the Langley Research Center where radiosondes were launched for comparison

143

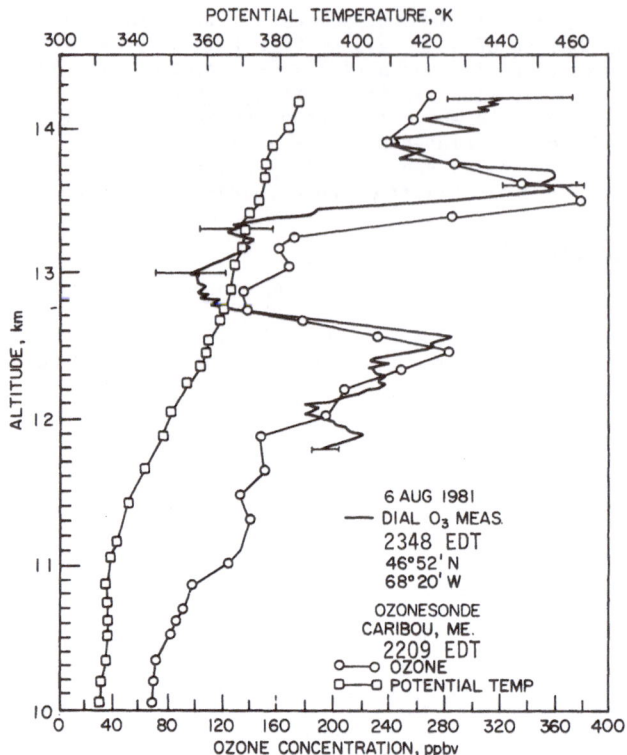

Figure 4. Comparison of DIAL and ozonesonde measurements of ozone layers in the vicinity of the tropopause

water vapor profiles. Operational parameters for the DIAL near-IR operation are given in Table 1. Due to the high precision wavelength control necessary for insuring coincidence of the laser wavelength with the narrow water vapor absorption line, a high-spectral resolution instrument was used to provide automatic servo-control of the dye laser wavelength with a precision of less than 0.3 pm.[20] A comparison is shown in Figure 5 of the water vapor profiles obtained by the airborne DIAL system and from radiosonde data on November 5, 1981, over the Langley Research Center. Also shown in the figure is the aerosol (off-line) lidar display along the flight path of the water vapor DIAL measurement. Several distinct layers can be seen in the aerosol picture. An abrupt decrease in the water vapor mixing ratio at a level corresponding to the top of the uppermost layer can be seen in both water vapor profiles. The 200 shot average represents a horizontal resolution of about 4 km. The water vapor concentration measured by the radiosonde below 1500 m altitude was within the standard deviation of the water vapor DIAL average profile. The slightly lower DIAL measurement between 1500 and 2250 m altitude was due to the horizontal inhomogeneity in the upper layer. Conditions in this region were changing rapidly due to a cold front passage in the vicinity of Langley. This horizontal inhomogeneity was seen in a sequence of water vapor profiles measured between Langley and Williamsburg, Virginia. Water vapor profiles such as these are important for investigations of latent heat flux, air mass modification over bodies of water, water vapor transport into the stratosphere, and initialization conditions for weather forecast models.[17]

144

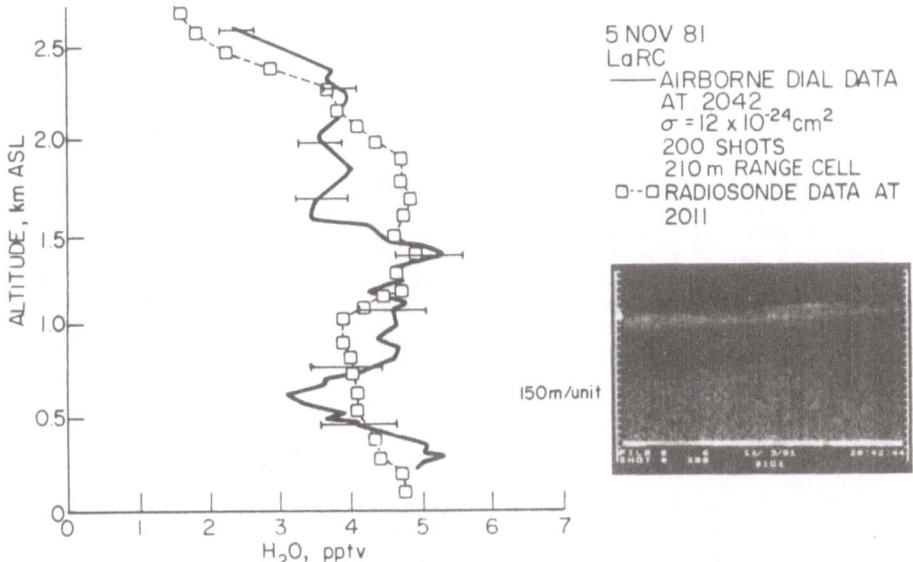

Figure 5.   Comparison of water vapor profiles measured by the DIAL system
and a radiosonde over the Langley Research Center.   Aerosol
display using DIAL off-line return for this region is shown

OTHER AIRBORNE DIAL SYSTEM APPLICATIONS

The NASA airborne DIAL system also has the capability to measure SO_2,
NO_2,[2] and simultaneous water vapor and temperature[9] (with the addition of a
third laser). Simulations have been conducted to assess the potential
accuracy for making these measurements in the troposphere. These simulations use the system parameters given in Table 1, and in all cases the
measurement uncertainties are indicative of random statistical errors. A
simulation of a DIAL SO_2 profile measurement near 300 nm indicated that at
an SO_2 concentration of 10 ppb, the measurement uncertainty was about 15
percent from an aircraft altitude of 3 km and for a vertical and horizontal
resolution of 500 m. This capability will permit the application of the
airborne DIAL system to investigations of SO_2 in power plant plumes and in
urban plumes.

The measurement of NO_2 with an airborne DIAL system is hindered by the
lack of strong NO_2 absorption lines in the 440 nm wavelength region and the
relatively low NO_2 concentrations to be measured. As an example, in the
rural environment with an ambient NO_2 concentration of 1 ppb, a measurement
of NO_2 over a 2 km vertical path would provide an optical depth of only
2 ppb km. Simulations indicate that NO_2 could be measured with an
uncertainty of about 50 percent or +0.5 ppb over 2 km. Likewise, in an
urban environment where the NO_2 concentration is 100 ppb, a 5 percent NO_2
DIAL measurement would be possible with a resolution of 500 m. Airborne
lidar investigations of NO_2 in a rural area are likely to be limited to a
column content assessment of NO_2 concentration so that the path distance
will be large when the NO_2 concentration is low.

By extending the current two wavelength DIAL system to a three
wavelength system, simultaneous DIAL measurements of tropospheric

temperature and water vapor profiles are possible.[9,10] Temperatures and humidities can be extracted, employing the method outlined by Mason[21] and described by Endemann and Byer[22], by choosing two laser frequencies that coincide with the center of two temperature dependent water vapor transitions with the third laser frequency in a non-absorbing region. A DIAL error analysis was conducted for appropriate water vapor absorption lines in the 730 nm wavelength region. These calculations indicate that temperatures with a relative accuracy of about 1°K and water vapor mixing ratios with uncertainties of approximately 2 percent can be obtained in the lower troposphere with a vertical resolution of 250 m from an aircraft altitude of 4 km. A 6 km horizontal resolution was assumed in this calculation. Potential benefits for simultaneous temperature and water vapor measurements occur in many areas of meteorology and environmental science; viz., weather prediction on all scales, transport of latent and sensible heat flux, cloud physics, severe storm genesis, and aerosol growth.

SUMMARY

This paper has discussed many of the recent DIAL measurements performed by the multipurpose airborne DIAL system developed at the NASA Langley Research Center. These have included the measurement of O_3 and aerosols in and above the boundary layer and in the tropopause/lower stratosphere region. In addition, the first water vapor profile measurements were discussed, and water profiles obtained near a cold front were presented. Simulation results for airborne DIAL measurements of SO_2, NO_2, and simultaneous water vapor and temperature were discussed and several potential tropospheric investigations were cited.

The results presented here demonstrate the capability of the airborne DIAL technique to make range-resolved measurements of a broad range of tropospheric gases. The application of lidar remote sensing to investigations of atmospheric phenomena is increasing as the DIAL technique becomes more accepted as an operational method for gathering data on tropospheric gases in local, regional, and global scale studies. Also, many of the lidar measurements that are now being made from the ground and aircraft will be made from space to open a new era in tropospheric investigations.

REFERENCES

1. Browell, E. V.; Wilkerson, T. D.; and McIlrath, T. J.: Water Vapor Differential Absorption Lidar Development and Evaluation. Appl. Opt., 18, 3474 (1979).
2. Browell, E. V.: Lidar Measurements of Tropospheric Gases. Opt. Eng.,21, 128 (1982).
3. Harris, J. E. and Browell, E. V.: Evolutionary Shuttle Atmospheric Lidar Program. Conf. Abs., Ninth International Laser Radar Conference, Munich, Germany, July 2-5, 1979.
4. Browell, E. V., ed.: Shuttle Atmospheric Lidar Research Program - Final Report of Atmospheric Lidar Working Group. NASA SP-433 (1979).
5. Greco, R. V., ed.: Atmospheric Lidar Multi-User Instrument System Definition Study. NASA CR-3303 (1980).
6. Browell, E. V.; Carter, A. F.; and Shipley, S. T.: An Airborne Lidar System for Ozone and Aerosol Profiling in the Troposphere and Lower

Stratosphere. Proceedings of the IAMAP International Quadrennial Ozone Symposium, NCAR, Boulder, CO, August 4-9, 1980.

7. Browell, E. V.; and Shipley, S. T.: Airborne Lidar Investigations of Ozone and Aerosols in the Nonurban Troposphere. Proceedings of the Second Symposium on the Composition of the Nonurban Troposphere, Williamsburg, VA, May 25-28, 1982.

8. Carter, A. F.; Browell, E. V.; Butler, C. F.; Mayo, M. N.; Hall, W. M.; Wilkerson, T. D.; and Siviter, J. H., Jr.: Remote Measurements of Tropospheric Water Vapor with an Airborne DIAL System. Conf. Abs., Eleventh International Laser Radar Conference, Madison, Wisconsin, June 21-25, 1982.

9. Browell, E. V.; Shipley, S. T.; Rosenberg, A.; Hogan, D.; and Wilkerson, T. D.: An Airborne Lidar for Simultaneous Measurements of Temperature and Water Vapor. Conf. Abs., IAMAP Third Scientific Assembly, Hamburg, Federal Republic of Germany, August 17-28, 1981.

10. Rosenberg, A. and Hogan, D. B.: Lidar Technique of Simultaneous Temperature and Humidity Measurements: Analysis of Mason's Method. Appl. Opt., 20, 3286 (1981).

11. Schotland, R. M: Some Observations of the Vertical Profile of Water Vapor by Means of a Laser Optical Radar. Proceedings of the Fourth Symposium on Remote Sensing of the Environment, Ann Arbor, Michigan, April 123-14, 1966.

12. Measures, R. M. and Pilon, G.: A Study of Tunable Laser Techniques for Remote Mapping of Specific Gaseous Constituents of the Atmosphere. Opt-Electronics, 4, 141 (1972).

13. Byer, R. L. and Garbuny, M.: Pollutant Detection by Absorption Using Mie Scattering and Topographic Targets as Retroreflectors. Appl. Opt., 12, 1496 (1973).

14. Schotland, R. M.: Errors in the Lidar Measurement of Atmospheric Gases by Differential Absorption. J. Appl. Meteorol., 13, 71 (1974).

15. Thompson, R. T., Jr.: Differential Absorption and Scattering Sensitivity Predictions. NASA CR-2627 (1976).

16. Inn, E. C. Y. and Tanaka, Y.: Ozone Absorption Coefficients in the Visible and Ultraviolet Regions. Advances in Chemistry, No. 21, American Chemical Society, Washington, D.C., 1959, p. 263.

17. Browell, E. V.; Carter, A. F.; and Wilkerson, T. D.: An Airborne Differential Absorption Lidar System for Water Vapor Investigations. Opt. Eng., 20, 84 (1981).

18. Wilkerson, T. D.; Schwemmer, G.; Gentry, B.; and Giver, L. P.: Intensities and N_2 Collision-Broadening Coefficients Measured for Selected H_2O Absorption Lines Between 715 and 732 nm. J. Quant. Spectrosc. Radiat. Transfer, 22, 315 (1979).

19. Browell, E. V.; Shipley, S. T.; Butler, C. F.; and Ismail, S.: Airborne DIAL Measurements of Ozone and Aerosol Profiles in the 1980 EPA PEPE/NEROS Field Experiment. NASA TN in preparation.

20. Wavelength Control Instrument Provided by M. L. Chanin of the Service d'Aeronomie du CNRS, Verriers-le-Buisson, France.

21. Mason, J. B.: Lidar Measurement of Temperature: A New Approach. Appl. Opt., 14, 76 (1975).

22. Endemann, M. and Byer, R. L.: Simultaneous Remote Measurements of Atmospheric Temperature and Humidity Using a Contiunuously Tunable IR Lidar. Appl. Opt., 20, 3211 (1981).

3.4 Pollution Monitoring Using Nd:YAG Based Lidar Systems

K. Frederiksson

National Swedish Environment Protection Board, Studsvik,
S-611 82 Nyköping, Sweden, and

S. Svanberg

Department of Physics, Lund Institute of Technology, P.O. Box 725
S-220 07 Lund, Sweden

Laser-radar techniques for remote measurements of atmospheric and hydrospheric conditions, especially pollution levels, are becoming increasingly important. Recently we have reviewed different aspects of laser remote sensing [1,2]. A program of laser-based pollution monitoring was initiated in 1975 within our research group at Chalmers University of Technology, Göteborg. A lidar system employing a 25 cm diameter Newtonian telescope and a dye laser (flashlamp- or N_2-laser-pumped) was constructed. A specially designed microcomputer was used for transient averaging and system steering. The system was used for technique developments, and several field tests on air-pollution monitoring and hydrospheric probing were performed. The lidar system and examples of atmospheric particle, NO_2 and SO_2 monitoring are presented in Ref. 3, which also contains references to more detailed laboratory reports.

Whereas mainly general laser-spectroscopy equipment from our basic research program was used in the early experiments, plans for a dedicated, mobile laser-radar laboratory were made. In early 1979 the lidar group received a grant from the Swedish Space Corporation for developing a mobile measuring system intended for research but also for routine monitoring. The project was supported by the Swedish Board for Space Activities, The National Swedish Environmental Protection Board and the Swedish Board for Technical Developments. The basic system was completed at the end of 1979. At this time the group was also split up. The mobile laboratory was transferred to the Environmental Protection Board in Studsvik for a 2 year critical evaluation period and the basic lidar research program was transferred to Lund Institute of Technology. We will now describe the construction of the mobile and the laboratory laser-radar systems and give examples of field and fixed-laboratory measurements.

A detailed description of the mobile lidar system was given in a recent paper[4]. The basic design of the system is shown in Fig. 1. The laser beam is emitted from a rotatable dome structure on the roof of the covered truck. The beam is deflected to the measurement area by a large mirror in the dome. The dome has a quartz window for the weather protection of the system. The backscattered light is deflected by the dome mirror into a Newtonian telescope, which has a fixed vertical position. The lasers, electronics and optics in the mobile laboratory are mounted with shock absorbers and vibration isolators that make it possible to drive the truck at its maximum speed (80 km/h) without any problems. The mobile system also includes a trailer carrying two motor generators.

The optical and electronic arrangement of the mobile lidar system is shown in Fig. 2. The laser equipment consists of a Nd:YAG laser and a dye laser. In some applications the Nd:YAG laser is used separately but it usually serves as a pump laser for the dye laser. In DIAL measurements the wavelength of the dye laser, operating at 10 Hz, is changed between each laser pulse. This rapid wavelength shift is achi-

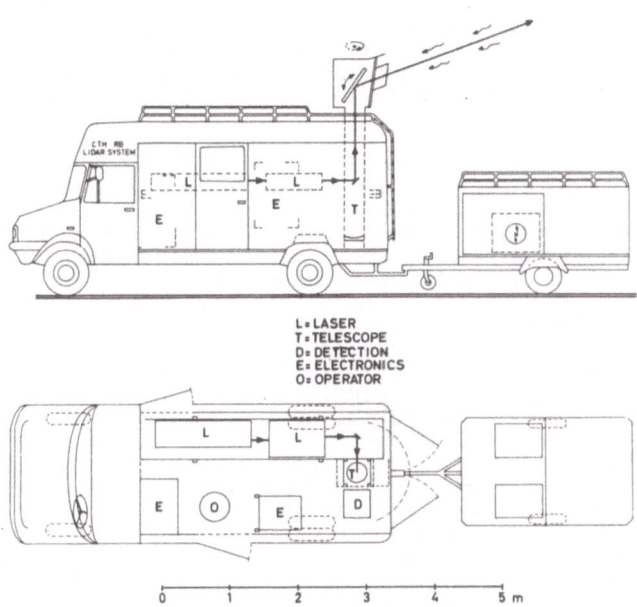

L = LASER
T = TELESCOPE
D = DETECTION
E = ELECTRONICS
O = OPERATOR

Fig. 1. Schematic of mobile lidar laboratory

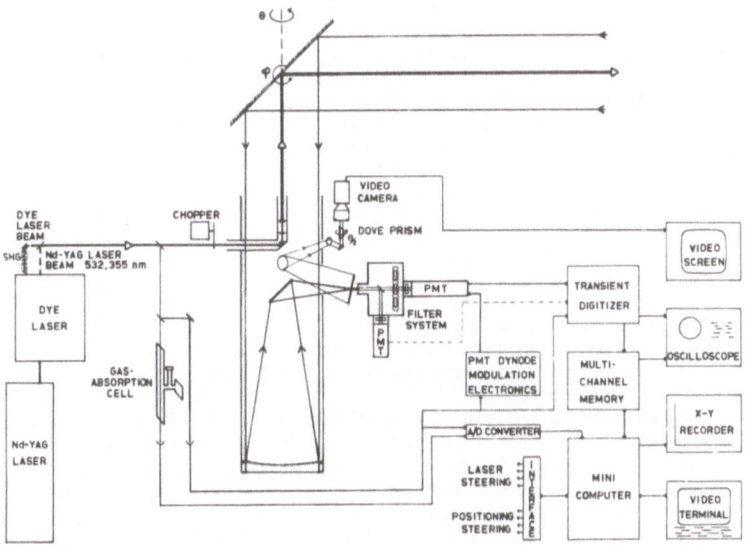

Fig. 2. Optical and electronic arrangements of mobile lidar system

eved by rocking the laser grating with the use of an excenter wheel, rotated by means of a stepper motor. Two frequency-doubling crystals, mounted in sequence, are used to generate laser light in a UV DIAL recording. The two crystals are then phase-matched to the absorption and reference wavelengths, respectively.

For measuring the absorption spectrum of the studied gas, a spectrometer with gas-absorption cells is used for accurate calibration of the laser wavelengths. A chopper is used for blocking of the laser beam, when background light and rf interference are recorded during a measurement cycle.

The detection Newtonian telescope has a 30-cm diam f/3.3 spherical mirror. The telescope mirrors and the dome mirror have UV-enhanced aluminum coatings. The field of view of the telescope is determined by an iris aperture. There are two alternative photomultipliers for recording the optical signal. One of these has a modulated dynode chain which enables close-range signals to be suppressed. Interference filters are used to suppress the sky light. A video camera with an optical arrangement including a dove prism is used to get an optical image of the telescope view. On the video screen, displaying the measurement area the orientation of the laser beam is seen as a sharp point.

The lidar signals are captured by a transient digitizer and transferred to a multichannel memory for averaging. The further processing of the data and the calculation of the pollution levels are performed by a minicomputer. The measurement data are stored on floppy discs. Besides the data recording the minicomputer controls the laser firings, the wavelength changes of the dye laser, the chopper, the choices of measurement directions, and the other parts in the system. Thus, the entire system is run automatically during the measurement cycles. The output media in the system are a video terminal, an oscilloscope screen, a plotter and a printer.

For the mobile system, the set-up time at a new measurement location is reduced to only few minutes. The system has proved to be reliable and well suited for field work.

Several field tests and measurements, where in particular the DIAL technique was utilized to measure SO_2, NO_2 and O_3, were done with the mobile lidar system for a period of almost two years. The possibilities for monitoring and charting of particle pollutants were also examined in field experiments. Industrial emissions and traffic pollution in urban areas were the main objects in this field work. The work will be summarized in half a year and the conclusions will then be a guidance for future measurements performed on a routine basis. Some examples of measurements from the field work[4,5] are given here.

An example of an SO_2 DIAL measurement is shown in Fig. 3. This measurement was made in a chemical factory area. The laser beam was directed horizontally along a local street at 3-5 m height above the ground. The lidar recordings at an absorption wavelength of SO_2 and at a reference wavelength are shown on a logarithmic scale. The divided DIAL curve is shown at top right in the figure. The lower curve shows the evaluated SO_2 concentration as a function of the distance from the monitoring system. The concentration of SO_2 in this case originates from diffuse emissions.

From a set of measurements of the type shown in Fig. 3 a horizontal charting of a gas concentration can be made for a certain area. Such an example is shown in Fig. 4. A number of probing directions were selected as shown by the lines on the chart. At the top of the figure a sketch of the measurement area, a paper mill, is shown. The directions as viewed from the location of the monitoring system are marked in the sketch. From the set of DIAL measurements performed, the concentrations in $\mu g/m^3$ are calculated and presented in a graphic form. On this occasion three areas with high concentrations of SO_2 were found. The measurement result shown in Fig. 4 is an average recording representing a two-hour period.

In Fig. 5 an example of the monitoring of NO_2 in industrial stack effluents is shown. The mobile system was at a distance of 1350 m from a smokestack of a saltpetre plant. The laser beam was directed immediately above the three-meter-diameter smokestack. In the

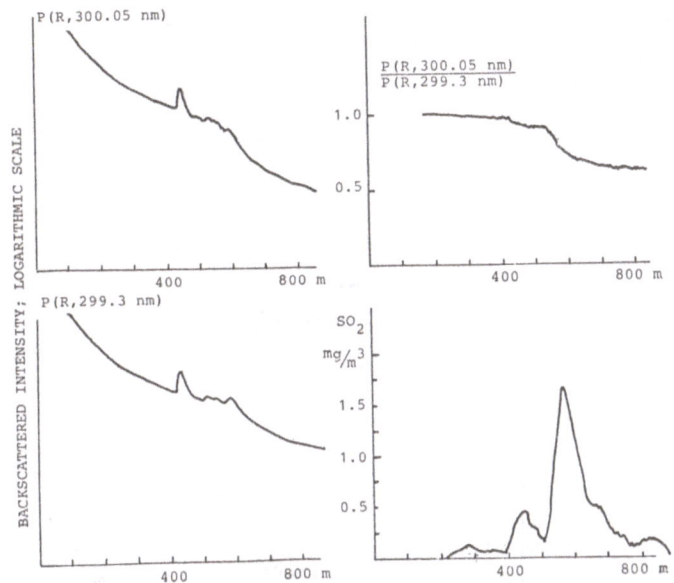

Fig. 3. SO$_2$ *DIAL measurement of the ambient air at a local street in a chemical factory area*

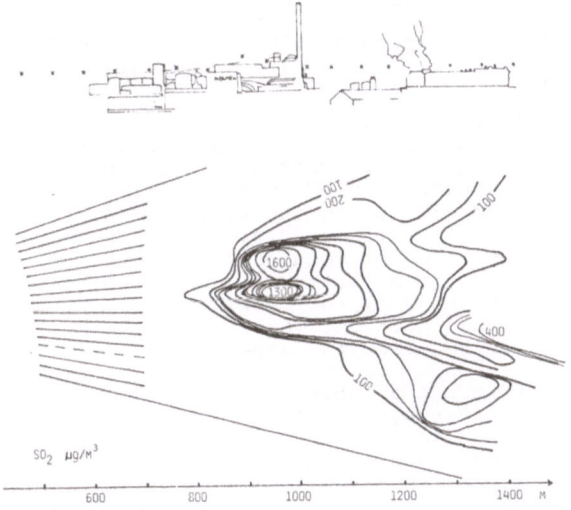

Fig. 4. Charting of SO$_2$ *concentrations in a paper mill area. Each level represents 100 μg/m^3. The scale in the figure represents the distance from the lidar system*

figure an example of the measurements is shown. The recordings at absorption and reference wavelengths of NO$_2$ are given on a logarithmic scale. Several plumes of particles in the industrial area are also displayed in the lidar recordings. In the DIAL curve at top right in the figure the particle dependence is eliminated and the dependence of the NO_2 is accentuated. From this curve it is obvious that it is sufficient in this case to measure the absorption and reference wavelengths one tenth of a second apart. In the lower right part of the figure the concentration of NO_2 is shown, integrated with the distance. The NO_2 concentration in the plume was calculated to be 122 ± 30 mg/^{3} using the known stack diameter.

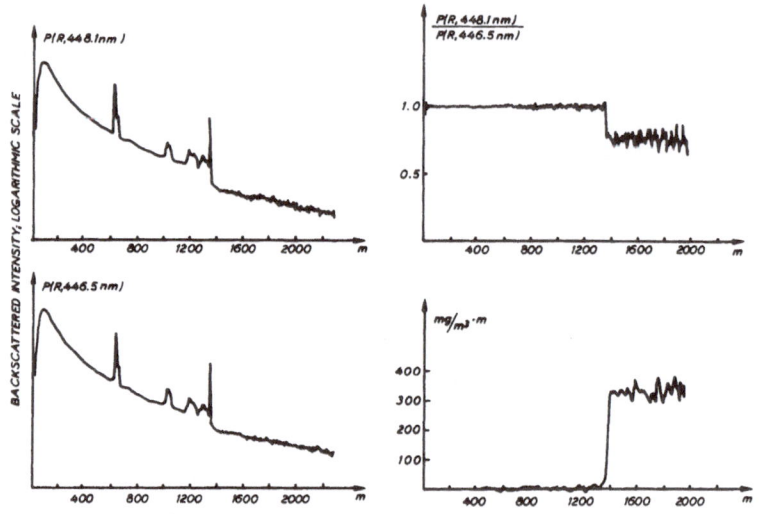

Fig. 5. *Monitoring of NO_2 in a plume from a salt-petre plant*

The DIAL technique is well suited for measurements of the total flow of a gas from an area or in a smoke-stack plume. An example is given here, whereas the technique is discussed more extensively in Ref. 5. The diagram in Fig. 6 shows the SO_2 concentration in a vertical section on the leeward side of an industrial area with fairly high emissions of the gas. A distributed plume due to the diffuse emission from a factory producing hydrochloric acid is displayed together with a dense plume coming from a stack. In addition to information about the distribution of the gas, the readings also give the total concentration in the vertical section. These numbers times the wind velocity yield the flows of SO_2 from the area.

For technique developments we use a laboratory lidar system employing a powerful Nd:YAG- and dye-laser system of the same kind as the one used in the mobile laboratory. An excimer- and dye-laser system can also be utilized. In frequency-doubling and mixing experiments a self-tracking crystal system is used. Short UV and near IR wavelengths can also be generated in a high-pressure cell for Raman shifting. The set-up used in a recent lidar experiment on atomic mercury is shown in Fig. 7 [6]. UV light at the 253.65 nm mercury resonance wavelength was generated by frequency-doubling of 567.06 nm dye-laser light and subsequent first anti-Stokes shifting by 4155 cm^{-1} in H_2. Typical pulse energies at the mercury wavelength were 0.7 mJ. Part of the light was used for reference purposes employing a mercury cell and power monitors. The main part of the UV light was transmitted into the atmosphere and back-scattered light was received by a 25 cm diameter telescope. The electronic detection system has been described in Ref. 3. In Fig. 8 lidar curves for a distant open-ended mercury vapour cell are shown.

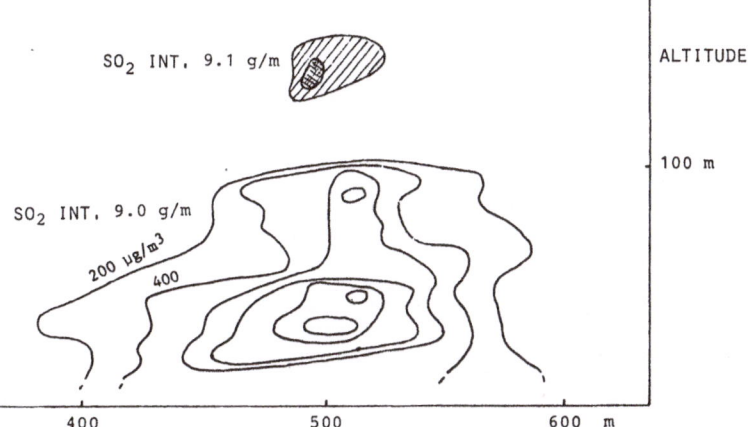

ALTITUDE

SO_2 INT. 9.1 g/m

100 m

SO_2 INT. 9.0 g/m

200 µg/m³

400

400   500   600 m

Fig. 6. Charting in a vertical section of the total
flow of SO_2 emitted in an industrial area

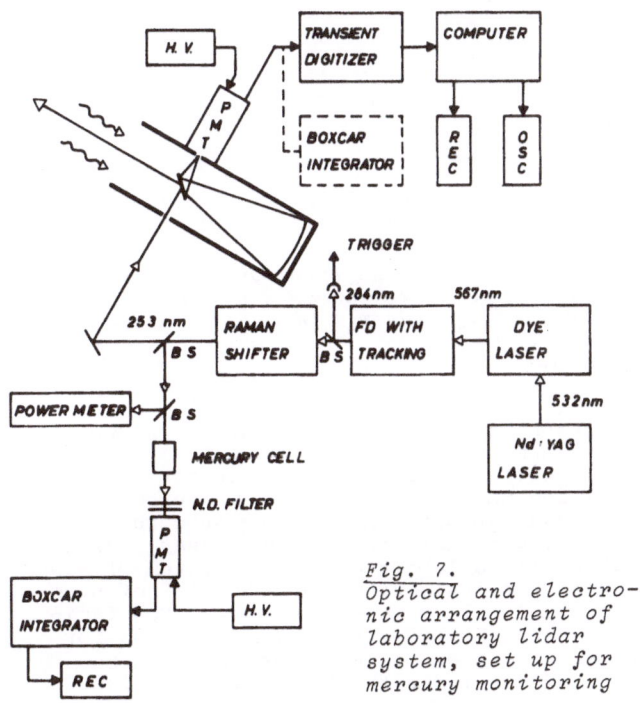

Fig. 7.
Optical and electro-
nic arrangement of
laboratory lidar
system, set up for
mercury monitoring

For the on-resonance curve both fluorescence and absorption are observed. In the divided DIAL representation the mercury absorption is
manifested in the distinct step in the curve. The absorption corresponds to a mean concentration of 100 µg/m³ in the 1.5 m long cell. Absorptions of a strength similar to the one shown in the figure are
expected at the mouth of refuse-burning-plant stacks. Experiments on
NO, HCHO and OH, all absorbing in the UV region, are planned.

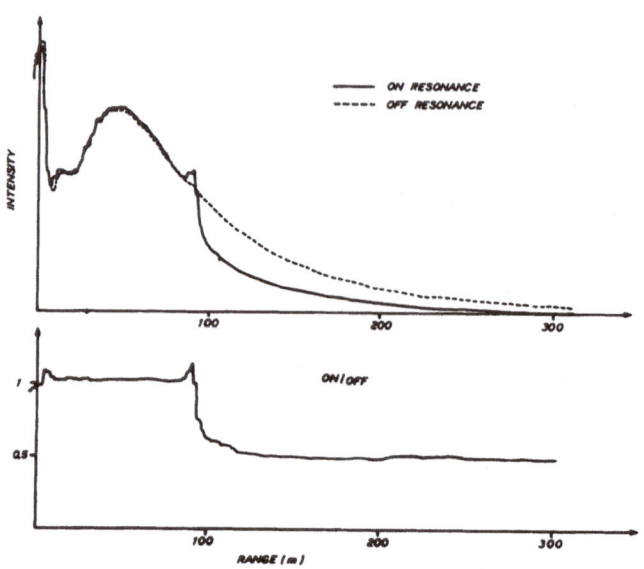

Fig. 8.
DIAL measurement
of the mercury
vapour content
in a distant
open-ended
chamber

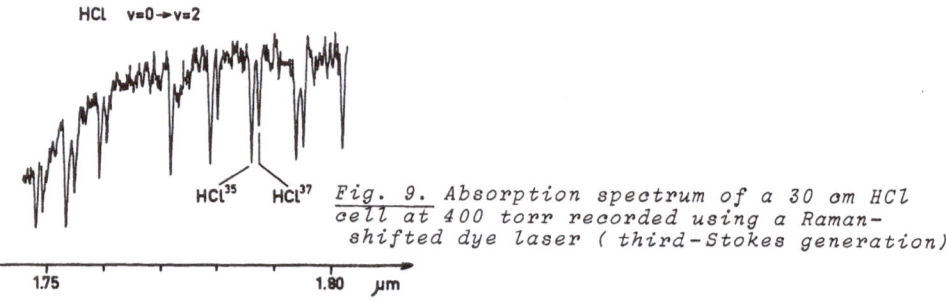

Fig. 9. Absorption spectrum of a 30 cm HCl
cell at 400 torr recorded using a Raman-
shifted dye laser (third-Stokes generation)

Several molecules have overtones or combination bands in the near
IR region, accessible with a Raman-shifted dye laser. An InAs detector is suitable in this spectral region. We have performed
spectral studies on CO, HCl and CH_4 and atmospheric measurements
are in preparation. In Fig. 9 an absoprtion overtone spectrum
for HCl is shown, recorded using a Raman-shifted dye laser[7]. After
initial experiments using our fixed-laboratory lidar system, field
experiments will be performed for evaluating the potential of practical monitoring of previously little studied molecules. For facilitating the transition between laboratory and field work and for
general strengthening of our effort on the pollution monitoring
area, a relocation of the Studsvik lidar facilities to Lund Institute of Technology is planned.

References

1. S. Svanberg, Lasers as probes for air and sea, Contemp. Phys. 21,
    541 (1980).

2. K. Fredriksson, Laser spectroscopy applied in studies of alkali-
    -atom structures and in environmental monitoring, Ph.D. disserta-
    tion, Göteborg 1980.

3. K. Fredriksson, B. Galle, K. Nyström and S. Svanberg, Lidar system applied in atmospheric pollution monitoring, Appl. Opt. 18, 2998 (1979).

4. K. Fredriksson, B. Galle, K. Nyström and S. Svanberg, Mobile lidar system for environmental probing, Appl. Opt. 20, 4181 (1981).

5. K. Fredriksson and H. Hertz, to be published.

6. M. Aldén, H. Edner and S. Svanberg, Remote measurement of atmospheric mercury using differential absorption lidar, Opt. Lett. in press.

7. M. Aldén, H. Edner and S. Svanberg, to be published.

Part 4

Atmospheric Propagation and System Analysis

4.1 Effects of Atmospheric Obscurants on the Propagation of Optical/IR Radiation

S.A. Clough, R.W. Fenn, F.X. Kneizys, J.D. Mill, and E.P. Shettle
Air Force Geophysics Laboratory, Hanscom Air Force Base, MA 01731, USA

Every optical remote sensor, whether looking at the natural atmospheric constituents or other obscurants or pollutants, has to be able to look through the atmosphere. It is therefore important that one understands and can reliably predict the propagation properties of the ambient atmosphere for these remote sensing systems.

Depending on the wavelength region (from visible through infrared to mm and cm waves) any of the constituents of the natural atmosphere from air molecules to haze and fog particles to rain and snow may become an obscurant to optical and LASER remote sensing systems (see e.g. Table 5-7 in reference 1).

The following figures illustrate the magnitude and importance of the effect which various natural atmospheric pollutants can have on the transmission of radiation at different wavelengths. In this example, we are considering a 10 Km long slant path from 1 Km altitude to the surface. The mid-latitude winter atmosphere contains either a rural, urban, or maritime aerosol, corresponding to 10 Km visibility (reference 2). The case of a fog is also considered with a 100 meter thick layer of radiation fog (reference 2) at the surface, corresponding to a visibility of 450 meters. Rain occurs at rates of 0.25 and 2.5 mm/hr.

The transmittance over this 10 Km slant path due to these atmospheric obscurants is calculated for the wavelength regions around 1 μm, 4 μm, 10 μm, and for millimeter wave frequencies from 0 to 900 GHz using the AFGL FASCODE Program (references 3 and 4).

In the 1 μm region (Figure 1) the water vapor continuum determines the upper limit in transmittance in the absence of any particulate extinction. Continental dry aerosols (at a 10 Km visibility) reduce the transmittance to approximately 20%. Larger maritime aerosol particules would reduce it to a few percent, and fog and rain to a fraction of one percent.

In the 4 μm window (Figure 2), the molecular continuum absorption gives the upper limit in transmittance; there is a weak N_2O absorption band whose lines can reduce the transmittance to 50-80% at the peak of an absorption line. Aerosol attenuation is only about 1/2 to 1/3 of that at 1 μm. Only fog and moderate rain would significantly limit systems performance in this example. In the 10 μm region (Figure 3), molecular line and continuum absorption again will limit transmittance in a particulate-free atmosphere. Only moderate rain will reduce the transmittance to less than 1% in this example.

In the mm-wavelength region, rain and snow are the major attenuators in addition to molecular absorption by water vapor and oxygen. The major transmission windows are at frequencies around 30 GHz (approximately 1 cm wavelength), 90 GHz (3 mm), and 130 GHz (2 mm) (Figure 4). Light to moderate rain seriously affects the transmittance. The attenuation by fog (over the 1 Km fog path) would be small compared to the molecular absorption over 10 Km. At shorter wavelengths in the submillimeter region (see Figure 5), molecular absorption becomes the major attenuation process (larger than 4 db Km^{-1}); over a 10 Km path the transmittance would drop to a fraction of one percent.

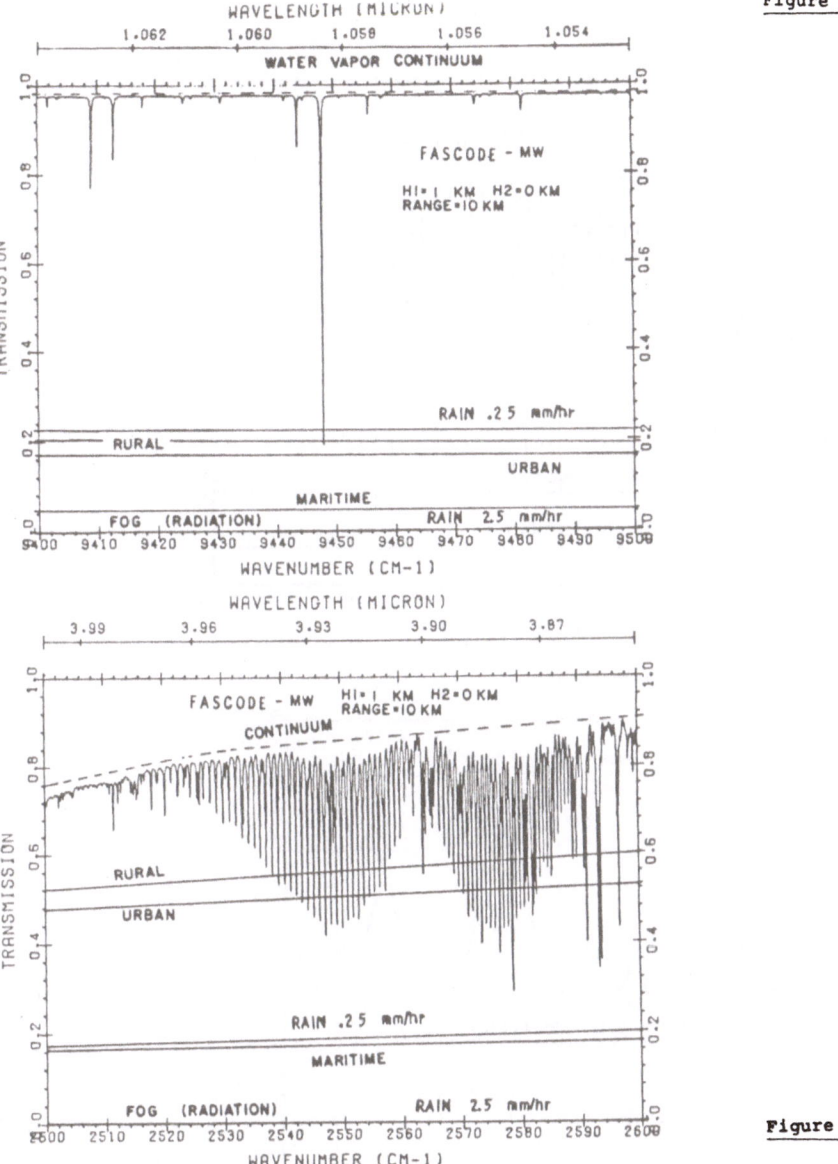

Figure 1

Figure 2

During the past years considerable interest developed in the attenuation effects of falling snow. Extensive experimental programs have been conducted during the winter of 80/81 and 81/82 in New England on IR and mm-wave propagation through falling snow. Previous measurements indicate that the extinction coefficient at wavelengths above 2 μm is from 1.2 to 1.45 times larger than at visible wavelengths (Figures 6 & 7). Whether this somewhat surprising relationship is the result of narrow angle forward scattering into the receivers at shorter wavelengths or due to the unknown scattering and absorption properties of the complex shaped snow flakes is not completely clear, but the former explanation is gaining favor at this point.

Figure 3

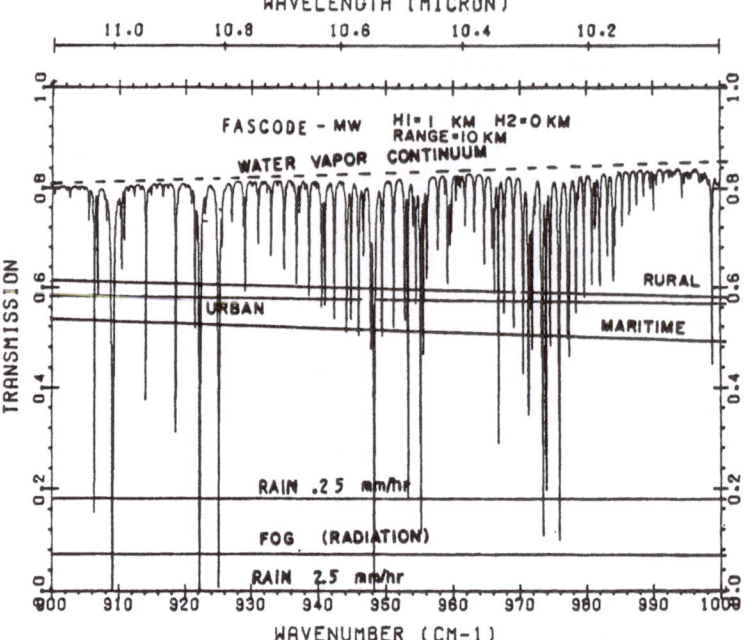

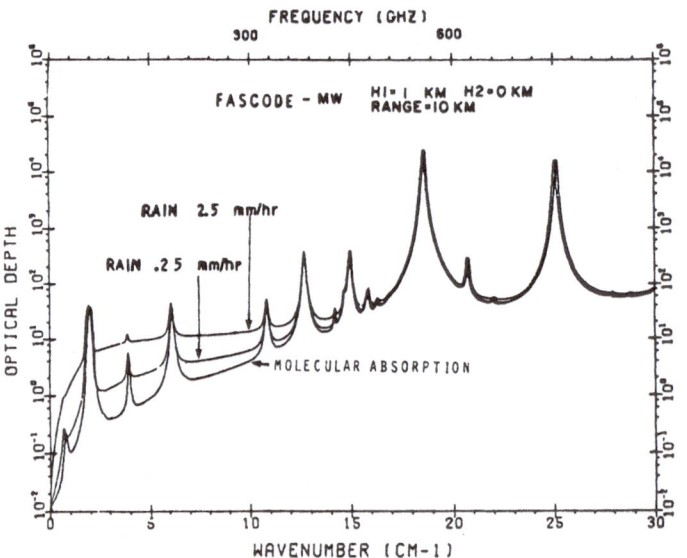

Figure 4

Theoretical studies conducted at the Canadian Defence Reseach Establishment by Tam (reference 5) on forward-angle single and multiple scattering show the importance of this effect in transmission measurements (Figure 8).

For a number of applications one is not only interested in the narrow angle forward scattering properties and its effect on the beam transmission, but also in the off-axis scattering for beam-detection.

160

Figure 5

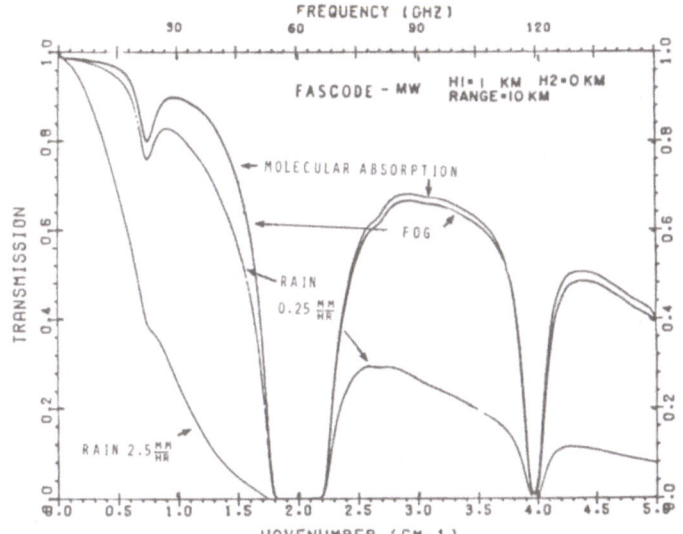

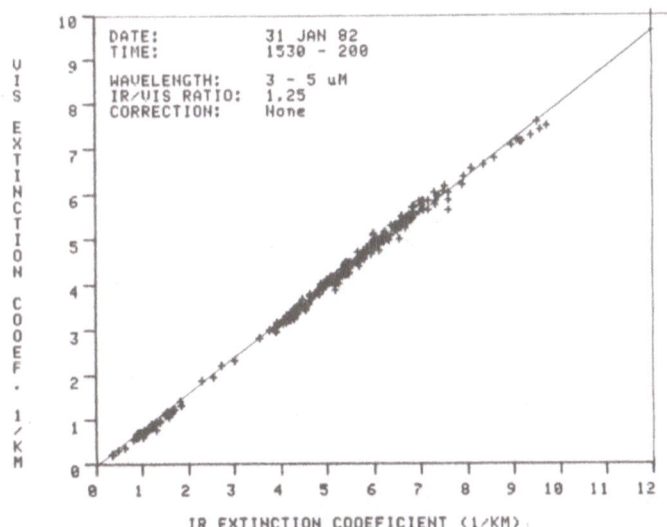

SNOW-ONE-A Visible vs IR Extinction

DATE: 31 JAN 82
TIME: 1530 - 200

WAVELENGTH: 3 - 5 uM
IR/VIS RATIO: 1.25
CORRECTION: None

Figure 6

Very often the assumption is made that in the infrared, especially in the 10 μm region the angular scattering from the natural hazy atmosphere becomes more or less Rayleigh like. Experiments and theory show that the influence of large aerosols in the size distribution is strong enough to cause considerably more forward than back-scattering. Figures 9 and 10 show a comparison of the scattering from haze and fog particles at different wavelengths.

Although atmospheric optical models can provide relationships between various parameters (e.g. visible versus IR transmittance), the variability of these relation-ships and their dependence on weather conditions can only be derived from extended field measurements. Figure 11 shows the variability of the relationship between

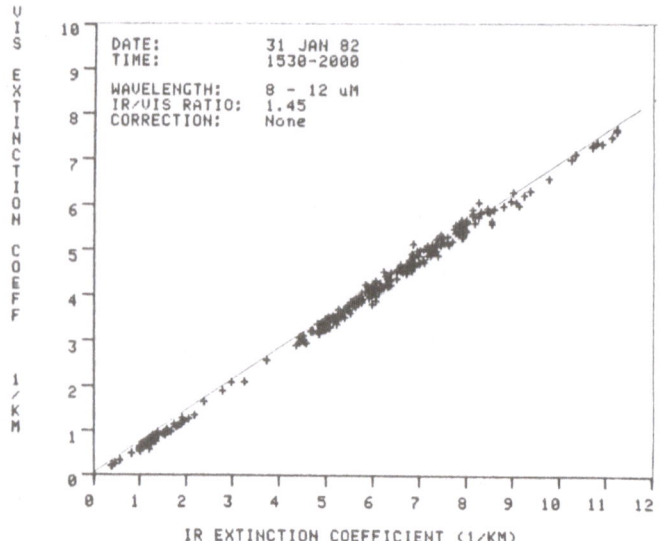

SNOW-ONE-A Visible vs IR Extinction

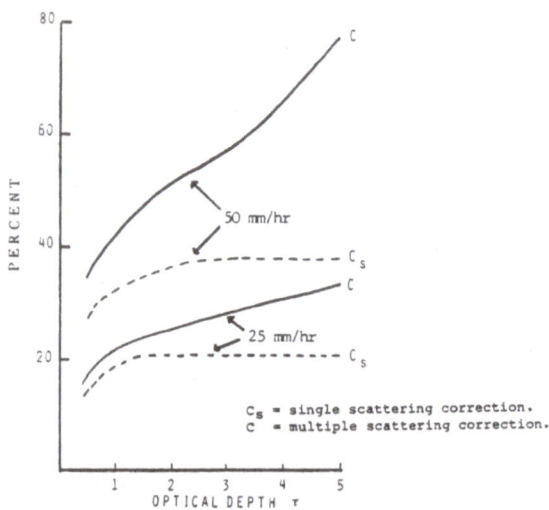

Figure 8 Correction factors versus optical depth for rain (from ref 5)

visual and 8-12 μm transmittance based on hourly measurements conducted over a one month period at a site in northern Germany.

In summary, natural atmospheric obscurants such as haze, fog, rain, or snow can significantly impede the performance of laser remote sensing systems, depending on laser wavelength, atmopheric conditions and, of course, path length. Existing models can provide generally reliable estimates of these effects. Some problems such as multiple scattering, propagation through snow, and correlation between optical effects and weather variability need further study.

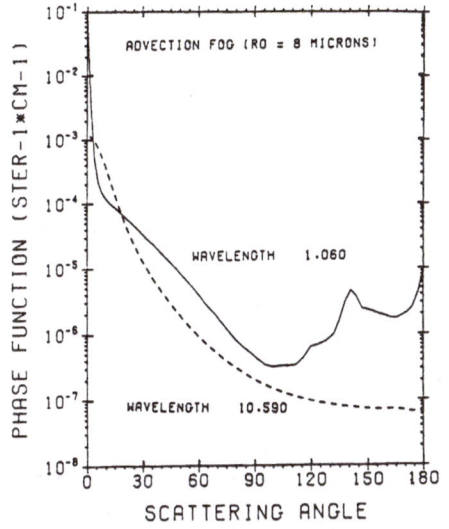

Figure 9

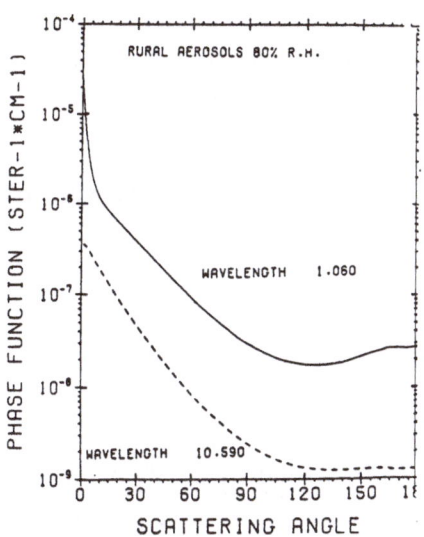

Figure 10

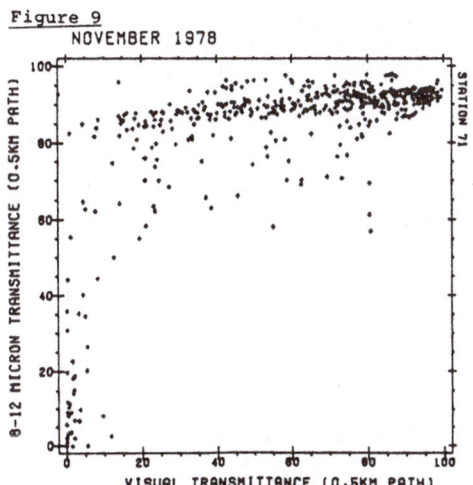

Figure 11

1. M.D. Kays, M.A. Seagraves, H.H. Monahan, R.A. Sutherland; Qualitative Description of Obscuration Factors in Central Europe, ASL Monograph No. 4, September 1980.

2. E. Shettle, R. Fenn: Models for the Aerosols of the Lower Atmosphere and the Effects of Humidity Variations on Their Optical Properties, AFGL-TR-79-0214, September 1979. NTIS ADA085951

3. V.J. Falcone, Jr., L.W. Abreu, E.P. Shettle: Atmospheric Attenuation of Millimeter and Submillimeter Waves: Models and Computer Code, AFGL-TR-79-0253, October 1979. NTIS ADA084485

4. S.A. Clough, F.X. Kneizys, L.S. Rothman, W.O. Gallery: Atmospheric Spectral Transmittance and Radiance: FASCODE 1B, in Proceedings of SPIE, The Intern. Soc. for Opt. Eng., Vol. 277, Atm. Transm., R.W. Fenn, Ed., April 1981.

5. W.G. Tam: "Multiple Scattering Corrections for Atmosphere Aerosol Extinction Measurements," Appl. Opt., Vol. 19, July 1980.

4.2 The Effects of Target-Induced Speckle, Atmospheric Turbulence, and Beam Pointing Jitter on the Errors in Remote Sensing Measurements

J. Fred Holmes

Oregon Graduate Center, 19600 N.W. Walker Road, Beaverton, OR 97006, USA

Introduction

Atmospheric turbulence and target-induced speckle can have a significant and deleterious effect on the performance of optical remote sensing systems The primary effect is to introduce a strong fluctuation in the intensity and a random perturbation of the phase of the fields at the optical receiver. This is equivalent to introducing a large noise source at the receiver and consequently multiple samples and/or clever processing techniques are needed to accurately estimate the magnitude of the received intensity. In addition, for coherent detection schemes, the random phase and finite transverse coherence length introduced by the speckle and turbulence will limit the size of aperture that can be used.

It would be very useful to be able to predict the effect of turbulence and speckle on the expected value of the mean square error for the parameters being remotely sensed in terms of the number of samples taken, path length, beam size, receiver aperture, strength of turbulence and wavelength. Unfortunately there are very few results of this type available. [1,2,3] However, the science base for speckle propagation through turbulence at least for the monochromatic case is now fairly complete[4-10] and awaits its application to the problems of determining the errors in remote sensing measurements. Consequently, it is expected that in the near future, solutions for the mean square error for many types of remote sensing systems will become available to aid in the design of these systems and to allow optimum processing of the received signal.

Speckle Effects

Before discussing the combined effects of speckle and turbulence, some background on speckle statistics will be presented to give the reader a better feel for the magnitude of the speckle problem. The statistical parameters that are important are the normalized variance, which is a measure of the degree of signal fluctuation, the covariance, from which the transverse correlation length can be derived, and the probability density function. For a monochromatic laser source, a diffuse target, and no turbulence the normalized variance of the received intensity is equal to unity and is defined as

$$\sigma_{I_N}{}^2 = \frac{\langle (I - \langle I \rangle)^2 \rangle}{\langle I \rangle^2}$$

A sample function for this type of signal is shown in Figure 1.

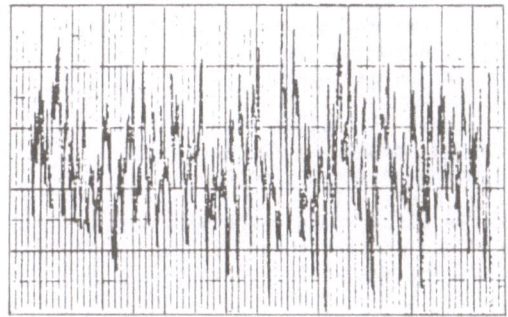

Fig. 1. Signal contaminated by speckle with unity contrast ratio

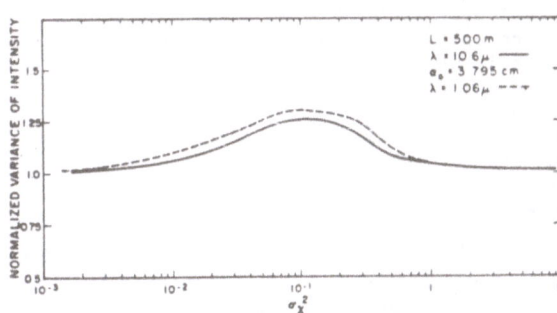

Fig. 2. Normalized variance of the received intensity versus the log-amplitude variance, focused beam, 500-m range, 1.06- and 10.6-μm wavelengths, α_0 = 3.795 cm

The normalized covariance of the received intensity is defined as

$$C_{I_N}(p) = \frac{\langle (I(\overline{p_2}) - \langle I \rangle)(I(\overline{p_1}) - \langle I \rangle) \rangle}{\sigma_I{}^2}$$

where $\overline{p} = \overline{p}_2 - \overline{p}_1$

and in the absence of turbulence can be expressed as

$$C_{I_N}(\overline{p}) = \exp\left\{ -\frac{p^2}{2}\left[\frac{1}{\alpha_o{}^2} + \left[\frac{k}{L}\alpha_o\left(1 - \frac{L}{F}\right)\right]^2 \right] \right\}$$

where α_o is the illuminating beam radius at the transmitter, p is the spacing between detectors, L is the path length, F is the distance at which the transmitter is focused and k is the wave number. It can be seen that for a focused transmitter, the speckle size back at the receiver is approximately the same size as the transmitted beam. As the focus distance F is increased, the spot size on the target grows and the speckle size back at the receiver decreases continuously.

165

The speckle scale size is important if aperture averaging is used to reduce the variance of the received signal. Each speckle can be thought of as being independent of all the other speckles. Consequently, what the collection aperture does then is add up the intensities from each speckle and since they are independent, the variance goes down as 1/N where N is the number of independent speckles. This effect can be estimated using the covariance scale size and the aperture size to find an approximate N; or an exact calculation can be performed.[11]

For a heterodyne detection system there is a limit to the size of aperture that can be used. This is caused by the fact that the received signal must be coherent with respect to the optical local oscillator. Consequently there will be an aperture size of the order of the speckle size beyond which very little additional signal will be gained and in fact there will be an optimum aperture size that maximizes the signal to noise ratio.[12]

The speckle scale size at the receiver is independent of the diffuse target. However, the actual pattern back at the receiver does depend on the target and as the spot on the target is moved, the new pattern increasingly decorrelates with the original pattern. This is important in systems such as DIAL where the ratio of two intensities is used and the speckle effect tends to cancel if the speckle pattern is not too different for the two measurements Consequently, such things as target motion, source motion and beam jitter which will cause the speckle pattern to change during the measurements are important considerations since they may degrade the beneficial effect of taking a ratio.

Changing the wavelength between measurements can have the same effect as moving the spot in that it changes the speckle pattern. This is a target dependent phenomenon; and in general if $\Delta W \ll \sigma_z$, where σ_z is the standard deviation of the target surface roughness, then the pattern does not change very much.

Another important parameter in evaluating the effect of speckle on the errors in remote sensing is the probability density function (PDF) of the received intensity or fields. The parameter that is being remotely sensed is in general a function of the received intensity or fields. Consequently, the PDF is needed to find the expected value of the mean square error. For a monochromatic source, a diffuse target and no turbulence at the receiver, the fields have a Gaussian distribution, the field amplitude has a Rayleigh distribution, and the intensity has an exponential distribution. As will be discussed in the next section, turbulence will modify these distributions. However, for integrated path turbulence corresponding to Rytov variances less than about 0.05, the above distributions could be used. It should be noted that system gains should be adjusted to take into account the strong fluctuation present in the signal in order not to clip. For instance, if it is desired to have not more than five percent of the samples (intensity) clipped, the gain must be set such that the mean signal is only one-third of the peak allowable (no aperture averaging).

Turbulence and Speckle Effects

When turbulence is present, an additional fluctuation is introduced in the signal at the receiver in addition to that caused by speckle. This is illustrated[6] in Figure 2 which shows the normalized variance of the received intensity versus the Rytov variance which is the normalized variance as observed by a point detector located at the receiver due to a point source at the target. As can be seen from the figure, the normalized variance increases up to a Rytov variance of around 0.1 and then decreases until it reaches unity again. The maximum increase is around 25 percent.

There is an additional effect that can increase the variance beyond that predicted in Figure 2. If the beam wanders on and off the target (or retroreflector) or wanders around on a target of greatly varying reflectivity, the variance can increase markedly.[13] This wander can be caused by the atmospheric turbulence or by beam jitter in the transmitter system.

As discussed in the previous section, there is a target—induced speckle scale size that for a given scenario is fixed. However, when turbulence is present, this scale size is dynamic and decreases as the strength of turbulence increases. This is illustrated[6] in Figures 3 and 4 for two different wavelengths.

There is an additional statistical quantity that is important when turbulence is present and that is the power spectral density. When there is a wind blowing, the turbulence is blown through the transmitter beam and the field of view of the receiver. This imparts a temporal fluctuation to the received signal and will cause a decorrelation of two received signals which are separated in time. Figure 5 shows a power spectral density function[4] that is valid up to a Rytov variance of about 0.05. The maximum time separation for insignificant decorrelation can be inferred from this figure or the more complete formulation for the autocorrelation function contained in Reference 6 can be used.

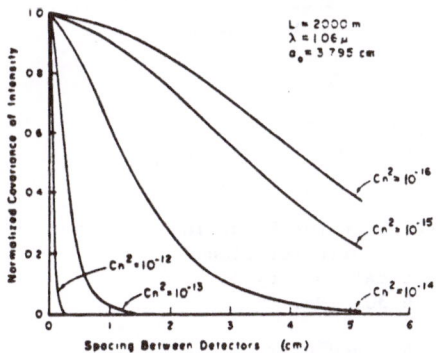

Fig. 3. Normalized covariance of the received intensity, focused beam, 2000-m range, 1.06-μm wavelength, α_o = 3.795 cm

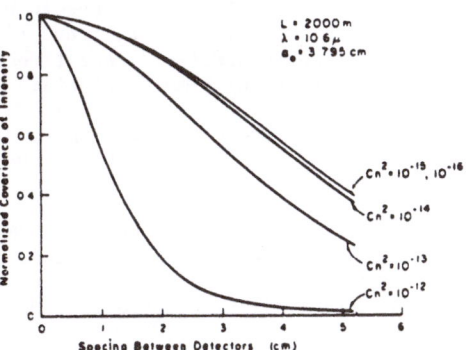

Fig. 4. Normalized covariance of the received intensity, focused beam, 2000-m range, 10.6-μm wavelength, α_o = 3.795 cm

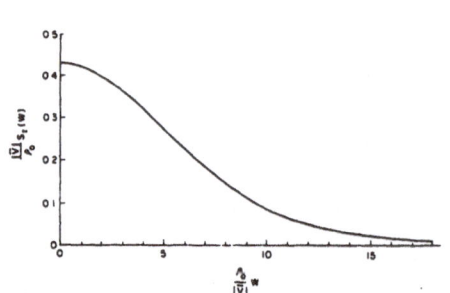

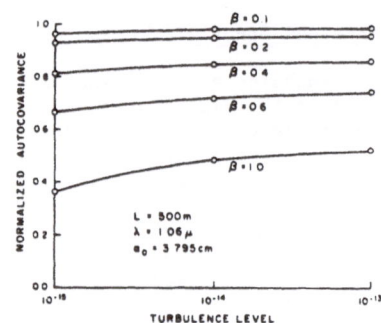

Fig. 5. Temporal, power spectral density, received intensity

Fig. 6. Normalized autocovariance of a randomly misaligned laser beam versus C_n^2. Focused beam, 500-m range, 1.06-μm wavelength, α_o = 3.795 cm

Beam jitter can also contribute both to temporal fluctuations and to the decorrelation between signal samples. This is illustrated[7] in Figure 6 which shows the crosscovariance between two time samples (pulses) versus the strength of turbulence for different amounts of beam jitter.

When turbulence is present, the PDF for the intensity for a monochromatic speckle pattern changes from an exponential to a K distribution.[9,10] Since the work that gives this result is not yet available, the form for the distribution will be given here.

$$P_I(\alpha) = 2\left(\frac{M}{\lambda}\right)^{\frac{M+1}{2}} \frac{\alpha^{\frac{M-1}{2}}}{\Gamma(M)} K_{M-1}\left(2\sqrt{\frac{M\alpha}{\lambda}}\right)$$

where K_{M-1} is the modified Bessel function of order M-1, λ equals $\langle I \rangle$, and M equals $2/(\sigma_{I_n}^2 - 1)$.

References

1. Dennis K. Killinger and Norman Menyuk, "Comparative Sensitivity of Dual-Laser and Single Laser Remote Sensing of Atmospheric Species," Annual Meeting of the Optical Society of America, Kissimmee, Florida, October 26-30, 1981.

2. N. Menyuk and D. K. Killinger, "Temporal correlation measurements of pulsed CO_2 lidor returns," Optics Letters 6, June 1981.

3. J. Fred Holmes and Myung Lee, "Estimation in the Error in Remote Crosswind Sensing Caused by Atmospheric Turbulence and Speckle," Annual Meeting of the Optical Society of America, Kissimmee, Florida, October 26 30, 1981.

4. Muung Hun Lee, J. Fred Holmes and J. Richard Kerr, "Statistics of Speckle Propagation Through the Turbulent Atmosphere," J. Opt. Soc. Am. 66, November 1976.

5. Philip A. Pincus, Michael E. Fossey, J. Fred Holmes and J.

R. Kerr, "Speckle Propagation Through Turbulence – Experimental," J. Opt. Soc. Am. 68, June 1978.

6. J. Fred Holmes, Myung Hun Lee, and J. Richard Kerr, "Effect of the Log-Amplitude Covariance Function on the Statistics of Speckle Propagation Through the Turbulent Atmosphere," J. Opt. Soc,. Am. 70, May 1981.

7. Myung Hun Lee, J. F. Holmes, "Speckle Intensity Crosscovariance Function for Two Misaligned Laser Beams in a Turbulent Atmosphere," J. Opt. Soc. Am. 70, May 1981.

8. J. Fred Holmes, V. S. Rao Gudimetla, "Variance of Intensity for a Discrete Spectrum, Polychromatic Speckle Field After Propagation Through the Turbulent Atmosphere," J. Opt. Soc. Am. 71, October 1981.

9. V. S. Rao Gudimetla and J. Fred Holmes, "Probability Density Function of the Intensity for a Laser Generated Speckle Pattern in the Turbulent Atmosphere," Annual Meeting of the Optical Society of America, Kissimmee, Florida, October 26-30, 1981.

10. V. S. Rao Gudimetla and J. Fred Holmes "Probability density function of the intensity for a laser generated speckle field after propagation through the turbulent atmosphere," submitted to J. Opt. Soc. Am.

11. J. C. Dainty, Laser Speckle and Related Phenomena (Springer-Verlag, 1975).

12. Steven F. Clifford and Stephen Wandzura, "Monostatic heterodyne lidar performance: The effect of the turbulent atmosphere," Appl. Opt. 20, 1 February 1981.

13. C. M. McIntyre, Myung Hun Lee and J. H. Churnside, "Statistics of irradiance scattered from a diffuse target containing multiple glints," J. Opt. Soc. Am. 70, September 1980.

4.3 Lidar System Analysis for Measurement of Atmospheric Species

R.T. Menzies

Jet Propulsion Laboratory, California Institute of Technology
Pasadena, CA 991109, USA

G. Mègie

Service d-Aeronomie du CNRS, F-91370 Verrieres-le-Buisson, France

Introduction

In this paper several important aspects of lidar are discussed, which pertain to a system analysis of the applicability of certain types of lidar to specific measurement objectives. The differential absorption lidar (DIAL) technique is emphasized. A brief treatment of the DIAL technique is followed by a discussion of the comparison of DIAL with other lidar measurement techniques. Sensitivity expressions are presented for DIAL with either direct or coherent (heterodyne) detection systems. A treatment of the fundamental similarities and differences in using the UV and IR spectral regions for species measurements, such as backscattering properties, spectral characteristics of the absorbing species, and factors relevant to the optimization of the DIAL measurement accuracy, is also included. Finally, results obtained from an existing ground-based DIAL system used to measure water vapor profiles are briefly discussed.

Summary Analysis

Applications of the differential absorption lidar (DIAL) technique for species measurements have been discussed previously [1-5], and the reader is encouraged to study those references for a more complete analysis of DIAL. The lidar equation is:

$$P_{\lambda R} = P_T \; \beta_{\lambda R} \left(\frac{ct_L}{2}\right) \left(\frac{A}{R^2}\right) \; \eta_2 \; \exp[-2(\tau^e_{\lambda R} + \tau^i_{\lambda R})], \tag{1}$$

where $P_{\lambda R}$ = the signal power incident on the detector, at wavelength λ, from the backscattering cell at range R and thickness ΔR;

P_T, t_L = the transmitted power and laser pulse duration, respectively;

$\beta_{\lambda R}$ = the backscattering coefficient at wavelength λ and range R;

A = receiver area;

η_2 = optical efficiency of the receiver;

$\tau^i_{\lambda R}$ = the integrated optical thickness due to absorption by the constituent i of interest, i.e.,

$$\tau^i_{\lambda R} = \int_0^R \sigma_i \; (\lambda,R) n_i (R) dR,$$

where σ_i (λ,R) is the absorption cross section, and

$n_i(R)$ is the number density of specie i; and

$\tau^e_{\lambda R}$ = the integrated optical thickness excluding that of the absorbing
constituent i.

By processing from different laser signals at two different wavelengths, λ,
on an absorption line and λ_2 off absorption, and from two successive cells
at ranges $R_1 = R$ and $R_2 = R + \Delta R$, one obtains the local optical thickness $\Delta\tau^i$
of the absorbing specie i in that range interval:

$$2(\Delta\tau^i_1 + \Delta\tau^i_2) = \ell n \left(\frac{P_{22}P_{11}}{P_{12}P_{21}}\right) + \ell n \left(\frac{\beta_{12}\,\beta_{21}}{\beta_{22}\,\beta_{11}}\right) - 2\,(\Delta\tau^e_1 - \Delta\tau^e_2), \quad (2)$$

where the first subscript refers to the wavelength and the second one to the
range, and $\Delta\tau^e$ is the local optical thickness due to other absorbers or
scatterers. It is important to choose two wavelengths such that the quantity
on the lhs of Eqn. (2) is maximized but the values of the second and third
terms on the rhs are minimized. If the magnitudes of these "undesirable"
terms cannot be ignored, then extra care must be taken to generate accurate
values for them. This can become a significant problem in both the UV [4]
and IR [5] spectral regions. In the ultraviolet, the problem arises because
of the comparatively broad spectral features, necessitating the use of widely
spread wavelengths. In the infrared, the spectra of the species of interest
are often sharper, but the most efficient laser sources are discretely tunable
at present, again resulting in the use of widely spaced wavelengths.

The noise processes in incoherent or direct detection (e.g., using a
photoemissive or photoconducting detector directly) can be described by using
Poisson statistics to characterize the fluctuations of the various sources of
current in the photodetector. Then the signal-to-noise ratio depends on the
ratio of the number of "signal" photoelectrons to the square root of the total
number of signal and background radiation induced photoelectrons and the "dark
current" electrons. Thus the signal-to-noise ratio for a single pair of pulses
(at the two wavelengths, respectively) can be quite high if the laser pulse
energy and receiver area are large enough. In contrast, the signal-to-noise
ratio when using coherent detection depends only on the number of "signal"
photoelectrons. (The background radiation and "dark" current are not important
in a properly designed heterodyne receiver.) However, it does not improve
matters to increase the number of received "signal" photons well above the
level needed to compete with quantum noise, because of speckle-induced shot-
to-shot fluctuations. If a collection of scatterers is illuminated coherently,
this collection will produce a speckle pattern in the "far field," which
produces significant received intensity fluctuations if the successive pulses are
not correlated. Thus it is normally necessary to average over several pulses
in order to attain a high signal-to-noise ratio. This implies that given a
fixed amount of available power or energy, it is better to use a high repetition-
rate laser than a high pulse energy, low repetition-rate laser for coherent
detection. Coherent detection for species measurements is of much more value
in the infrared than in the visible or ultraviolet regions, because the back-
ground-induced and intrinsic sources of current fluctuation in infrared detectors
are several orders of magnitude larger than the quantum noise limit.

What are the advantages of DIAL when compared with other lidar techniques for species measurements? The prospect of having to process the "difference" between returns at two wavelengths may seem to be a disadvantage. However, it is often necessary to operate fluorescence and Raman scattering lidars at multiple wavelengths also, in order to reduce the effects of spurious signals. The DIAL technique is superior when a spectral match can be made between the laser and the absorber of interest, and when a suitable backscattering medium is available. Consequently, the comparative advantages of DIAL are more evident in the lower atmosphere, especially in the planetary boundary layer, where the aerosol is much more dense. In comparison, the DIAL "signal" can be approximately expressed as the product of the backscatter coefficient and the local optical thickness due to the absorber of interest, i.e.,

$$\text{DIAL "signal"} \sim \beta \Delta \tau^1 \sim 10^{-5} \text{ to } 10^{-7}$$

for a case in which we are using a range interval of 100 meters and the constituent is present with a mixing ratio of about one part-per-million. The "signal" from a fluorescence lidar under the same conditions in the lower atmosphere is

$$\text{Fluorescence "signal"} \sim \sigma \ (\Gamma_{rad}/\Gamma_{coll})(\Delta f/f) \sim 10^{-6} \text{ to } 10^{-8},$$

(Where σ is the excitation cross section, the Γ's are the decay rates, and ($\Delta f/f$) is the branching ratio to the desired lower level.)

for cases in which the excited specie radiates an appreciable fraction into a small number of narrow spectral regions (i.e., the branching ratio, $\Delta f/f$, into the spectral region for which the receiver is sensitive, is ~ 0.1). The return signals from Raman scattering are much lower,

$$\text{Raman "signal"} \sim 10^{-11}$$

for measurement conditions as stated above. Under certain circumstances the fluorescence lidar sensitivity can exceed that of DIAL. The detection of OH in the ambient troposphere or stratosphere is an example [6]. For most cases, however, especially in the lower troposphere, the DIAL technique is more feasible. The Raman scattering approach does not require a tunable laser (a definite advantage in simplicity), but the very low signal levels make it relatively impractical for "remote" sensing.

Backscatter Efficiencies and Relative Factors

The backscattering efficiency for a DIAL system is wavelength dependent. In the UV ($\lambda \sim 300$ nm) the processes involved are both Rayleigh scattering by the major atmospheric gases and Mie scattering by aerosol particles, whereas in the IR only the latter is effective due to the rapid decrease of the Rayleigh scattering cross section with increasing wavelength. As the altitude of observation increases, the aerosol backscatter decreases relative to the Rayleigh backscatter, because the scale height for Rayleigh scattering is 7-8 km and the corresponding scale height for aerosol backscatter, on the average, is 1-2 km [7]. The relative total backscatter efficiencies are plotted in Figure 1, for representative wavelengths, as a function of altitude [5]. These relative efficiencies, especially at the longer wavelengths, are strongly model dependent, as they are influenced by meteorological conditions.

The relative enhancement in lidar efficiency at UV wavelengths due to higher backscatter coefficients is partially compensated by the relative

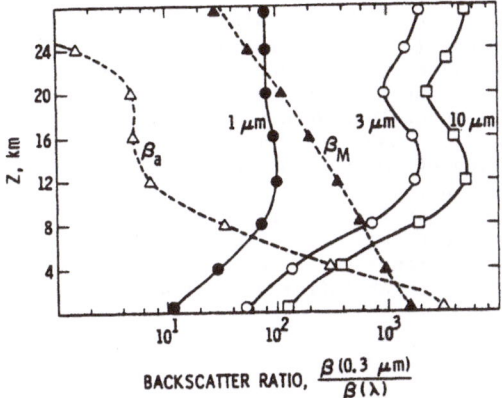

BACKSCATTER RATIO, $\dfrac{\beta\,(0.3\ \mu m)}{\beta\,(\lambda)}$

Fig. 1.   Relative efficiencies of the Rayleigh + Mie backscatter coefficient
at three IR wavelengths (1, 3, and 10 µm) as compared with the total
backscatter coefficient at 0.3 µm in the UV. The value of β (0.3 µm)
for moderate visibility conditions is about $2.5 \times 10^{-5}\ m^{-1}\ sr^{-1}$

efficiencies of the available laser systems in the UV and IR. The overall
efficiency of a system using the second harmonic (0.53 µm) of a pulsed YAG
laser to pump a dye for operation in either the 300 nm or 700 nm regions
is of the order of 10^{-4}. On the other hand, a discretely tunable TEA CO_2
laser in the 9-11 µm region has an efficiency of > 1%. The development of
the rare gas halide excimer lasers in the near UV with efficient methods
for shifting their "pump" wavelengths, has the potential for producing
overall efficiencies which are within a factor of 5 of CO_2 laser devices.

The transmittance through the lower atmosphere, and its spectral
dependence, is an important factor for long range applications or for

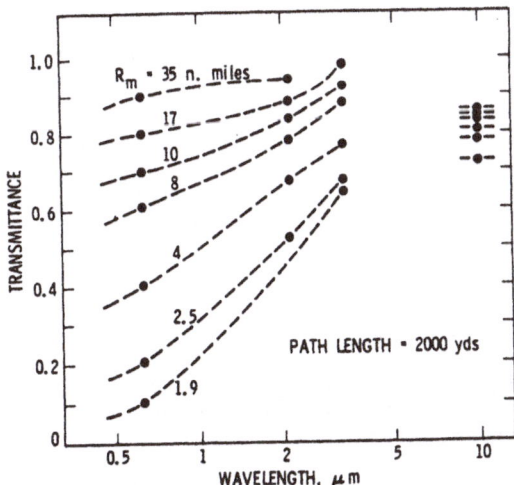

Fig. 2.   Transmittance of a 2000-yd horizontal path in haze, at visible
and infrared wavelengths, for various values of the meteorological
range, R_m

low visibility conditions. In Fig. 2 are shown data obtained from the extensive investigations of Gebbie, et al. [8], for transmission over a horizontal path over water, with a nearly constant amount of integrated water vapor (1.7 cm precipitable water) along the path for each haze (visibility) condition. The visibility conditions were characterized by values of the meteorological range, R_m in nautical miles. These data were taken in "window" regions, free of absorption bands of major atmospheric absorber gases. The IR region between 4 μm and 9 μm is heavily affected by bands, hence the data gap for this region. The absorption in the 10-μm region is almost entirely due to the water vapor continuum. For low visibility conditions, the haze affects transmittance in the UV and visible much more than at infrared wavelengths longer than 3 μm.

Tropospheric Water Vapor Measurements

An example of the type of data obtained with a ground-based DIAL system operating in the near IR is shown in Fig. 3, along with corresponding data obtained from ballon-borne soundings. The transmitter was a dye laser, pumped with the second harmonic of a YAG laser, producing a pulse energy of 80 mj in the 720 nm region, with a linewidth of 1.5 pm, and a repetition rate of 10 Hz. The receiver telescope diameter was 36 cm, the optical bandwidth was 2.4 nm, and the photomultiplier quantum efficiency was 4.5%. A transient digitizing recorder with a 10 MHz rate and 10 bit resolution was used for the sampling of the return signal vs range. Agreement between the laser measurements and the soundings is quite good. The lidar was able to generate data for a vertical profile up to 3 km in a 5-minute period. The ability to produce range-gated remote measurements in a short amount of time is a demonstration of the unique capability of lidar instruments when applied to species detection. Although commplexity is a drawback, future developments should mitigate this problem.

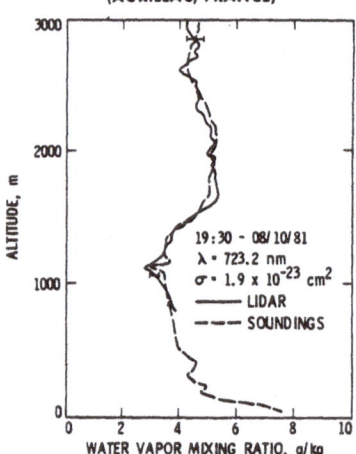

DIAL WATER VAPOR PROFILES
(AURILLAC, FRANCE)

19:30 - 08/10/81
λ = 723.2 nm
σ = 1.9 x 10⁻²³ cm²
— LIDAR
--- SOUNDINGS

ALTITUDE, m

WATER VAPOR MIXING RATIO, g/kg

Fig. 3. Tropospheric water vapor profiles obtained with a ground-based near-IR DIAL system. Reference 9

References

1. R. M. Schotland, in Proceedings of the Fourth Symposium on Remote Sensing of the Environment (U. Michigan, Ann Arbor, 1966).

2. R. L. Byer and M. Garbuny, Appl. Opt., 12, 1496 (1973).

3. T. Kobayaski and H. Inaba, Opt. and Quantum Electron., 7, 319 (1975).

4. E. E. Remsberg and L. L. Gordley, Appl. Opt., 17, 624 (1978).

5. G. Megie and R. T. Menzies, Appl. Opt., 19, 1173 (1980).

6. D. K. Killinger, C. C. Wang and M. Hanabusa, Phys. Rev. A, 2145 (1976).

7. E. J. McCartney, Optics of the Atmosphere (J. Wiley, New York, 1976).

8. H. A. Gebbie, Proc. Roy. Soc., A206, 87 (1951).

9. C. Cahen, G. Megie, and P. Flamant, "Lidar Monitoring of the Troposphere
    Water Vapor Mixing Ratio" , Proceedings of the IAMAP Symposium on
    'nowcasting' , K. Browning, ed., ESA Publication SP-165, p. 149 (1981).

4.4 CO$_2$ DIAL Sensitivity Studies for Measurements of Atmospheric Trace Gases

P. Brockman, R.V. Hess, and C.H. Bair

NASA Langley Research Center, Hampton, VA 23665, USA

Introduction

There is great need for atmospheric trace constituent measurements with higher resolution than attainable with passive radiometers. Infrared (IR) differential absorption lidar (DIAL), which depends on Mie scattering from aerosols, has special advantages for tropospheric and lower stratospheric applications and has great potential importance for measurements from Shuttle and aircraft (refs. 1,2,3,4). DIAL data reduction involves comparing large amplitude signals which have small differences. The accuracy of the trace constituent concentration inferred from DIAL measurements depends strongly on the errors in determining the amplitude of the signals. Thus, the commonly used SNR expression (signal divided by noise in the absence of signal) is not adequate to describe DIAL-measurement accuracy and must be replaced by an expression which includes the random coherent (speckle) noise within the signal (refs. 5,6). A comprehensive DIAL computer algorithm (ref. 7) is modified to include heterodyne and direct detection with speckle noise. Results of studies for monitoring vertical distribution of O$_3$, H$_2$O, and NH$_3$ from ground aircraft and Shuttle platforms are extended to include measurements of ground level plumes with larger than ambient densities, from ground and aircraft platforms. Results are shown for plumes of NH$_3$ but these results can be generalized in terms of the plume optical depth.

Analysis of DIAL Sensitivity

The expectation value P of the number of measured photons from a scattering cell at range R of length Δr including near field correction (ref. 8) is

$$P = \frac{\eta \Gamma E \Delta r A \beta \quad \exp \int_0^R -2(\xi + \sigma\rho)dr}{h\nu R^2 \left[1 + \left(\frac{A\nu}{Rc}\right)^2 \left(1 - \frac{R}{f}\right)^2\right]} \tag{1}$$

where η = detector efficiency, Γ = optical efficiency, E = laser energy, β = 180° backscatter coefficient per length per steradian, Δr = cell length = $\frac{cT}{2}$, T = integration time, ν = frequency, A = receiver area (= transmitter for heterodyne system with single detector), ξ = extinction coefficient (total minus that of measured gas), σ = absorption coefficient of measured gas, and ρ = density of gas being measured. The receiver is focused at range f.

In differential absorption, measurements are made at two frequencies selected to maximize signal and differential absorption of the species being measured while minimizing interference effects. The double ratio of signals at adjacent scattering cells, at two frequencies, yields information about absorption in the region between the scattering cells.

$$2\int_{R_1}^{R_2} (\rho\Delta\sigma)dr = \ln\left(\frac{P_{22}}{P_{12}}\frac{P_{11}}{P_{21}}\right) + \ln\left(\frac{\beta_{12}}{\beta_{11}}\frac{\beta_{21}}{\beta_{22}}\right) - 2\int_{R_1}^{R_2} (\Delta\xi)\,dr \qquad (2)$$

where P_{ij} = expectation value of measured return from cell j at frequency i, β_{ij} = backscatter coefficient from cell j at frequency i, $\Delta\sigma$ = absorption coefficient difference between frequencies 1 and 2, $\Delta\xi$ = extinction coefficient difference between frequencies 1 and 2 (not including gas being measured), i = 1 or 2 frequencies on or off absorption line, respectively, and j = 1 or 2 for distances R_1 and R_2, respectively. For scattering cells of equal length the resolution length $R_2 - R_1$ equals the scattering cell length Δr. For narrow plumes of density ρ_p length L < $(R_2 - R_1)$ and $\sigma_{on} >> \sigma_{off}$.

$$2\int_{R_1}^{R_2} (\rho\Delta\sigma)dr = 2\,\rho_p L \sigma_{on}$$

The $\ln\dfrac{\beta_{12}\,\beta_{21}}{\beta_{22}\,\beta_{11}}$ term is an error term due to changes in the frequency dependence of backscattering across scattering cells. For range resolved measurements this term is zero unless the frequency dependence of β changes with range, i.e., changes in aerosol chemistry or size distribution or transitions from atmospheric aerosol to ground or cloud. This term can be important for column content measurements where the integrals are taken from the transceiver to the scattering cell and ratios are only taken across the two wavelengths.

The $2\int_{R_1}^{R_2} (\Delta\xi)dr$ term is a bias term due to interferent species. It is an error term for that part of the interferent differential absorption that cannot be predicted. This term should not be important for plume measurements assuming that the atmosphere is homogeneous except for the plume and that base line atmospheric returns can be compared to measurements through the plume. Both these terms are small for closely spaced frequencies.

The random error calculated in this paper is solely the error occuring in the measurement process. Errors caused by lack of knowledge of laser frequency or variation of the absorption coefficient of the species being measured, due to uncertainties in temperature and pressure, are not included. Effects of turbulence on the measured signals are neglected. These effects are small for short range coaxial coherent systems (ref. 9,10) and are reduced by aperture averaging for direct detection systems. Turbulence effects can also be reduced by ratioing nearly simultaneous pulse pairs (ref. 11,12).

The random error in $2(\rho_p L \sigma_{on})$, which is calculated assuming that signal plus background are measured during each pulse and that background is measured between pulses and subtracted, will depend on the random error in the measurement of P_{ij}. The uncertainty in the two way plume optical depth is given by

177

$$\delta(2\rho_p L\sigma_{on}) = \left[\frac{1}{N} \sum_{i=1}^{2} \sum_{j=1}^{2} \frac{1}{(SNR)_{ij}^2} \right]^{1/2} \qquad (3)$$

where N is the number of pulse pairs per measurement and SNR is the single pulse signal-to-noise ratio. For heterodyne detection, the major errors in P_{ij} are due to quantum noise in the local oscillator and speckle noise in the return signal.

The heterodyne signal-to-noise ratio for a single coherence volume $SNR = \frac{P_{ij}}{P_{ij} + BT}$ is limited to 1.0 due to the speckle noise in the return signal. The number of coherence lengths per scattering cell is BT. The number of coherence areas viewed by the detector is M, where M is the ratio of receiver to transmitter area. For heterodyne detection an individual detector is required for each coherence area. The single pulse signal-to-noise ratio for determining the single from a scattering cell is then

$$SNR = \frac{P_{ij}}{P_{ij} + MBT} \sqrt{BTM} \qquad (4)$$

where BTM is the number of statistically independent samples from a scattering cell. Although some progress has been made on heterodyne detector arrays for the purpose of this paper a single detector is considered. The post-detection bandwidth B is constrained by matching with pulse duration T_p, with $B \leq 1/T_p$, and by the width of atmospheric spectral lines. The integration time T is constrained by spatial resolution requirements. The random error in two way optical depth for heterodyne detection with bandwidth B, integration time T, N pulse pairs, and with one detector is

$$\delta(2\rho_p L\sigma_{on}) = \left[\frac{1}{N} \sum_{i=1}^{2} \sum_{j=1}^{2} \left(\frac{P_{ij} + BT}{P_{ij} \sqrt{BT}} \right)^2 \right]^{1/2} \qquad (5)$$

The minimum error for a single heterodyne detector at $P_{ij} >> BT$ is $\delta(2\rho_p L\sigma_{on}) = \frac{2}{\sqrt{BTN}}$ for a range resolved measurement. For a column content measured using the ground as a reflector, the minimum error is $\frac{\sqrt{2}}{\sqrt{N}}$ since BT = 1 for a pulse reflected from a solid target and since only two signals are ratioed for a column content measurement.

Direct detection can have advantages over single-detector heterodyne detection when signal levels are high since direct detection allows averaging over multiple coherence areas with a single detector. The major disadvantage of direct detection is background noise, which is limited by optical filters

178

($\sim 10^{11}$ Hz) compared with the electronics filters (10^6 to 10^9 Hz) for heterodyne detection, and thermal noise which can be reduced by cooling below 77° K (ref. 13). Reducing the background noise by reducing the field of view will result in an increase in speckle noise.

The single pulse direct detection signal-to-noise ratio can be written as a function of the number of coherence areas M, the direct detected signal P_{ij} and the detector thermal noise figure D_T^*. The SNR for direct detection is

$$(SNR)_{DD} = \frac{P_{ij} \sqrt{BT}}{\left[\frac{n^2 A_D BT^2}{(hvD_T^*)^2} + P_{ij} BT + (GFMT)BT + \frac{P_{ij}^2}{M} \right]^{1/2}} \tag{6}$$

Equation (6) includes a detector noise term $\frac{n^2 A_D BT^2}{(hvD_T^*)^2}$, a speckle term P_{ij}^2/M,

a Poisson term $P_{ij}BT$, and a background term (GFMT)BT. In equation (6), G is the background signal in detected photons/sec/Hz per coherence area, M is the number of coherence areas, F is the bandwidth in Hz of the optical filter, and A_D is detector area.

The single pulse SNR for direct detection will exceed that of heterodyne detection when P_{ij} is greater than the background and thermal noises. The upper limit is $\sqrt{BTM}$ for distributed measurements and $\sqrt{M}$ for ground reflections. Thus, with high signal return, single pulse direct detection accuracies can exceed single pulse heterodyne accuracies by $\sqrt{M}$. An additional set of independent statistics can be achieved for direct detection by operating multimode and transmitting M' frequencies for the purpose of examples in this paper M = 1000 and M' = 1. The assumption that the SNR improves with $\sqrt{BT}$ implies that there are no major nonuniformities across a scattering cell. The assumption that the measurement error is reduced by $\sqrt{N}$ implies that the range distribution of scatterers does not vary during the measurement period. This assumption will remain valid if the measurement period is short compared to time scales of atmospheric nonuniformities.

Conditions of Simulations

A comprehensive computer algorithm is used to calculate the expectation values of P_{ij} for various measurement conditions. Pressures, temperatures, and gas species densities are input from a midsummer midlatitude atmospheric model. Trace gas species densities can be modified using card inputs. Line absorption parameters are accessed from a comprehensive data base. Sources for line data are given in reference 14. At each altitude, molecular absorption at v_1 and v_2 is calculated for each species by summing contributions from absorption lines in the vicinity of the laser frequency. Lorentz, Voigt, or Doppler line shapes are used at appropriate altitudes. Water vapor continuum absorption is added to the line absorption. Extinction due to particulate and molecular scattering is summed with molecular absorption at each altitude to give the total loss in each scattering cell. The integrated two-way loss is calculated by summing contributions from altitude layers between the laser and the cell being considered. The backscattering coefficient β is calculated by combining a Rayleigh term, which is small at infrared frequencies, and a Mie term. Mie backscatter and extinction for the cases shown here are calculated using parameters of Deirmendjian's Haze L size distribution (ref. 15). The vertical aerosol distribution is based on reference 16 with a ground level concentration (350 particles/cm^3) corresponding to a 23-km visibility. At the CO_2 laser frequency used, β at ground level is 2.6×10^{-8} m^{-1} ster^{-1}.

For all simulations of plume measurements, the optical efficiency is 20% and quantum efficiency is 50%. Telescope area is .01 m^2 and is focused at infinity and the optical bandwidth is 1.5×10^{11} Hz. The "on" and "off" frequencies have been selected to be at the R(30) (1084.63 cm^{-1}) and R(28) (1083.479 cm^{-1}) normal isotope CO_2 laser lines. The "on" absorption is 48 (atm cm)$^{-1}$ and the "off" absorption is .3 (atm cm)$^{-1}$. For these frequencies the only significant interference is from CO_2, with ground level absorption coefficients of 6.27×10^{-5}/m at R(30) and 7.31×10^{-5}/m at R(28) and from water vapor with ground level absorption coefficients at 14 torr water vapor of 2.71×10^{-4}/m at R(30) and 2.67×10^{-4}/m at R(28). Laser pulse energies are .01 to 10 Joule/pulse for horizontal ground based measurements, and .01 to 1 Joule/pulse for vertical measurements. M = 1 for heterodyne measurements and 1000 for direct detection measurements. The range to the plume center is 3000 m for horizontal cases and the range to the ground is 3500 m for vertical cases. The resolution length is 500 m for horizontal case, with T = 3.3×10^{-6} sec. The error in the measurements of the two way optical depth ($2\rho_p L \sigma_{on}$) of a plume will be proportional to the reciprocal of the square root of the resolution length for the plume length less than the resolution length. Pulse width is set at 10^{-7} and bandwidth at 10^7 Hz based on commercially available lasers.

Results and Discussion

Figures 1 and 2 show sensitivities of plume measurements for horizontal range resolved and vertical column content measurements. The percent

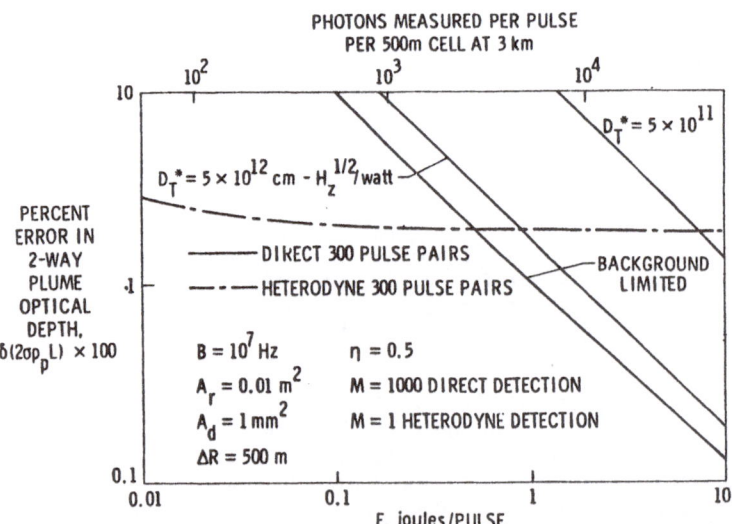

Fig. 1. Sensitivity of horizontal range resolved plume measurements

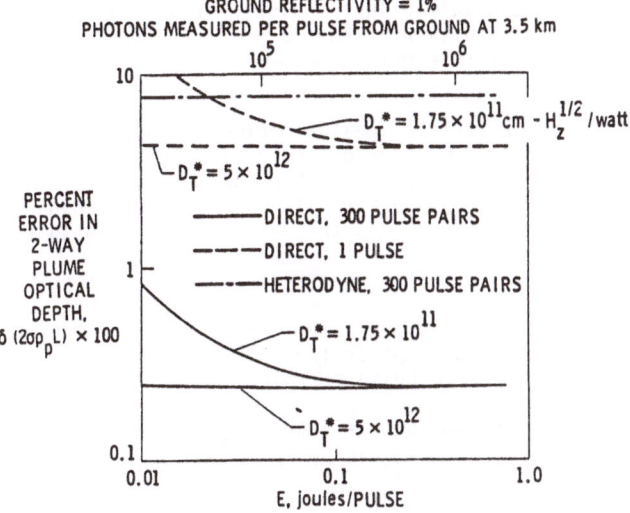

Fig. 2. Sensitivity of vertical column content plume measurements

error in measuring the two way optical depth $100 \times \delta \ (2\rho_p L\sigma_{on})$ of the plume is plotted vs laser energy and measured photons per range cell for heterodyne detection and direct detection. The figures are based on NH_3 measurements where a 2% two way optical depth error would imply an error in $(2\rho_p L)$ of 4.2×10^{-4} atm cm or a NH_3 density error of ~ 21 ppb in a 100 m plume. The results can be extended to other chemicals if the absorption coefficients are known. The results cannot be extended to plumes with high optical densities due to the loss of signal from behind the plume.

Figure 1 displays measurement accuracy for horizontal range resolved measurements using 300 pulse pairs for both heterodyne and direct detection for a plume at 3000 m range with a resolution of 500 m. The measurement error for heterodyne detection shows slight improvement from .01 J/pulse to .1 J/pulse and then remains constant at the maximum heterodyne sensitivity of $\dfrac{2}{\sqrt{BTN}}$, 2% in this case. The strategy for a heterodyne detection system would be to operate between .01 and .1 Joule/pulse with a high repetition rate. Heterodyne sensitivity also can be improved by operating with shorter pulses and higher bandwidths. A particular advantage of heterodyne detection is that Doppler wind velocity measurements can be combined with DIAL measurements. Although Doppler measurements usually are performed with pulse lengths greater than or equal to the inverse of the Doppler frequency shift, recent studies (ref. 4,17) have indicated the possibility of Doppler measurements using shorter pulse durations compatible with DIAL measurements.

For direct detection with M = 1000, three D_T^*'s are used: 5×10^{11}, 5×10^{12} 5×10^{13} $\dfrac{cm - Hz^{\frac{1}{2}}}{Watt}$. For the first two, the system is thermal noise limited and for the third background limited. For the horizontal range resolved direct detection examples shown, the SNR is limited by thermal detector and background noise, rather than speckle noise. The SNR is thus independent of bandwidth and pulse duration. For $D_T^* = 5 \times 10^{11}$ laser energies of ~ 10 Joules/per pulse would be required to exceed heterodyne sensitivity. For the higher D_T^* values, laser energies of ~ 1 Joule/pulse would be sufficient to exceed the heterodyne SNR. However, these D_T^* values would require detectors cooled below LN$_2$ temperatures. It should be noted that the required laser pulse energies are dependent on receiver area and atmospheric attenuation. Thus larger area receivers, expecially if-focused at the range of interest, and lower humidity will reduce the required laser energy.

Figure 2 shows sensitivities for vertical measurement of plumes using ground returns from an altitude of 3500 m. The ground reflectivity is set at 1%. Heterodyne detection calculations are for 300 pulse pairs. For ground reflections $T = T_p$ and $BT = 1$. Since for all laser energies shown $P_{ij} >> 1$ the heterodyne sensitivity remains constant at 8% corresponding to $\frac{\sqrt{2}}{\sqrt{N}}$, the maximum sensitivity for a heterodyne column content measurement. Heterodyne detection is sensitive enough in a ground reflections mode for CW operation with a few milliwatts of laser power (ref. 18) or quasi CW high pulse repetition rate, operation with pulse energies well below 10^{-3} J.

Direct detection sensitivity is shown for $N = 1$ and 300 and $M = 1000$. At $D_T^* = 1.75 \times 10^{11} \frac{cm - Hz^{\frac{1}{2}}}{Watt}$ a value available at LN$_2$ temperatures a sensitivity of $\frac{\sqrt{2}}{\sqrt{NM}}$ is achieved for laser energies above 0.1 Joules/pulse. For $D_T^* = 5 \times 10^{12} \frac{cm - Hz^{\frac{1}{2}}}{Watt}$ a sensitivity of $\sim \frac{\sqrt{2}}{\sqrt{NM}}$ is achieved at $\sim .01$ Joule/pulse. Since high horizontal resolution is required to measure narrow plumes from flight platforms the $\sim 4\%$ sensitivity for single pulse pair operation direct detection appears attractive.

DIAL measurements at long ranges can benefit from operating at rare isotope laser frequencies and utilizing collision broadening in the laser to tune the frequencies across atmospheric absorption lines. The use of closely

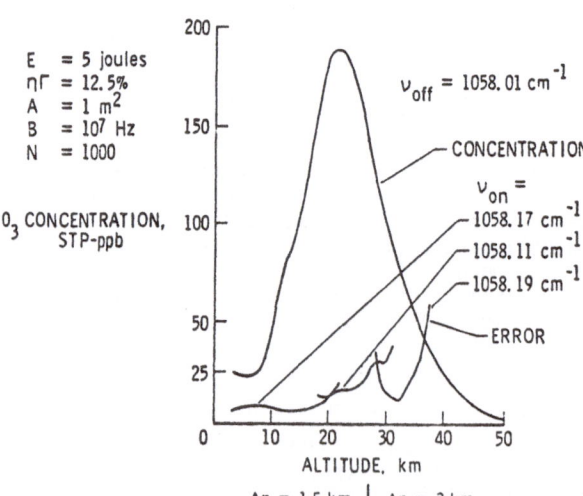

Fig. 3. Sensitivity of range resolved ozone measurements from 250 km

spaced on and off frequencies also minimizes errors due to frequency variation in backscattering and to interferring species. One example is figure 3 (ref. 1) which displays O_3 concentration in STP-ppb and measurement error in the same units versus altitude for measurement from Shuttle at 250 km. The pulse energy is 5 Joules per pulse with 1000 pulse pairs per measurement. Telescope area is 1 m^2, bandwidth is 10^7 Hz, quantum efficiency is 50%, optical efficiency is 25%, resolution is 1.5 km below 30 km and 3 km above 30 km. Simulations are for an "off" frequency of 1058.01 cm^{-1} and "on" frequencies of 1058.17, 1058.11, and 1058.19 cm^{-1}. It should be noted that a fine tuning of the "on" frequency results in a sharp variation of altitude at which the best measurement can be accomplished.

1. Brockman, P; Hess, R.V.; Staton, L.D.; and Bair, C.H.: DIAL with Heterodyne Detection Including Speckle Noise: Aircraft/Shuttle Measurement of O_3, H_2O and NH_3 with Pulsed Tunable Lasers. Heterodyne Systems and Technology Part II, p. 557 NASA CP 2138, March 1980.

2. Hardesty, R.M.: A Comparison of Heterodyne and Direct Detection CO_2 DIAL Systems for Ground Based Humidity Profiling. NOAA Technical Memorandum ERL WPL-64, October 1980.

3. Megie, G. and Menzies, R.T.: Complementarity of UV and IR Differential Absorption Lidar for Global Measurement of Atmospheric species. _Applied Optics_, Vol 19, p. 1773, 1980.

4. Hess, R.V.: CO_2 Lidar for Measurements of Trace Gases and Wind Velocities. SPIE 25th Annual Technical Symposium, Physics and Technology of Coherent Infrared Radar. Paper 300-05, August 1981.

5. Elbaum, M. and Teich, M.C.: Heterodyne Detection of Random Gaussian Signals in the Optical and Infrared: Optimization of Pulse Duration. _Opt. Commun._, Vol 27, no. 2, pp. 257-261, Nov. 1978.

6. Rye, B.J.: Differential Absorption Lidar System Sensitivity with Heterodyne Reception. _Appl. Opt._, Vol. 17, no. 24, pp. 3862-3864, Dec. 1978.

7. Remsberg, E.E. and Gordley, L.L.: Analysis of Differential Absorption Lidar from the Space Shuttle. _Appl. Opt._, Vol. 17, no. 4, pp. 624-630, Feb. 1978.

8. Sonnenschein, C.M. and Horrigan, F.A.: Signal-to-Noise Relationships of Coaxial Systems that Heterodyne Backscatter from the Atmosphere. _Appl. Opt._, Vol. 10, p. 1600, July 1971.

9. Clifford, S.F. and Wandzura, S.: Monostatic Heterodyne Lidar Performance, The Effect of the Turbulent Atmosphere. _Appl. Opt._, Vol 20, p. 514, Feb. 1980.

10. Hardesty, R.M.; Keeler, R.J.; Post, M.J.; and Richter, R.A.: Characteristics of Coherent Lidar Returns from Calibration Targets and Aerosols. _Appl. Opt._, Vol. 20, p. 3763, Nov. 1981.

11. Killinger, D.K. and Menyuk, N.: Remote Probing of the Atmosphere Using a CO_2 DIAL System. IEEE Jl. Quantum Electronics, p. 1917, Sep. 1981.

12. Kjelaas, A.G.; Nodal, P.E.; and Bjerkestand A.: Scintillation and Multi-wavelength Coherence Effects in a Long Path Laser Absorption Spectrometer. _Appl. Opt._, Vol. 17, p. 277, 1978.

13. Blouke, M.M.; Burgett, C.B.; and Williams, R.L.: Sensitivity Limits for Extrinsic and Intrinsic Infrared Detectors. Infrared Physics, Vol. 13, pp. 61-77, 1973.

14. Park, J.H.: Optical Measurement in the Middle Atmosphere. Pure & Appl. Geophys., Vol. 117, no. 3, pp. 395-429, 1978-1979.

15. Deirmendjian, D.: Electromagnetic Scattering on Spherican Polydispersions. American Elsevier Publishing Co., New York, NY, 1969.

16. McClatchey, Robert A. and Shelby, John E. A.: Atmospheric Attentuation of Laser Radiation from 0.76 to 31.25 μm. AFCRL-TR-74-0003, U.S. Air Force, Jan. 1974.

17. Hardesty, R.M.: Feasibility of Combined DIAL - Doppler Coherent Lidar Systems for Moisture Flux Measurement. Topical Meeting on Coherent Laser Radar for Atmospheric Sensing, Aspen, July 1980.

18. Menzies, R.T. and Chahine, M.T.: Remote Atmospheric Sensing with an Airborne Laser Absorption Spectrometer. Appl. Opt., Vol. 13, no. 12, Dec. 1974.

4.5 Signal Averaging Limitations in Heterodyne- and Direct-Detection Laser Remote Sensing Measurements*

N. Menyuk and D.K. Killinger

Lincoln Laboratory, Massachusetts Institute of Technology,
Lexington, MA 02173, USA

C.R. Menyuk **

Laboratory for Plasma and Fusion Energy Studies, University of Maryland,
College Park, MD 20742, USA

Introduction

The use of laser remote sensing to determine the concentration of molecular species in the atmosphere by differential-absorption LIDAR (DIAL) requires the measurement of the average transmission of laser signals through the atmosphere at two or more different wavelengths. The accuracy with which the average transmission value, and hence the molecular concentration, may be determined is limited in many cases by the presence of large atmosphere-induced pulse-to-pulse fluctuations in the LIDAR returns.

Examples of such LIDAR fluctuations are shown in Figs. 1a and 1b. These results were obtained using direct (noncoherent) detection of successive LIDAR return signals from normalized CO_2 laser pulses backscattered from a retroreflector and from a diffusely reflecting target located 2.7 km from the laboratory. The magnitude of the fluctuations is seen to depend strongly upon the nature of the reflector. Over this range typical normalized standard deviation values, σ, of the pulse-to-pulse fluctuation of return signals from the retroreflector, where glint forms the major reflective component, is 60-80%. For the diffusely reflecting target, where speckle effects play the dominant role, aperture averaging effectively reduces the standard deviation to typical values of the order of 20%.[1]

When a heterodyne (coherent) detection system is used to measure the backscattered LIDAR return signals, aperture averaging cannot be employed, and the pulse-to-pulse fluctuations are larger than those shown in Figs. 1a and 1b. The normalized standard deviation has been found to be approximately 100%, which is consistent with the expected negative exponential probability distribution of LIDAR return signals due to speckle.[2,3]

The normalized standard deviation serves to define the uncertainty of the LIDAR measurements. Over intermediate ranges (< 4 km), the signal-to-noise ratio is frequently limited by this measurement uncertainty[4] rather than by instrumental dark-current or shot-noise limitations. To improve measurement accuracy, a standard procedure is to increase the number of measurements and take the average value. For N measurements, the standard deviation of the mean, σ_N, is expected to decrease as $N^{-1/2}$, assuming all the measurements are independent.

*The Lincoln Laboratory portion of this work was supported by the National Aeronautical and Space Administration and the Air Force Engineering and Services Center.
**The University of Maryland portion of this work was supported by the Department of Energy.

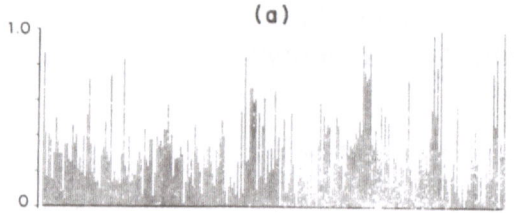

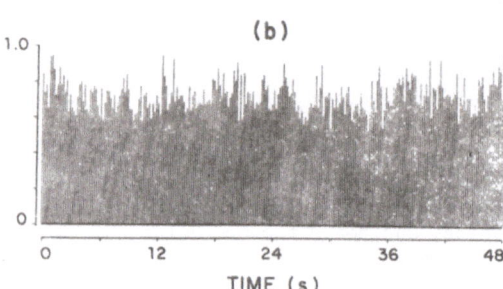

Fig. 1.
Backscattered LIDAR returns of successive normalized CO_2 laser pulses from a target at a range of 2.7 km; (a) returns from a 1-inch retroreflector, and (b) returns from a diffusely-reflecting target.

TIME (s)

In this paper we describe the results of a direct investigation of the effect on measurement uncertainty of averaging increasing numbers of successive pulse return signals. Our measurements have found σ_N to be much more weakly dependent on N than $N^{-1/2}$. The departure from $N^{-1/2}$ behavior indicates that the assumption that successive LIDAR signal returns are independent is invalid.

A theoretical analysis of the effect of temporal correlation on σ_N is described in the following section. Experimental results for both direct-detected and heterodyne-detected LIDAR signals are given in Section III and are shown to be in excellent accord with the analysis. The results indicate that the departure from $N^{-1/2}$ behavior can be explained by the presence of small but long-term temporal auto-correlation caused by time-varying atmospheric effects. These long-term correlation effects are shown to severely limit the improvement available through signal averaging, and imply a limitation to the improvement in the associated signal-to-noise ratio. Conclusions based on these results are discussed in Section IV. They are shown to have a direct bearing on the question of the relative merits of heterodyne- and direct-detection in DIAL systems, and are in general accord with recent theoretical studies.[5,6]

II. Theoretical Analysis

In order to describe analytically the statistical and temporal characteristics of the LIDAR returns, the pertinent parameters must be defined. First, define $I_k = I(t_k)$ as the normalized deviation of the kth pulse return (occurring at time t_k) from its mean value, $\overline{P}$, over the full data set. Then,

$$I_k = (P_k - \overline{P})/\overline{P} \tag{1}$$

where P_k is the kth LIDAR signal. The normalized variance of the full set of individual signals is defined as

$$\sigma^2 = \langle (I_k)^2 \rangle = \frac{1}{\Gamma} \sum_{k=1}^{\Gamma} I_k^2 \qquad (2)$$

where Γ is the total number of pulses in the set. The temporal auto-correlation is then defined as

$$\rho_j = \frac{1}{\sigma^2} \langle I(t_k) I(t_k + j\tau) \rangle = \frac{1}{\sigma^2(\Gamma - j)} \sum_{k=1}^{\Gamma - j} I_k I_{k+j} \qquad (3)$$

for a delay time $j\tau$, where τ is the time interval between pulses. It has been shown that, to a good approximation, the relationship between the standard deviation of the mean after averaging over N pulses, σ_N, and the temporal correlation is given by[7]

$$\sigma_N = \frac{\sigma}{\sqrt{N}} \left[1 + 2 \sum_{j=1}^{N-1} (1 - j/N) \rho_j \right]^{1/2} . \qquad (4)$$

In the absence of temporal correlation ($\rho_j = 0$), Eq. (4) predicts the ex-pected $N^{-1/2}$ behavior.

III. Experimental Results

A dual-hybrid-TEA CO_2 laser DIAL system was employed for the experiments. Through the use of a beam-splitter and two independent detection units, the system is capable of obtaining simultaneous heterodyne and direct-detection measurements of the same backscattered LIDAR signal. A schematic of the system is shown in Fig. 2; a more detailed description will be given elsewhere.[8] For the auto-correlation study described here, only one of the hybrid TEA CO_2 lasers was used.

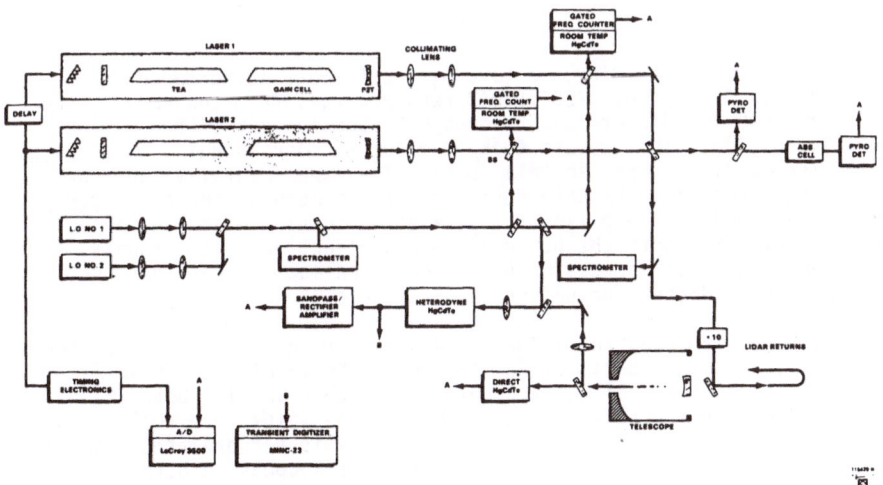

Fig. 2. Schematic of dual hybrid-TEA CO_2 laser differential-absorption LIDAR system.

Table I. Percentage standard deviation of signal-averaged LIDAR returns.

N	DIRECT DETECTION		HETERODYNE DETECTION	
	MEASURED	CALCULATED	MEASUREMENT	CALCULATED
1	22.1		99.3	
2	17.1	17.1	76.5	76.5
4	13.4	13.4	59.0	59.0
8	10.7	10.7	46.0	46.5
16	8.7	8.8	36.4	37.5
32	7.3	7.3	29.2	30.1
64	6.4	6.2	24.2	24.8
128	5.8	5.7	20.6	21.5
256	5.2	5.3	17.9	19.3
512	4.7	4.9	15.9	17.1
1024	4.0	4.3	14.2	13.9

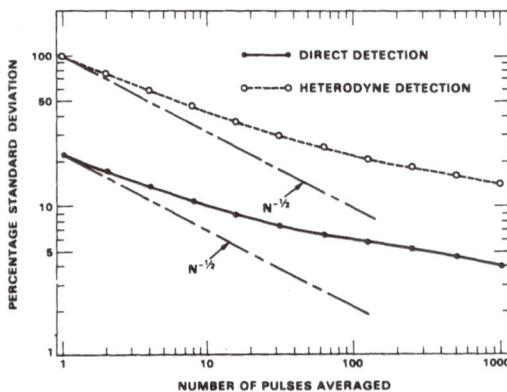

Fig. 3.
Measured percentage standard deviation of both direct- and heterodyne-detected LIDAR return signals from a diffuse reflector at a range of 2.7 km as a function of N, the number of pulses averaged. The dashed lines correspond to a $\sqrt{N}$ dependence.

Laser beam pulses were directed to a diffusely reflecting target at a range of 2.7 km. A total of 12288 LIDAR return pulses from the target were recorded from both the direct- and heterodyne-detection units [r = 12,288 in Eqs. (2) and (3)]. The process took 20 minutes, corresponding to a pulse repetition frequency of approximately 10 Hz. The experimental values obtained for the normalized standard deviation of the mean of the LIDAR signal returns averaged over N pulses, σ_N, are given in Fig. 3 and in Table I as a function of N for both the heterodyne- and direct-detected signals. The reduction of σ_N with increasing N is seen to be much slower than the $N^{-1/2}$ dependence predicted for uncorrelated signals.

The data were also used to evaluate the correlation coefficients ρ_j for $j=1,2,4,8....$ 1024 on the basis of Eq. (3), with the results shown in Fig. 4. Significant temporal auto-correlation is observed for both the direct- and heterodyne-detected signal returns. The values are smaller for the heterodyne-detected returns, but the difference is not dramatic. Small positive correlation is seen to persist for tens of seconds in both cases.

To establish if the experimental temporal correlation values are sufficient to explain the observed departure of σ_N from $N^{-1/2}$ behavior, the values

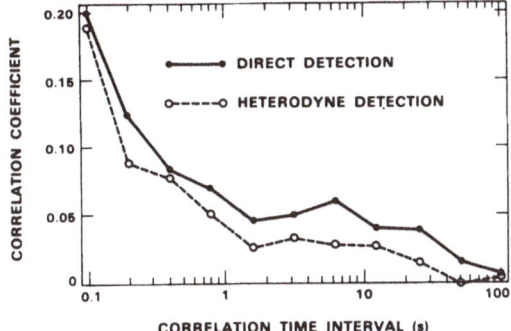

CORRELATION COEFFICIENT

— DIRECT DETECTION

o----o HETERODYNE DETECTION

CORRELATION TIME INTERVAL (s)

Fig. 4.

Auto-correlation coefficients of both direct- and heterodyne detected LIDAR return signals from a diffuse reflector at a range of 2.7 km as a function of time delay between signals based on a 10 Hz pulse-repetition rate.

of σ_N were calculated on the basis of Eq. (4) using the values of ρ_j shown in Fig. 4 and assuming a linear interpolation for all other values of ρ_j. The calculated results are also given in Table I, and are seen to be in excellent agreement with the measured values of σ_N. This agreement serves to validate the assumptions made in the derivation of Eq. (4), and illustrates the ability of small but persistent temporal correlation to severely limit the improvement, σ_N/σ, available through signal averaging of LIDAR returns from hard targets.

IV. Discussion

The above results, which show a limitation in the ability of signal averaging to reduce the standard deviation of the mean, are a direct consequence of the fact that the atmospheric transmission is varying during the measurement period with time constants from a few milliseconds upward.[9] As seen in Fig. 3 and Table I, the effect upon both heterodyne- and direct-detected signals is nearly the same, with approximately seven fold and six-fold improvements in σ_N/σ obtained in the two cases respectively after averaging over 1024 pulses. In addition, the curves of σ_N vs N indicate that additional improvement by further increasing the number of pulses averaged will be small. It may be added that the above results are consistent with other experimental measurements[4] which have, in general, shown a limitation in σ_N/σ of approximately 10 or less under various atmospheric conditions.

In view of these results, assuming a ten fold improvement in σ_N/σ, signal averaging of LIDAR returns from a hard target at a range of the order of a kilometer can not be expected to yield σ_N values below ~ 10% for the case of heterodyne detection, while the comparable limit for direct detection is ~ 2%. Therefore, if both types of detection system are capable of performing a given experiment, and the normalized standard deviation of the measurement is the limiting factor defining the signal-to-noise ratio[10,11] (as opposed to detector noise limitations), a direct-detection system appears to be preferable by a significant margin. The advantage of the direct-detection system in this instance can, of course, be reduced by using multiple detectors in conjunction with the heterodyne system to achieve additional signal averaging.

However, it should be noted that for a given laser energy, the ratio of the mean return signal value to the mean detector noise value[11] is much higher for a heterodyne-detection system than for a direct-detection system. It is therefore capable of making measurements at much greater ranges than are achievable with a direct-detection system. Therefore, both the increased detection range of the heterodyne-detection DIAL system and the greater accuracy at shorter ranges of the direct-detection system must be considered when designing for the overall system requirements.

REFERENCES

1. N. Menyuk and D. K. Killinger, "Temporal Correlation Measurements of Pulsed Dual CO_2 LIDAR Returns," Opt. Lett. 6, 301 (1981).

2. P. A. Pincus, M. E. Fossey, J. F. Holmes and J. R. Kerr, "Speckle Propagation Through Turbulence:Experimental," J. Opt. Soc. Am. 68, 760 (1978).

3. R. M. Hardesty, R. J. Keeler, M. J. Post and R. A. Richter, "Characteristics of Coherent LIDAR Returns from Calibration Targets and Aerosols," Appl. Opt. 20, 3763 (1981).

4. N. Menyuk, D. K. Killinger and W. E. DeFeo, "Laser Remote Sensing of Hydrazine, MMH, and UDMH Using a Differential-Absorption CO_2 LIDAR," Appl. Opt. 21, 2275 (1982).

5. R. M. Hardesty, "A Comparison of Heterodyne and Direct Detection CO_2 DIAL Systems for Ground-Based Humidity Profiling," NOAA Tech. Memo. ERL WPL-64, (October 1980).

6. P. Brockman, R. V. Hess, L. D. Staton and C. H. Bair, "DIAL with Heterodyne Detection Including Speckle Noise: Aircraft/Shuttle Measurements of O_3, H_2O, and NH_3 with Pulsed Tunable CO_2 Lasers," NASA Tech. Paper 1725, (August 1980).

7. N. Menyuk, D. K. Killinger and C. R. Menyuk, "Limitations of Signal Averaging Due to Temporal Correlation in Laser Remote Sensing Measurements," Appl. Opt. (to be published, September 15, 1982).

8. D. K. Killinger, N. Menyuk and W. E. DeFeo, (to be published).

9. A. G. Kjelaas, P. E. Nordal and A. Bjerkestrand, "Scintillation and Multiwavelength Coherence Effects in a Long-Path Laser Absorption Spectrometer," Appl. Opt. 17, 277 (1978).

10. J. H. Shapiro, B. A. Capron and R. C. Harney, "Imaging and Target Detection with a Heterodyne-Reception Optical Radar," Appl. Opt. 20, 3292 (1980).

11. When the measurement accuracy limitation due to the fluctuations of the signal returns is significantly greater than the limitations due to the detection capability of the system, as is the case in the measurements presented here, the signal-to-noise ratio is defined as the mean value of the LIDAR return signals divided by the standard deviation of those signals. This is equivalent to the inverse of the normalized standard deviation values used throughout this paper. The ratio of the mean value of the signal to the mean value of the noise (defined as the carrier-to-noise ratio in Ref. 10) is an equally important factor in establishing system capabilities. In general, for a given energy level, this ratio for a heterodyne-detection system can be expected to be 3 to 4 orders of magnitude greater than for a direct-detection system.

Part 5

UV-Fluorescene Remote Sensing

5.1 Rayleigh and Resonance Sounding of the Stratosphere and Mesosphere

Marie Lise Chanin

Service d'Aéronomie du CNRS, BP3,
F-91370 Verrieres-Le-Buisson, France

Lidar measurements of atmospheric species and physical parameters have been performed by our group from the Haute Provence Observatory (44°N, 6°E) since the early seventies. Until 1975, the main emphasis was placed on studying the upper atmosphere, and more precisely the alkali metals present in the atmosphere, between 80 and 100 km : Na[1], K[2] and more recently Li[3]. In the last 6 years we have extended our lidar facility to study the middle and lower atmosphere using Rayleigh and Mie scattering as well as range-resolved differential absorption technique. In this presentation I shall limit myself to present the state of the art in Lidar measurements in the stratosphere and mesosphere using resonance and Rayleigh scattering. As the detection of Rayleigh scattering up to 90 km came as a consequence of successive refinements on the resonant detection of the alkalis, both processes will be presented in their chronological order as a source of geophysical data.

It should be mentioned that the different technical approaches which will be presented here to answer specific questions should not be generalized for another altitude range or to measure other parameters : we are very aware of the inadequacy of a unique multipurpose Lidar to sound with the best performances the all atmospheric range from a few 100 meters to 100 km and to answer all requirements.

RESONANCE SCATTERING

Several metals from meteoritic origin are present in the form of atoms or ions in the upper atmosphere. As some of them have their atomic resonant transitions in the visible range, the large value of the resonance cross-sections ($\sim 10^{-12} cm^2$) made their detection one of the first achievements of the Lidar technique. Most of the work was concerned with the observation of sodium at nighttime. The purpose of my paper is not to present a review of that work, but only to describe our two more recent developments which could be considered as the state of the art in this domain : the nighttime detection of lithium (as its concentration is about 1000 time less than sodium) and the daytime detection of sodium in a background 10^6 time stronger than at nighttime[4].

Experimental system

Since 1970 the Lidar system has been in permanent evolution mainly if one looks at the laser performances. Until recently, the lasers used for resonant scattering were the results of our laboratory developments[5][6] : they are flash-pumped dye lasers emitting an energy of 1 Joule/pulse at 1 Hz repetition rate. The divergence of the emitted beam is 2 mrad which can be reduced to 0,2 mrad by using a collimating telescope.

A large telescope (ϕ = 81 cm) of optical quality ($\frac{\lambda}{4}$) has been used for all our mesospheric work ; its field of view can be adjusted from 3 to 0,15 mrad, but its reduction to 0.3 mard compatible with the laser divergence requires a quality of parallelism between emitter and receiver better than 0.1 mrad. A study of the geometrical form factor (ratio of the effective intersection volume of the two beams to a full overlapping volume) has been conducted [7] and the error due to misalignment should be taken into account in estimating the accuracy.

The absolute measurements of the concentration has also gained in accuracy by a better knowledge of the effective resonance cross-section. The fine tuning of the laser emitted wavelength and its monitoring by a Fizeau interferometer[8] insure a stability with respect to the reference wavelength better than 0.5 pm for a emitted linewidth of 6 pm (usually used in those experiments) insuring then a maximum error of 5 % on σ_{eff}.

At last, and in order to allow daytime measurements, the receiver spectral bandwidth can be reduced from 0,5 nm to 20 pm by a Fabry-Perot interferometer. This reduction implies a tempe-- rature controled system insuring a perfect matching between emitter and receiver central wavelengths within a few pm.

Experimental results

The reduction of the sky background obtained by limiting the field of view to 0.4 mrad has led to the observation of atomic lithium even though its concentration in the atmosphere varies between 0,1 and 1 atom.cm^{-3}. Fig. 1 presents an example of 3 hourly profiles obtained on November 14 1979.

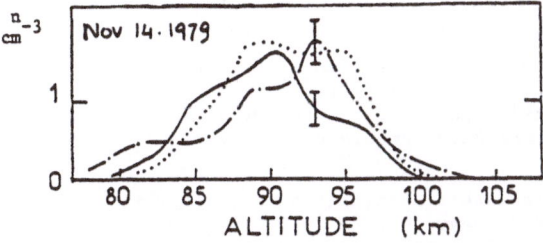

Fig. 1. Consecutive profiles of atomic lithium integrated over one hour

The reduction of both the field of view and the receiver spectral bandwidth was necessary to observe the sodium during daytime with a signal to noise ratio up to 50 [4]. Fig. 2 shows how the sodium is observed for a 16 seconds integration. The time resolution used to study diurnal variation is more usually 1 hour.

More recent laser developments and the use of $Nd^{3\pm}$ Yag pumped dye laser with 10 Hz repetition rate and a reduced energy/ pulse (100 mJ) do not affect the type of observations described above for which the energy/pulse is more important than the repetition rate at constant average power. We are aiming towards the development of flash pumped dye laser (1J/pulse) with improved

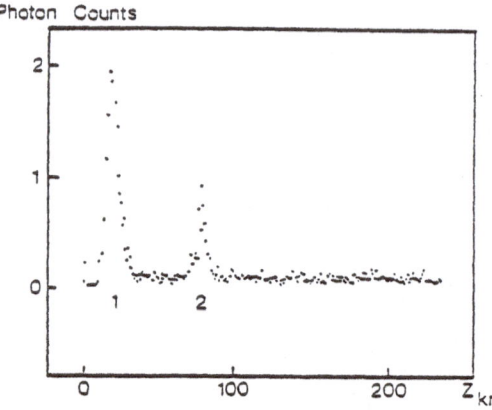

Photon Counts

Fig. 2. Realtime photon counts integrated over 16 laser shots. The signal 1 corresponds to the Rayleigh scattering above 20 km and the signal 2 refers to the scattered signal from the atomic sodium layer

repetition rate (10 Hz) to lower our detection level. But the gain in repetition rate has already proved its interest to study short-term varation of nighttime sodium, as the integration time can then be reduced to 1 minute. Furthermore such a laser is able to provide an emission line of 1.5 pm very precious for other applications.

Geophysical impact

Let us summarize briefly what has been learned from these Lidar observations of alkali metals about the behaviour of those species and,when using them as tracers, about the atmosphere it-self. The seasonal variations of the 3 alkalis we have been able to observe (Na, K, Li),and the impossibility for existing models to explain those variations,triggered in our group the development of a new model [9] which is now able to explain, under the combined influence of photo-chemistry and dynamics, all the long term variations which had been observed. An extension of this model is also in progress to explain the short term variations which are observed from the day and night survey of the Na behaviour.

The sodium (as the most abundant of the alkalis in the atmosphere) has been used as tracer of temperature[10], eddy diffusion coefficient[1] and waves[11] and we expect to use it in the near future to measure simultaneously temperature and winds between 80 and 100 km.

Further developments of dye lasers at different wavelengths are undertaken now in order to study heavy metals in their neutral and ionized forms : Ca, Ca^+, Fe^+... This programme is mostly related to the understanding of the history of meteoritic debris entering into our atmosphere. Another topic related to cosmology is pushing us to improve the detection of lithium in order to measure the isotopic ratio of lithium in comets.

RAYLEIGH SCATTERING

The absolute calibration of the alkali density profile is obtained by reference with neutral density measured from Rayleigh scattering between 30 and 40 km, which was then recorded simulta-neously. The detection of very small lithium concentration led to a reduction of the background which then allowed the detection of

Rayleigh scattering echos as high as 90 km. The first density profiles were then obtained at the Na and Li resonance wavelengths
while observing those alkalis. In the last year the measurements
have been made using the second harmonic of a Nd-Yag laser emitting
300 mJ at 532 nm with a 10 Hz repetition rate. The efficiency has
then been increased by more than an order of magnitude.

Principle - Limits and accuracy

The method has been described in several articles[12][13] and
will be only briefly recalled. In the absence of any aerosol layer
the backscattered light is proportional to atmospheric density. The
presence of aerosols, checked by two-wavelengths Lidar sounding,
usually limits the measurements to altitudes above 30 km. The upper
limit is limited by the signal to noise ratio and is right now 80 km
(Fig. 3). The temperature profile is then deduced from the density
assuming that the atmosphere obeys the perfect gas law and is in
hydrostatic equilibrium. It is important to notice that, even though
the density measurement is relative, the temperature derived from
it is absolute. Fig. 4 represents a temperature profile deduced from
the density measurements of Fig. 3. The performance of the method
has been improved continuously in the last 2 years through several
steps (more adequate choice of wavelength, higher repetition rate,
reduction of sky background) and the accuracy as for today is given
in Fig. 5. The method until now has been limited to nighttime but
the reduction of the receiver spectral bandwidth, which proved to be
satisfactory for sodium daytime observations, will allow to extend
the measurements during the day to a range ob about 60 km.

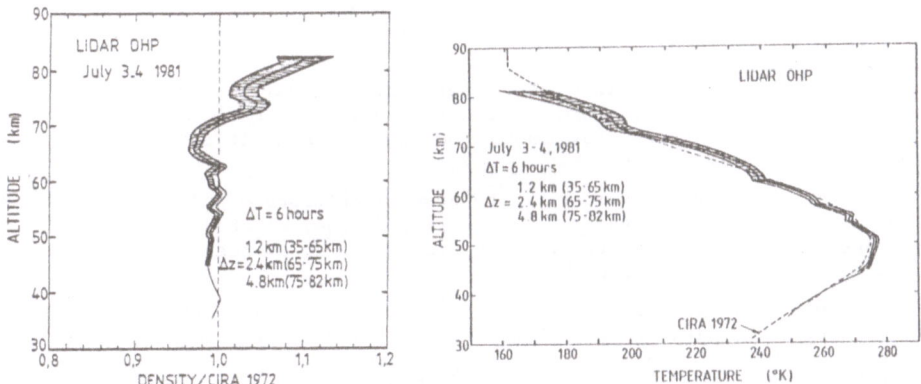

Fig. 3. Density profile normalized to the CIRA 1972 model. The shaded area
correspond to ± 1 standard deviation

Fig. 4. Temperature profile for the night of July 3-4 1981. The dotted line
is the CIRA 1972 model. The shaded area corresponds to ± 1 standard deviation

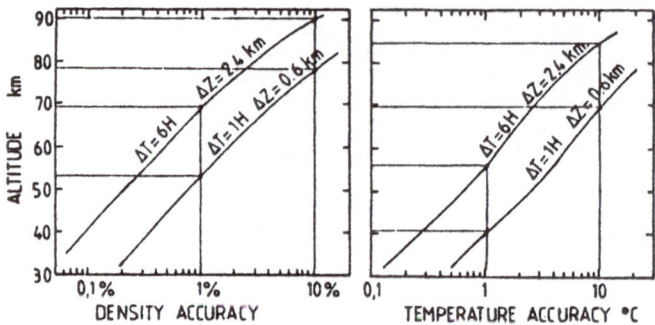

Fig. 5. Density and temperature accuracy as experimentally obtained in July 1981

Geophysical results

This method has been used on a routine basis for a full year
and a certain number of data of poorer quality are available since
1979. From this set of data we can conclude to a large variability
of density and temperature with different time scales :

- density fluctuations with amplitudes of 2% at 50 km are
often observed to last for several hours and a spectral analysis in
time and space of these fluctuations is being conducted now on a
large number of cases. It looks premature to-day to conclude about
their nature.

- when integrated over more than 6 hours, those structures
disappear and the density and temperature profiles compared with a
model (CIRA 72 for example) indicate a surprisingly good agreement
during summer periods (Fig. 4). This agreement is less obvious in
other seasons.

- Nighttime cooling of the stratosphere and mesosphere can be
observed when the amplitude of short-term fluctuations does not hide
the effect.

- Variations of density and temperature from day to day are
observed and seem to be related to planetary wave propagation (as
this propagation is blocked by easterly winds in summer it does not
appear at that time).

- Study of the behaviour.of the upper stratosphere and the
mesosphere has been studied in detail during 3 major stratospheric
warmings and indicate that the warming propagates downwards from
the mesosphere in about 20 days [14].

- Detection of a general trend in the long term variation of
stratospheric temperature is within the possibility of the technique
as the amplitude of variation induced by environmental pollution could
be of the order of 0.3 K/year and should be measurable.

Future developments

We already mentioned the possibility of daytime operation.
Improvement of the time resolution in order to have access to short-
term fluctuations would also be worth some laser developments.

Without any further instrumental improvements, we think that these observations should be soon performed from 2 lidar stations (one being mobile) to have access to the horizontal extension of the density fluctuations and we also have in mind the organisation of a worlwide network of such Lidars.

CONCLUSION

We are convinced that the Lidar is a unique tool to study, during day and night and from the ground to 100 km, a certain number of species and several basic parameters, and more precisely in the high stratosphere and mesosphere which is not accessible to balloons and radars and where satellite data do not provide enough spatial resolution .

ACKNOWLEDGEMENTS

The work described in this paper reflects the activity of the whole team whom I have been working with and I am pleased to offer my warmful thanks to each of its member.

REFERENCES

[1] Megie G and Blamont J.E.
Laser sounding of atmospheric sodium : interpretation in terms of global atmospheric parameters,
Planet Space Sci. 25 1039, 1977

[2] Megie G., Bos F., Blamont J.E. and Chanin M.L.,
Simultaneous nightime Lidar measurements of atmospheric sodium and potassium
Planet Space Sci. 26, 27 1978

[3] Jegou J.P., M.L. Chanin, G. Mégie, J.E. Blamont,
Lidar measurements of atmospheric Lithium
Geophys. Res. Lett. 7, 995, 1980

[4] Granier C. and G. Mégie,
Daytime Lidar measurements of mesospheric sodium layer
Planet Space Sci. (accepted) 1981

[5] Loth C. and G. Mégie
High spectral luminance dye amplifier
Journal of Physics E 7 80, 1974

[6] Allain J.Y.,
Description of high energy tunable dye lasers for atmospheric sounding
Applied Optics 18, 287, 1979

[7] Lefrère J.
Etude par sondage laser de la basse atmosphère
Thèse de 3ème cycle, University of Paris, 1982

[8] Cahen C, Jegou J.P., Pelon J., Gildwarg P and Porteneuve J.,
Wavelength stabilization and control of the emission of pulsed dye lasers by means of a multibeam Fizeau interferometer,
Rev. Phys. Appl. 16, 6, 353, 1981

[9] Jegou J.P.,
    Etude expérimentale et théorique des métaux alcalins dans
    la haute atmosphère 70 - 110 km
    Thesis University of Paris, 1982

[10] Blamont J.E., M.L. Chanin and G. Mégié,
    Vertical distribution and temperature profile of the night
    time atmospheric sodium layes obtained by laser bacscatter,
    Ann. Geophys. 28, 833, 1972

[11] Juramy P, M.L. Chanin, G. Mégie, G.F. Toulinov and Y.P. Doudolado,
    Lidar sounding of the mesospheric sodium layer at high
    latitude,
    J. Atmos. Terr. Phys. 43, 209, 1981

[12] Hauchecorne A. and M.L. Chanin
    Density and temperature profiles obtained by Lidar between
    30 and 80 km
    Geophys. Res. Lett. 7, 565, 1980

[13] Chanin M.L. and A. Hauchecorne,
    Lidar observation of gravity and tidal waves in the stratosphere
    and mesosphere
    J. Geophys. Res. 86 9715, 1981

[14] Hauchecorne A. and M.L. Chanin
    Mid latitude ground based lidar study of stratospheric warming
    and planetary waves propagation,
    submitted to J. Atmos. Terr. Phys., 1982.

5.2 High-Resolution Lidar System for Measuring the Spatial and Temporal Structure of the Mesospheric Sodium Layer

C.S. Gardner, C.F. Sechrist, Jr., and J.D. Shelton

Aeronomy Laboratory, Department of Electrical Engineering, University of Illinois, Urbana, IL 61801, USA

One of the first remote sensing applications of tunable dye lasers was in the study of mesospheric sodium. The layer, which is believed to be of meteoric origin, is confined to altitudes between approximately 80 and 100 km. The layer has been investigated using a variety of techniques since its discovery in the late 1930s. A review of early measurements and the theory of sodium layer chemistry is contained in Brown (1973). Before dye laser based lidar systems were developed, sodium measurements were largely restricted to studying resonantly scattered sunlight. Ground-based measurements of this type were able to define seasonal variations in column abundance, but the sharp layer boundaries were not revealed until rocket-borne dayglow measurements were made (Hunten and Wallace, 1967). Lidar observations of the vertical structure of the sodium layer were first made in England (Bowman et al., 1969). Since then, similar measurements have been reported from a variety of locations including France (Blamont et al., 1972), Brazil (Kirchoff and Clemesha, 1973), California (Hake et al., 1972), Illinois (Rowlett et al., 1978) and at high latitudes (Franz Joseph Land, USSR), (Juramy et al., 1981). Because the daytime sky produces very high background photocount rates in most sodium lidar receivers, the majority of the lidar observations of the layer have been restricted to nighttime. During the past decade, improvements in dye laser output and stability and the development of effective signal processing techniques, have significantly enhanced the quality and resolution (spatial and temporal) of sodium lidar measurements. This paper describes the design of the University of Illinois lidar system and presents some of the sodium measurements obtained with it.

The peak concentration of atomic sodium in the layer is only on the order of 10^9–10^{10} m^{-3}. However, because the resonant backscatter cross-section is large, adequate signal levels can be obtained when the laser is tuned to the D_2 resonance line at 589.0 nm. If we neglect convolutional effects of the laser pulse shape and sodium radiative lifetime, the photon counting rate due to the backscattered laser energy is given by

$$\lambda_s(t) = \frac{\eta A_R C}{2h\nu} J\gamma(t) \tag{1}$$

$$\gamma(t) = \frac{\beta(r)}{r^2}\exp\left[-2\int_0^r dz\alpha(z)\right], r = \frac{ct}{2} \tag{2}$$

where

η = overall receiver efficiency
A_R = receiver aperture area
h = Planck's constant
ν = laser frequency
$\beta(r)$ = volume backscatter coefficient (m^{-1}) at range r

$\alpha(r)$ = volume extinction coefficient (m^{-1}) at range r
c = velocity of light
J = laser pulse energy

The sodium density varies with time and position and can be determined by estimating the signal counting rate $\lambda_S(t)$. Estimates λ_S are obtained by counting the number of detected signal photons in contiguous time intervals or range bins. If the sodium density does not change appreciably within a range bin, the expected signal photocount is given by

$$<x(t)> \simeq \frac{\eta A_R CT}{2h\nu} J\gamma(t) \tag{3}$$

where T is the counting interval width. For the UI lidar system, T can be as short as 1 μsec which corresponds to a range resolution of 150 m. In practice, however, range resolution is usually limited by signal shot noise.

Table I is a summary of the major system parameters for the UI lidar. The telescope was designed and constructed using a 1.22 m (48 in) diameter acrylic Fresnel lens. The dye laser was manufactured by Candela Corporation and is tuned using a grating and etalon. The output wavelength is monitored by directing a small portion of the beam to a grating spectrograph and a sodium cell. Coarse tuning to the D2 line is monitored with the spectrograph, while fine tuning is accomplished by observing resonant scattering in the sodium cell.

TABLE I

University of Illinois Lidar System Parameters

Laser	
Type	Tunable Flashlamp Pumped Dye Laser
Output Energy	50 mj/pulse
Pulse Repetition Frequency	10 Hz
Pulse Width	2 μs FWHM
Wavelength	589.0 nm
Linewidth	1 pm FWHM
Beam Divergence	1 mrad

UI Telescope	
Type	f/1.56 Fresnel Lens Objective
Diameter	1.22 m (48 in)
Area	1.17 m^2
Bandwidth	5 nm FWHM
Field-of-View	3 mrad

Goddard Telescope	
Type	Cassegrain System, f/25 at Coudé focus
Diameter	1.22 m
Bandwidth	5 nm FWHM
Field-of-View	100 μrad
Elevation Angle	60°
Azimuthal Range	60°
Azimuth Scan Time	15 min/60° (signal strength limited)

Prior to firing the laser, the gain of the PMT is reduced by 20 dB to prevent overloading by the strong Rayleigh and aerosol scattering in the troposphere. A laser pulse detector is used to synchronize the photon counter with the firing of the laser to provide accurate range timing.

Typically, the number of detected signal photons per laser pulse varies between approximately 5 to 100 depending upon the atmospheric conditions and laser output. Since approximately 1000 photocounts are required to construct a good quality sodium profile, the photocounts from 100 to 250 laser shots are integrated and then stored on floppy disk for later processing. Because the laser pulse frequency is 10 Hz, the integration time per profile varies between 10 and 30 sec.

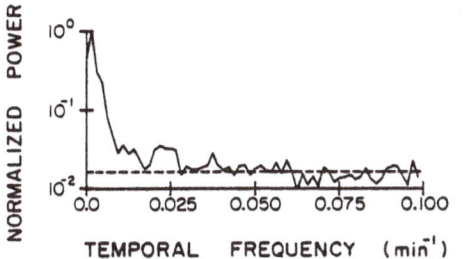

Figure 1. Plot of the normalized average temporal periodogram of data collected on March 16-17, 1981

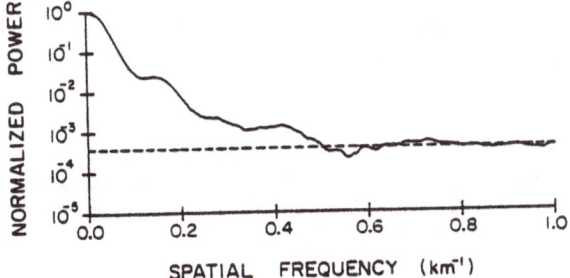

Figure 2. Plot of the normalized average spatial periodogram of temporally filtered data collected on March 16-17, 1981. The temporal filter cutoff was at 0.0308 min^{-1}

The system has been operated continuously from sunset to sunrise obtaining up to 1000 individual profiles during a single night's operation. The spatial and temporal resolution of the data can be enhanced considerably by employing 2-D filtering techniques to process the raw photocounts (Gardner and Shelton, 1981). Figures 1 and 2 are plots of the average temporal and spatial periodograms (power spectra) of the photocount data which were collected on March 16-17, 1981. The estimated sodium profiles, obtained by spatially and temporally filtering the photocount data, are plotted in Figure 3. The observations span a period of more than 9 hours and illustrate the highly dynamic nature of the layer. The wave-like features and oscillations in the topside boundary are believed to be caused by internal gravity waves and tidal effects. The dotted lines in Figures 1 and 2 are the shot noise levels. In the spatial periodogram, the signal drops into the noise at approximately 0.5 km^{-1} indicating that the spatial resolution is limited to about 2 km for these data. The temporal resolution is approximately 30 min.

Steerable lidar measurements have been used by some groups to study horizontal density variations in the sodium layer. The first steerable measurements were made between September 1975 and January 1976 at Winkfield, U.K.

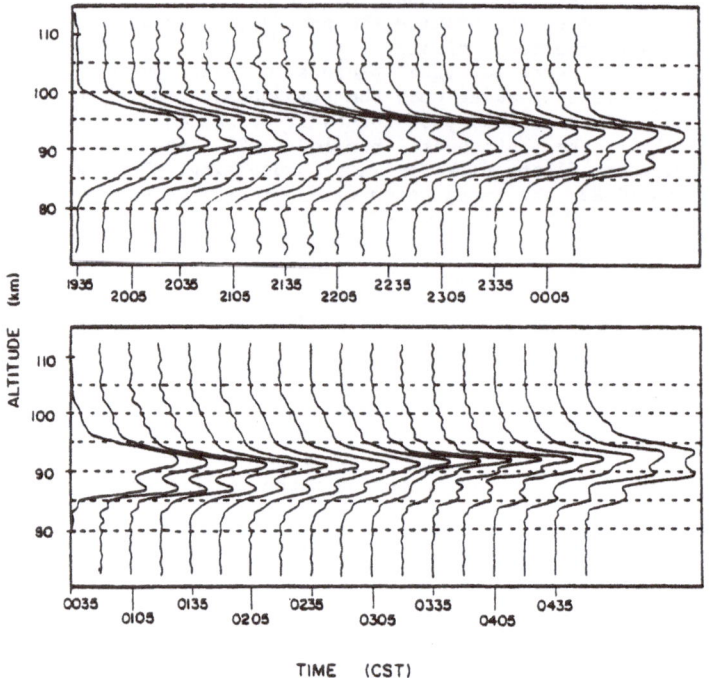

MARCH 16-17, 1981

Figure 3.   Time history of the estimated altitude profiles of sodium
             density observed on March 16-17, 1981. The spatial and temporal
             filter cutoffs were 0.476 km^{-1} and 0.0308 min^{-1}, respectively.
             Estimated profiles are plotted every 15 minutes

(Thomas et al., 1976). Some of these observations indicated changes in the
height distribution of sodium occurring over small horizontal separations
(approximately 15 km) in short time periods (approximately 2 to 3 minutes).
More extended observations have been made in the Southern Hemisphere by Clemesha
et al. (1981) at San Hose dos Campos, Brazil. In June and October, 1981, the
48-inch astronomical telescope located at the Goddard Space Flight Center
Optical Test Site was used in conjunction with the components of the University
of Illinois lidar system to make steerable measurements of the sodium layer.
The telescope facility consists of a 48-inch Cassegrain telescope with Coude'
focus and supporting equipment which enables this instrument to be used in a
variety of laser ranging experiments. The pointing of the telescope is computer
controlled and therefore it is easily and rapidly positioned at the desired
azimuth and elevation angles. The laser head was mounted directly on the telescope to ensure that the beam tracked the telescope motions accurately. The
laser beam was also expanded and collimated to reduce the divergence to about
100 μrad. In June measurements were obtained by scanning the telescope between
three points separated horizontally by about 45 km. Figure 4 is a plot of the
sodium profiles obtained at Goddard on June 17-18 by pointing the telescope at
zenith. The profile measured at 2348 EDST is replotted in Figure 5 along with
the range corrected photocount data. Similar profiles were also obtained at the
two off-zenith points. In October, additional measurements were made at Goddard
by upgrading the receiving system computer to permit azimuthal scanning of the

202

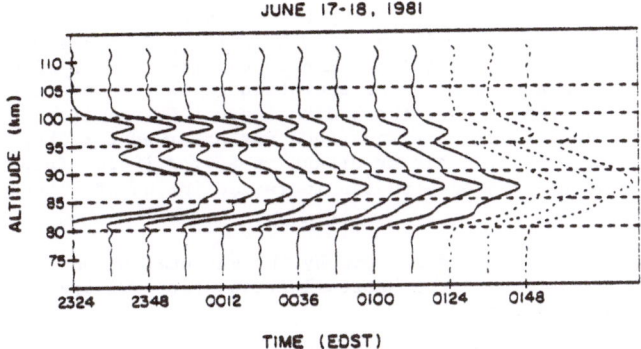

JUNE 17-18, 1981

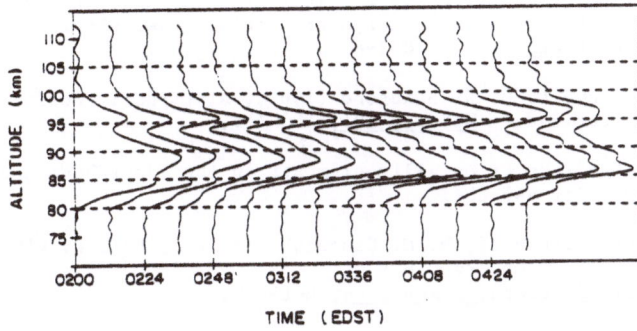

Figure 4. Altitude profiles of sodium density observed at the Goddard Space Flight Center on June 17-18, 1981. The spatial and temporal filter cutoffs were 0.4 km⁻¹ and 0.033 min⁻¹ respectively. Profiles are plotted at 12 minute intervals. Interpolated profiles are indicated by dashed curves

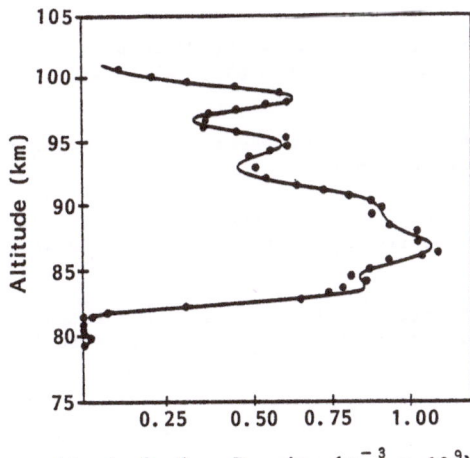

Figure 5. Plot of the sodium density profile observed at 2348 EDST on June 17-18, 1981. The dots are range corrected photocount data and the smooth curve is 2-D filtered photocount data

telescope. Measurements were obtained by pointing the telescope 30° off zenith and repeatedly scanning over a range of 60° in azimuth in 1° or 2° increments. The 60° azimuthal scan required approximately 15 min to complete. Over 3000 profiles were collected in 3 separate nights of operation revealing interesting features about the horizontal structure of the sodium layer. Presunrise enhancement of the column abundance of more than 100% was observed corroborating earlier measurements at Goddard in June and at Urbana in Spring 1981. This enhancement is believed to be associated with the increased influx of meteoric dust in the early morning hours.

Acknowledgement. This work was supported in part by the National Science Foundation under grant ATM 79-20726 and the National Aeronautics and Space Administration under grant NGR 14-005-181.

REFERENCES

Blamont, J. E., M. L. Chanin and G. Megie (1972), Vertical distribution and temperature profile of the nighttime atmospheric sodium layer obtained by laser backscatter, Ann. Geophys., 28, 833-838.

Bowman, M. R., A. J. Gibson and M. C. W. Sandford (1969), Atmospheric sodium measured by a tuned laser radar, Nature, 221, 456-457.

Brown, T. L. (1973), The chemistry of metallic elements in the ionosphere and mesosphere, Chem. Rev., 73, 645-667.

Clemesha, B. R., V. W. J. H. Kirchhoff, D. M. Simonich and P. P. Batista (1981), Mesospheric winds from lidar observations of the atmospheric sodium, J. Geophys. Res., 86, 868-870.

Gardner, C. S. and J. D. Shelton (1981), Spatial and temporal filtering techniques for processing lidar photocount data, Opt. Lett., 6, 174-176.

Hake, R. D., Jr., D. E. Arnold, D. W. Jackson, W. E. Evans, B. P. Ficklin and R. A. Long (1972), Dye-laser observations of the nighttime atomic sodium layer, J. Geophys. Res., 77, 6839-6848.

Hunten, D. M. and L. Wallace (1967), Rocket measurements of the sodium dayglow, J. Geophys. Res., 72, 69-79.

Juramy, P. M., L. Chanin, G. Megie, G. F. Toulinov and Y. P. Doudoladov (1981), Lidar sounding of the mesospheric sodium layer at high latitudes, J. Atmos. Terr. Phys., 43, 209-215.

Kirchhoff, V. W. J. H. and B. R. Clemesha (1973), Atmospheric sodium measurements at 23° S, J. Geophys. Res., 77, 6839-6848.

Rowlett, J. R., C. S. Gardner, E. S. Richter and C. F. Sechrist, Jr. (1978), Lidar observations of wave-like structure in the atmospheric sodium layer, Geophys. Res. Lett., 5, 683-686.

Thomas, L., A. J. Gibson and S. K. Bhattacharyya (1976), Spatial and temporal variations of the atmospheric sodium layer observed with a steerable laser radar, Nature, 263, 115-116.

5.3 Remote Sensing of OH in the Atmosphere Using the Technique of Laser-Induced Fluorescene

Charles C. Wang

Engineering and Research Staff, Research, Ford Motor Company, Dearborn, MI 48121, USA

INTRODUCTION

The hydroxyl radical (OH) is a reactive species which controls many of the chemical processes operative in the atmosphere. OH is important in ozone chemistry because it relates to the process of photochemical smog formation in the troposphere and to the partitioning of the odd-nitrogen and odd-chlorine compounds in the stratosphere. Various estimates [Weinstock and Niki, 1972; Liu, 1977; Chameides, 1978; Hameed et al., 1979] place the global yearly averaged OH concentration at between 10^5 molecules/cm^3 and 10^6 molecules/cm^3, depending to a large extent on the assumed perturbations to the natural atmosphere. However, measurements of OH in the troposphere have been scarce, and results of these measurements [Wang et al., 1975; Davis et al., 1976; Perner et al., 1976; Campbell et al., 1979] have been less than satisfactory. For example, the absorption technique [Perner et al., 1976] should in principle provide accurate OH measurements but remains to be developed [Killinger and Wang, 1977]. The isotope tracing technique [Campbell et al., 1979] offers an interesting alternative for OH monitoring, but remains to be calibrated, with possible systematic errors associated with the technique identified and remedied. The technique of laser-induced fluorescence [Wang and Davis, 1974a, b; Killinger et al., 1976] is probably the most sensitive, but it has so far been marred by a multitude of interferences [Wang et al., 1976; Hanabusa et al., 1977; Wu et al., 1976; Wang and Davis, 1975] induced by the probing laser beam and by inaccuracies arising from the uncertainties in the value of the parameters, such as fluorescence efficiency, linewidth, etc., in data reduction. Recently, we have obtained more reliable values for these parameters [Selzer and Wang, 1979; Wang and Killinger, 1977; Killinger and Wang, 1979; Wang et al., 1980; Wang and Huang, 1980; Wang and Killinger, 1979; Wang et al., 1981] and we have redesigned our OH-measuring instrument so that laser-induced interferences were reduced to a negligible level. This instrument was flown on a NASA aircraft in 1979 [Wang et al., 1981] and was used near Niwot Ridge, Colorado, both in 1979 and in 1981. In each case, reliable measurements of OH were demonstrated.

DETECTION SCHEME

The use of laser-induced fluorescence has been described in detail elsewhere [Wang and Davis, 1974a; Killinger et al., 1976]. This technique involves excitation of the OH radical, using one of the rotational-electronic lines in the $^2\pi(v'' = 0) \to {}^2\Sigma^+(v' = 1)$ transitions and observing the fluorescence emission associated with the $^2\Sigma^+(v' = 1) \to {}^2\pi(v'' = 1)$ transitions near 3145 Å. With the availability of high power laser sources and with the occurrence of fluorescence sufficiently red shifted from the exciting radiation, this technique promises unprecedented sensitivity and selectivity. However, it should be noted that a successful application of this technique is predicated upon knowledge of the fluorescence spectrum and many physical and spectroscopic parameters involved in the absorption and reemission processes. It required elaborate laboratory studies, the results of which are briefly summarized in the following paragraphs.

There are a number of considerations in the choice of excitation wavelength. At ambient temperatures, most of the OH populations resides in the lowest two rotational levels of the ground electronic state. However, the rotational transition probability for any given type of transition increases with increasing rotational quantum number; there are also close-by satellite structures for some of the rotational-electronic lines, which make it difficult to interpret the results obtained through wavelength tuning (see discussion in a later section). These considerations together dictate that the appropriate line for fluorescence excitation be the $P_1(2)$ line near 2825.8 Å or the $Q_1(2)$ line near 2820.7 Å.

With the laser excitation indicated above, the spectral distribution of the fluorescence emission is found to be pressure dependent at low pressures but becomes independent of pressure at pressures above 40 torr of nitrogen [Killinger et al., 1976]. This pressure independence simplifies the relation between the total fluorescence emission and the fluorescence intensity observed over a finite spectral width under ambient conditions. For OH fluorescence emitted under ambient conditions, approximately 70% of the total intensity is associated with the (0,0) transitions, with the remainder appearing in the (1,1) transitions.

With the laser light tuned in resonance with one of the above transitions, the OH fluorescence signal is given by

$$(\text{OH signal}) = A[\text{OH}] \frac{\sigma_o(\text{OH})}{\Delta\nu} \eta \ (\Delta n/n) \quad , \tag{1}$$

where [OH] is the concentration of OH, $(\Delta n/n)$ is the fraction of the OH population residing in the rotational level from which the exciting transition originates, $\sigma_o(\text{OH})$ is the integrated absorption cross section for the particular rotational-electronic transition, $\Delta\nu$ is the effective linewidth for the transition, η is the overall fluorescence efficiency under ambient conditions, and A is a constant determined by the overall excitation and collection efficiencies. This constant can be evaluated most conveniently by noting that the Raman-scattered signal owing to nitrogen molecules occurs around 3025 Å, for which the excitation configuration and collection efficiencies are the same for all practical purposes. This N_2 Raman-scattered signal is given by

$$(N_2 \text{signal}) = A[N_2]\sigma_R(N_2) \quad , \tag{2}$$

where $[N_2]$ is the nitrogen concentration in air, and $\sigma_R(N_2)$ is the cross section for Raman scattering of nitrogen. Elimination of A in (1) and (2) gives

$$[\text{OH}] = \frac{\text{OH signal}}{N_2 \text{ signal}} \times \frac{[N_2]\sigma_R(N_2)\Delta\nu}{\sigma_o(\text{OH})\eta \ (\Delta n/n)} \quad . \tag{3}$$

TABLE 1. Values for the Parameters of Equation (3) Used in OH Measurements Under Sea Level Conditions

Parameter	Uncertainty	Parameter Values
$\sigma(N_2)$	(8%)[a]	7.4×10^{29} cm2/molecule sr
$\sigma_o(OH)$, $Q_1(2)$	(15%)[b]	0.72×10^{-16}cm2-cm-1
$\sigma_o(OH)$, $P_1(2)$	(15%)[b]	0.54×10^{-16}cm2-cm-1
$\Delta n/N$ (computer value)[c]		0.197
η	(30%)[d]	0.001
Doppler width (computed value)		0.11 cm-1
Laser linewidth[e]		0.1-0.2 cm-1
Collision-broadened width[f]		0.14 cm-1

The appropriate values for the aprameters that occur in the above equations are summarized in Table 1. Among these parameters the Raman cross section of nitrogen has been known to ±8% [Hyatt et al., 1973]. Values for the fluorescence efficiency were scattered over an order of magnitude, but recent work by Selzer and Wang, [1979] has narrowed the range of uncertainties considerably; a thorough discussion on the value for this parameter also may be found there. The oscillator strength (and hence the integrated absorption cross section) of OH has been measured recently along with its rotation and spin dependences [Wang and Killinger, 1977; Wang and Huang, 1980]. The fractional population, ($\Delta n/n$), within a given rotational level is determined in the usual manner [Killinger et al., 1976; Killinger and Wang, 1979] with the partition function summed over all the rotational (and vibrational) levels, as well as the two spin sublevels and the Λ components for each rotational level. The effective linewidth involves convolution of the laser linewidth with the homogeneous and inhomogeneous linewidths of the OH transition [Killinger et al., 1976; Killinger and Wang, 1979; Wang et al., 1980].

For OH measurements at 33,000 feet, use of the appropriate values for the parameters listed in Table 1 lead to the following numerical value for the deduced OH concentration:

$$[OH] = 3 \times 10^9 \frac{(OH\ signal)}{(N_2\ signal)} . \qquad (4)$$

Similar values can be obtained easily for other altitudes and for varying degrees of water content, although the quenching owing to the latter is generally small. To the extent that water quenching is negligible, the numerical constant in the above equation varies as the inverse square of the nitrogen pressure.

Some of the values are indicated for a thermalized distribution at 300°K. For other temperatures, see Killinger and Wang [1979] and Wang et al. [1980]. The number in parenthesis indicates the uncertainty in the corresponding parameter value.

[a]From Hyatt et al. [1973].
[b]See Wang and Killinger [1979], and references cited therein.

cSee Killinger et al. [1976].
dSee Selzer and Wang [1979]. A multiplicative factor of 0.7 should apply if fluorescence from the (0,0) transitions only is detected.
eThe range of values given is primarily a result of jitter in the laser frequency.
fFor dry air only. See Wang et al. [1980], for the temperature-dependent contribution from water molecules.

Although most of the parameters used in deducing OH concentrations from fluorescence measurements are reasonably well determined, it would still be desirable to calibrate the fluorescence technique described above against some known source of OH under ambient conditions. One possible means for such a calibration is to make simultaneous measurements of OH in ambient air using both the fluorescence technique and the absorpion technique [Perner et al., 1976; Killinger and Wang, 1977]. Since the absorption technique can be calibrated against a thermal source of OH at high temperatures [Killinger and Wang, 1979], it serves as a secondary standard against which the fluorescence instrument can be calibrated. However, since the absorption technique is inherently much less sensitive, such a calibration can be realized only when high OH concentrations exceeding 10^7 OH/cm^3 are encountered.

DESCRIPTION OF EQUIPMENT

Figure 1 depicts the experimental setup used for OH measurements. The output from a doubled Nd-YAG laser operating near 5300 Å was used to

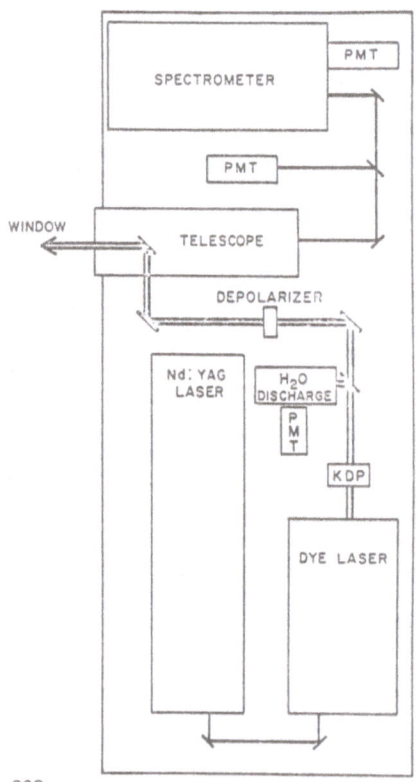

Fig. 1. Schematic of the experimental setup. The single line indicates the fundamental beam, and the double line indicates both the fundamental and the second harmonic beams

pump a tunable dye laser system consisting of an oscillator and two amplifiers. Next the output from this dye laser system was sent through a crystal of KDP (potassium dihydrogen phosphate) to generate second harmonic radiation near 2825 Å. The unfocused second harmonic radiation was expanded, recollimated, and then sent along the axis of a Cassegrainian telescope, through an aircraft window, to excite resonance fluorescence of OH in the outside air. The fluorescence light emitted in the backward direction was collected through the telescope and was imaged into the entrance slit of a spectrometer for processing. The output near 3090 Å from the spectrometer was then detected by a high gain photomultiplier and was processed by photon-counting apparatus with 10 ns resolution.

Under normal conditions the lasers operated at a repetition rate of 10 pps, delivering a maximum output of 5 mJ/pulse near 2825 Å with a linewidth of 0.1 cm^{-1} and a beam diameter expanded to 2.5 cm or larger. This radiation was tuned on and off the OH resonance by pressure tuning the air-spaced etalon inside the dye laser.

The use of a telescope for light collection in a coaxial configuration is advantageous for at least two reasons. First of all, it eliminates the need for air sampling and does not constrain measurements to the immediate proximity of the measurement platform. It also allows the excitation beam to be expanded so that interference owing to laser-induced generation of OH may be reduced to a negligible level. (See next section.) Estimates based on geometric optics indicate that the collection efficiency in a coaxial configuration is comparable to that obtainable with f/1 optics under actual measurement conditions. The telescope employed in our instrument was an eight inch Cassegrainian with f/4 primary and f/16 effective. The reflecting optics have been dielectric coated to provide about 95% reflectivity at 3090 Å. The focal distance was adjusted to be about 30 feet for aircraft operations.

For light processing a Spex model 1401 spectrometer equipped with a 3600 line/mm holographic grating was used. The spectrometer had a focal length of 0.85 m, with f/8.5 and a slit opening of 5 mm x 2 cm. To attenuate further the Rayleigh-scattered light near 2825 Å and N_2 Raman-scattered light near 3025 Å, a liquid filter consisting of 1,2,4,5 tetrachlorobenzene dissolved in cyclohexane also was employed.

Wavelength calibration for on-resonance excitation was accomplished by directing part of the exciting light into a water discharge and by analyzing the spectrum of the induced fluorescence emitted by the high OH concentration present therein [Selzer and Wang, 1979].

Note that the N_2-Raman-scattered light is polarized, whereas the OH fluorescence is completely depolarized. A depolarizer was thus used to depolarize the exciting beam so that both the N_2-Raman-scattered light and the OH fluorescence appeared to be completely depolarized. This procedure in effect eliminated any systematic error in the N_2 normalization arising from polarization-dependent efficiency of the grating.

Under most operating conditions, solar background and/or nonresonant fluorescence owing to other absorbing species may be excessively large. To obtain an OH signal in the presence of comparatively much larger background, it was necessary to tune periodically the exciting radiation on and off the OH resonance and to gate the detection electronics for a fixed duration both during and after the laser excitation. Let A_0 and B_0

be the number of events recorded during and sometime after the laser excitation with the laser on-line, respectively, and let A_F and B_F be the corresponding number of events with the laser tuned off resonance. Then the OH signal and its associated uncettainty for 2N laser shots are given by

$$\text{(OH signal)} = \sum_N [(A_0 - B_0) - (A_F - B_F)] \pm \left[\sum_N (A_0 + B_0 + A_F + B_F)\right]^{1/2} \qquad (5)$$

where the square root of the sum of the recorded events represents the statistical fluctuation in the deduced OH signal. It follows that in the absence of other uncertainties, the shot noise associated with these gated background counts determine the detection limit for our OH measurements.

TYPICAL RESULTS

With the equipment described in the preceding section, it has been possible to reduce to a negligible level the effects of absorption saturation [Killinger et al., 1976] and ozone interference [Wang et al., 1975]. By terminating the line of sight of the collection telescope, it has also been possible to reduce the solar background to a manageable level. Under normal operating conditions, this solar background is responsible for about 200 counts per 1,000 laser shots in the detected signal; this may be compared to a signal count of 14 per 1,000 laser shots due to an OH concentration of 10^6 molecules/cm^3, and a nonresonant fluorescence background which varies between 100 and 1,000 counts per 1,000 laser shots depending on meteorological and other conditions of the ambient. Our detection limit is determined by the shot noise associated with these background counts, by possible systematic changes in the property of the laser beam accompanying on-off tuning, and by the temporal variations of the nonresonant fluorescence background. To date, our best detection limit is about 0.7×10^6 OH/cm^3, but under adverse conditions the detection limit could be degraded by as much as an order of magnitude. With certain planned improvements in our detection and tuning schemes, we are confident that a working detection limit in the neighborhood of 1×10^6 OH/cm^3 can be maintained routinely.

This research has been supported in part by the U. S. Department of Energy, by National Science Foundation through the University of Michigan, and by National Aeronautics and Space Administration through Wayne State University.

REFERENCES

Campbell, M. J., J. C. Sheppard, and B. F. Au, Measurement of hydroxyl concentration in boundary layer air by monitoring CO oxidation, Geophys. Res. Lett. 6, 175-178 (1979).

Chameides, W. L., The photochemical role of tropospheric nitrogen oxides, Geophys. Res. Lett. 5, 17-20 (1978).

Hameed, S., J. P. Pinto, and R. W. Stewart, Sensitivity of the predicted CO-OH-CH$_4$ perturbation to tropospheric NO$_x$ concentrations, J. Geophys. Res. 84, 763-768 (1979).

Hanabusa, M., C. C. Wang, J. Japar, D. K. Killinger, and W. Fisher, Pulse width dependence of ozone interference in the laser fluorescence measurement of OH in the atmosphere, J. Chem. Phys. 66, 2118-2120 (1977).

Hyatt, H. A., J. M. Cherlow, W. R. Fenner, and S. P. S. Porto, Cross section for the Raman effect in molecular nitrogen gas, J. Opt. Soc. Am. 63, 1604-1606 (1973).

Killinger, D. K., and C. C. Wang, Absorption measurements of OH using a cw tunable laser, Chem. Phys. Lett. 52, 374-376 (1977).

Killinger, D. K., and C. C. Wang, Direct measurements of the Gibbs free energy of OH using a cw tunable laser, J. Chem. Phys. 71, 1582-1584 (1979).

Killinger, D. K., C. C. Wang, and M. Hanabusa, Intensity and pressure dependence of resonance fluorescence of OH induced by a tunable uv laser, Phys. Rev. A 13, 2145-2152 (1976).

Liu, S. C., Possible effects on tropospheric O_3 and OH due to NO emissions, Geophys. Res. Lett. 4, 325-328 (1977).

Perner, D., D. H. Ehhalt, H. W. Patz, U. Platt, E. P. Roth, and A. Volz, OH radicals in the lower troposphere, Geophys. Res. Lett. 3, 466-468 (1976).

Selzer, P. M., and C. C. Wang, Quenching rates and fluomescence efficiency in the A^2 + state of OH, J. Chem. Phys. 71, 3786-3791 (1979).

Streit, G. E., C. J. Howard, and A. L. Schmeltekopf, Temperature dependence of $O(^1D)$ rate constants for reactions with O_2, N_2, CO_2, O_3, and H_2O, J. Chem. Phys. 65, 4761-4764 (1976).

Wang, C. C., and L. I. Davis, Jr., Measurement of hydroxyl concentrations in air using a tunable uv laser beam, Phys. Rev. Lett. 32, 349-352 (1974a).

Wang, C. C., and L. I. Davis, Jr., Ground-state population distribution of OH determined with a tunable uv laser, Appl. Phys. Lett. 25, 34-35 (1974b).

Wang, C. C., and L. I. Davis, Jr., Two-photon dissociation of water: A new OH source for spectroscopic studies, J. Chem. Phys. 62, 53-55 (1975).

Wang, C. C., and C. M. Huang, Accurate determination of the band oscillator strength for the (0,0) ultraviolet transitions of OH, Phys. Rev. A 21, 1235-1236 (1980).

Wang, C. C., and D. K. Killinger, Simultaneous determination of rotational and translational temperatures of $OH(^2II)$ in a gas discharge, Phys. Rev. Lett. 39, 929-932 (1977).

Wang, C. C., and D. K. Killinger, Effect of rotational excitation on the band oscillator strength of OH, Phys. Rev. A 20, 1495-1498 (1979).

Wang, C. C., L. I. Davis, Jr., C. H. Wu, S. Japar, H. Niki, and B. Weinstock, Hydroxyl radical concentrations measured in ambient air, Science 189, 797-800 (1975).

Wang, C. C., L. I. Davis, Jr., C. H. Wu, and S. Japar, Laser-induced dissociation of ozone and resonance fluorescence of OH in ambient air, Appl. Phys. Lett. 28, 14-16 (1976).

Wang, C. C., D. K. Killinger, and C. M. Huang, Rotational dependence in the linewidth of the ultraviolet transitions of OH, Phys. Rev. A 22, 180-185 (1980).

Wang, C. C., M. T. Myers, and D. Zhou, Observation of competition of rotational effects in the intensity of ultraviolet bands of OH, Phys. Rev. Lett. 47, 490 (1981).

Wang, C. C., L. I. Davis, Jr., P. M. Selzer, and R. Munoz, Improved airborne measurements of OH in the atmosphere using the technique of laser-induced fluorescence, J. Geophys. Res. 86, 1181 (1981); 12,156 (1981).

Weinstock, B., and H. Niki, Carbon monoxide balance in nature, Science 176, 290-292 (1972).

Wu, C. H., C. C. Wang, S. M. Japar, L. I. Davis, Jr., M. Hanabusa, D. Killinger, H. Niki, and B. Weinstock, Hydroxyl radical measurements in a photochemical reactor by laser-induced fluorescence, Int. J. Chem. Kinetics 8, 756-776 (1976).

5.4 Use of the Fraunhofer Line Discriminator (FLD) for Remote Sensing of Materials Stimulated to Luminesce by the Sun

William R. Hemphill[1] and Arnold F. Theisen[2]

U.S. Geological Survey

Robert D. Watson[3]

Dallas, Texas

INTRODUCTION

Luminescence has been little used as a remote sensing tool in mineral exploration because artificial excitation sources are relatively low powered. Their effective range is on the order of a meter for hand-carried ultraviolet lamps, a few tens-of-meters for cathode-ray systems, and a few hundred meters for laser systems. The work must be performed at night with hand-carried lamps and cathode-ray systems in order to avoid obscuring the low intensity luminescence by bright sunlight. Luminescence stimulated by laser sources may be measured in daylight, but the luminescence signal at visible and near-visible wavelengths must compete with reflected daylight and background luminescence stimulated by the Sun.

The Fraunhofer line discriminator (FLD) is an electro-optical instrument operated from an aircraft which overcomes the problems mentioned above and permits the detection of solar stimulated luminescence several orders of magnitude less than the intensity detectable with the human eye. The FLD uses the Sun as an excitation source and permits detection of luminescing materials during daylight, thus avoiding both the power and distance limitations of artificial sources and the awkwardness of nighttime operations. The rationale for using the Fraunhofer line-depth method for measuring luminescence, and a detailed description of luminescence measurements with an FLD operated as a nonimaging radiometer, have been described previously by Hemphill and others (1969) and Watson and Hemphill (1976).

Rhodamine WT is an artificial dye which is used as a luminescence standard in FLD operations. The FLD is sensitive to concentrations as low as 0.1 parts per billion (ppb) of rhodamine WT dye in water. In order to assess optimum wavelength and sensitivity requirements before conducting an airborne survey, the luminescence of a sample target material may be measured with a laboratory fluorescence spectrometer and expressed in terms of rhodamine dye equivalency. For example, we may say that a material in sunlight luminesces at an intensity equivalent to a certain level of rhodamine dye. Watson and others (1974), and Hemphill and Watson (1975, p. 118-120) describe the apparatus and procedures for performing these measurements in the laboratory and the instrumentation and solar corrections that must be applied.

This paper describes recent airborne work in which the FLD has been modified to operate as an imaging system. The images acquired by the FLD depict the areal distribution of a variety of luminescent materials, notably phosphate rock and gypsum in the Sespe Creek area northeast of Santa Barbara, California, and a marine oil seep in the Santa Barbara Channel. The concept for an improved airborne imager, which could also be operated aboard the space shuttle, is introduced.

---

1   Reston, Virginia
2   Flagstaff, Arizona
3   Formerly with the U.S. Geological Survey.

During the early 1970's, development of the FLD, as well as associated laboratory and field studies, were jointly supported by the Advanced Applications Flight Experiments (AAFE) Program of the National Aeronautics and Space Administration (NASA) (NASA Order L-58,514) and the U.S. Geological Survey. Since 1975, the laboratory and field effort has been supported by the Geological Survey, with occasional funding and/or personnel support provided for specific tasks by other agencies (Environmental Protection Agency, the Bureau of Land Management of the Department of the Interior, the Science and Education Administration of the Department of Agriculture, and the Department of Defense).

INSTRUMENTATION

Fraunhofer Line Discriminator (FLD)

The FLD, shown in figure 1, was made by Perkin-Elmer[1] of Norwalk, Conn., and consists of an optical head, electronic console, and light collector (Plascyk, 1975). The main components in the optical head are two telescopes, one Earth-looking and one sky-looking; a rotating optical chopper wheel; three interchangeable optical filter sets; and a photomultiplier with its power supply.

Sunlight and skylight falling on the diffuse surface of the light collector are reflected by a mirror into the sky-looking telescope. The Earth-looking telescope receives radiation from the target whose reflectivity and luminescence are to be measured. Light from the two telescopes is sequentially routed through two different paths by the rotating chopper wheel.

In one path, light passes through a filter which is centered at a specific Fraunhofer line but whose bandwidth is an order of magnitude wider than the Fraunhofer line. This signal constitutes the light intensity measured on the solar continuum adjacent to the Fraunhofer line and is designated signal $\underline{a}$ in the sky-looking channel and signal $\underline{d}$ in the Earth-looking channel. In the other path, a Fabry-Perot interference filter, with a spectral bandwidth at the half power point of less than 0.07 nanometer (nm), passes light coincident with the intensity of the central part of the Fraunhofer line; this signal is designated signal $\underline{b}$ in the sky-looking channel and signal $\underline{c}$ in the Earth-looking channel.

These signals are processed by a hard-wired arithmetic processor that generates luminescence and reflectance values by solving the following equations: $\rho = (d-c)/(a-b)$; and $\epsilon = (d/a)-\rho$, where ρ is reflectance and ϵ is luminescence. Both ρ and ϵ are displayed as four-digit numbers (from 0000 to 9999 counts) on the front panel. Inasmuch as displayed and recorded counts are only proportional to luminescence or reflectance for each substance measured, a luminescence or reflectance count is referenced to the count from a standard (such as dye sample or a photographer's gray card) permitting relative luminescence measurements to be made. An FLD count of 100 is approximately equivalent to the luminescence of 1 ppb rhodamine WT dye. Three sets of Fabry-Perot filters are available, permitting measurement of luminescence of Fraunhofer wavelengths of 486.1, 589.0, and 656.3 nm.

FLD imaging system

The FLD imager (figure 2) is based on a simple optical-mechanical arrangement which uses two front surface mirrors, one of which is fixed at $45°$ with respect to the FLD Earth telescope, and the second of which is oscillated about an axis which is

---

[1] Use of trade names in this paper is for descriptive purposes only and does not constitute an endorsement of the produce by the U.S. Geological Survey

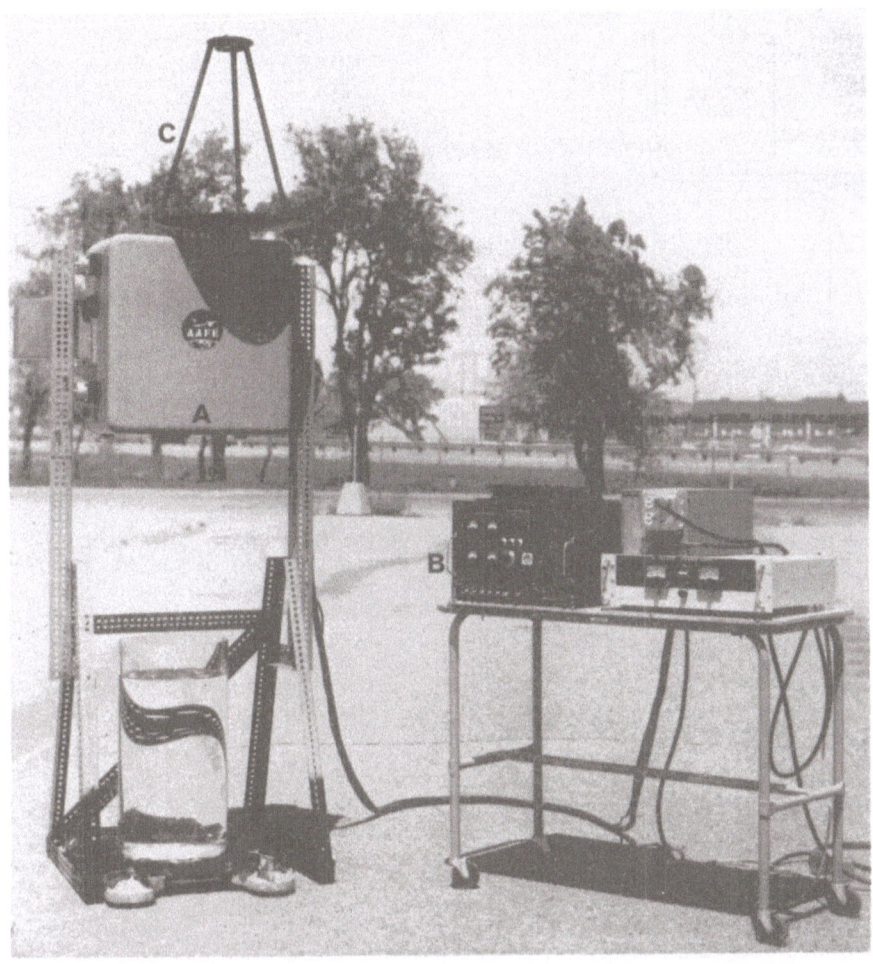

Figure 1. Fraunhofer line discriminator (FLD), showing the optical head (A), electronic console (B), and light collector (C). A pyrex jar beneath the optical head contains a solution of rhodamine WT dye in distilled water

parallel to the fixed mirror. The amount of oscillation provides a total sweep of $\pm18.5°$ with respect to the axis of the optical system, or a total scan angle of $37°$. The linearity and oscillation angle is controlled by a precision cam, based on an Archimedes spiral design. The rate of oscillation is determined by an electronically controlled precision DC motor with tachometer feedback. The feedback signal is monitored for accurate speed control of one complete sweep ($37°$) per second. A timing disk is mounted on the motor shaft to provide a beginning-of-sweep trigger, generated by an electro-optical coupler. Ground resolution is determined by the instantaneous field of view ($1°$) of the FLD. Maximum ground coverage is obtained when the aircraft speed matches the ground resolution of the FLD. For example, with the present system, an aircraft speed of 45 meters per second and an altitude above terrain of 2,382 m, complete coverage would be provided by a square 45 m on a side. However, because the instantaneous field of view is a circle 45 m in diameter,

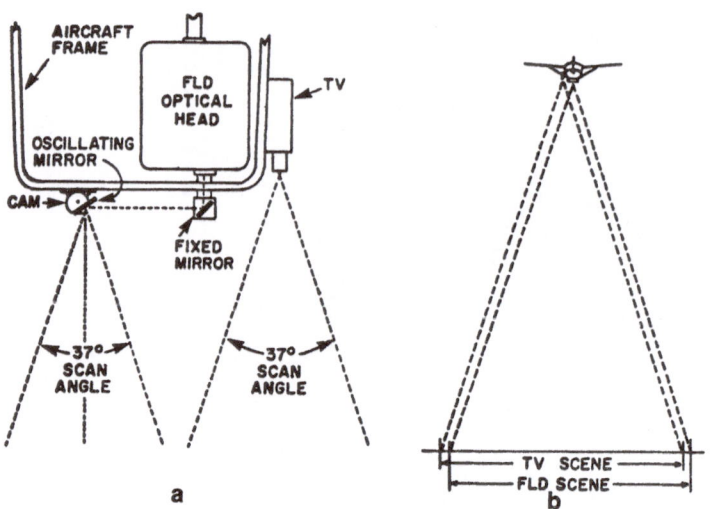

Figure 2. FLD imaging system (a) diagram of the FLD imager showing components; (b) schematic overview of ground coverage of the FLD and tele vision system

the actual ground coverage is only 80 percent. A bore sight television system permits identification of surface features and thus correlation of these features with measured luminescence and reflectance.

Data acquisition and analysis

The heart of the data acquisition system is a Motorola M6800 micro-processor, having 10,000 words of memory. The system uses a 1-megahertz clock and has an interface board specifically designed for use with the FLD. The M6800 is programed to organize FLD data in blocks of 15 sweeps with 36 resolution elements per sweep. The FLD provides data at 40 hertz (40 resolution elements per second) plus a data valid pulse. The beginning-of-sweep trigger, the data valid pulse, and the FLD image data are fed into the computer, where the beginning-of-sweep trigger produces both a beginning-of-block and a beginning-of-sweep mark, updates block and sweep counters, and sets the computer to accept FLD data valid pulses and data. Both block and sweep counts and the beginning-of-sweep mark are transferred to a video monitor and recorder. These three items plus the beginning-of-block mark (produced every 15 sweeps) are recorded on digital cassette tape. The data valid pulse causes the computer to transfer to the digital tape the luminescence and reflectance values for 36 resolution elements per sweep. The remainder of the time in which four additional resolution elements would have been acquired is used for the oscillating mirror to return to the beginning-of-sweep position.

FLD data, recorded on digital tape, are transferred to computer memory, where either luminescence or reflectance data for each resolution element can be analyzed and assigned a character representing either a specific value or a range in values. These characters are then reproduced by a printer with a capability of producing 94 distinct characters plus a blank. The computer is programed to produce gray-scale slicing with preselected window for luminescence or reflectance.

LUMINESCENT MATERIALS

Phosphate rock - Sespe Creek area, California

To assess potential use of the FLD in prospecting for phosphate rock, the luminescence of 10 phosphate rock samples from the United States, Colombia, and Brazil were measured with the laboratory fluorescence spectrometer at the 486.1 nm Fraunhofer line. These measurements, reported by Hemphill and others (1975) and by Watson and Hemphill (1976), showed a luminescence intensity as large as 8.0 ppb rhodamine WT equivalency; all samples luminesced within the sensitivity range of the FLD, suggesting the need for a field test.

The possibility of commercial phosphate deposits is particularly promising in a southeast-trending belt of the Santa Margarita Formation (late Miocene) in southern California (Gower and Madsen, 1964; p. D84). This belt includes phosphate rocks in the Sespe Creek area of western Ventura County, about 40 km northeast of Santa Barbara. Laboratory spectral measurements confirmed that samples of phosphate rock and associated gypsum from the Santa Margarita Formation in this area luminesce withinin the sensitivity range of the FLD.

In field experiments in November 1974 and May 1975, the FLD was mounted in a helicopter and operated in the nonimaging radiometer mode. Luminescence was measured during helicopter traverses (figure 3) and hovers above the Santa Margarita Formation as well as in what was presumably detrital material downslope from that formation. Further work showed that although luminescence of both the phosphate rock and gypsum were similar, the two materials could be distinguished because reflectance of the gypsum exceeded that of phosphate rock by as much as a factor of five. Details of the 1974-75 helicopter work are described by Hemphill and others (1975) and by Watson and Hemphill (1976).

On November 6, 1979, the FLD imaging system, operating at 486.1 nm from a fixed-wing aircraft, was used to acquire an image (figure 4) of the same area overflown previously with the helicopter. Also shown in figure 4 is the geologic map of the Sespe Creek area. High luminescence values, shown in black (>3.9 ppb rhodamine dye equivalency), correlate well with the phosphatic Santa Margarita Formation. Lower values (2.5-3.8 ppb rhodamine dye equivalency) south of outcrops of the Santa Margarita Formation are believed to be detrital material (including phosphate rock and gypsum) washed downslope from that formation. Luminescence values greater than 2.5 ppb rhodamine dye equivalency are rare south of Sespe Creek where phosphate rocks do not crop out. Watson (1981, p. 31) shows a more detailed, color version of this image.

Oil seep - Santa Barbara Channel, California

Watson and others (1974, p. 1964) used a laboratory fluorescence spectrometer to determine luminescence in terms of rhodamine dye equivalency of 29 samples of crude oil (from the United States, Canada, Venezuela, Trinidad, Denmark, Nigeria, Libya, and the Middle East) and 20 samples of refined oil. All but four samples appear to be within the sensitivity range of the FLD. Watson also showed that (1) lighter crudes (<0.84 specific gravity) tend to luminesce more strongly than heavier crudes, and (2) a good correlation exists between thickness of oil films (up to 50 μm) and luminescence of lighter crudes (<0.83 specific gravity).

Natural fractures in the ocean floor of the Santa Barbara Channel, California, permit oil of varying densities to seep to the surface. Throughout the channel, patches of oil film can be seen at various times with the naked eye, depending upon sea state, viewing direction with respect to the Sun, and other factors. At times, west of Santa Barbara, heavy crude oil forms narrow filaments that extend for several kilometers parallel to the shore.

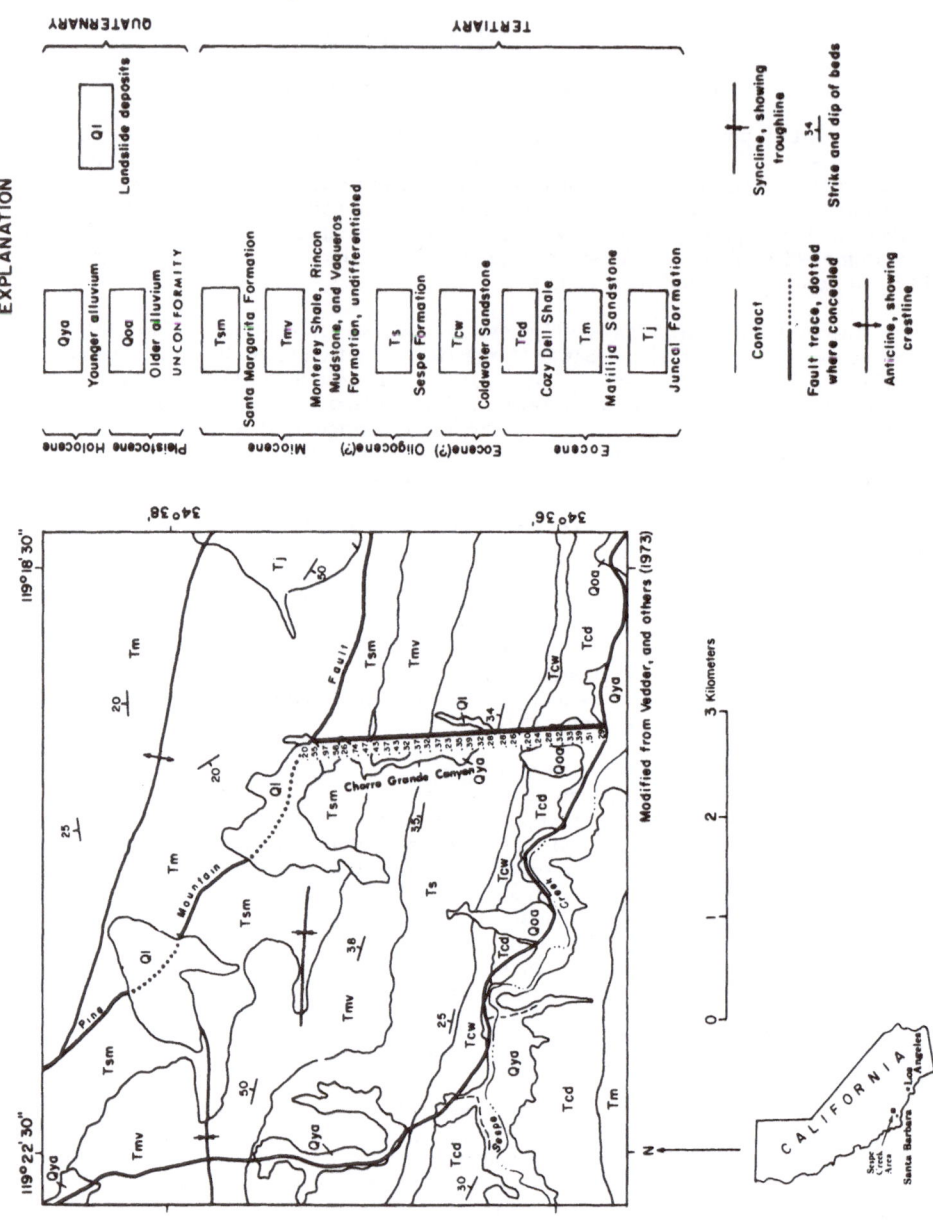

Figure 3. Geologic map of part of the Sespe Creek area, Ventura County, California, showing location of a helicopter traverse (solid line east of Chorro Grande Canyon). Numbers along traverse are FLD luminescence measurements in terms of rhodamine dye equivalency. Luminescence is markedly higher in the traverse segment across the outcrop of the phosphatic Santa Margarita Formation (Tsm)

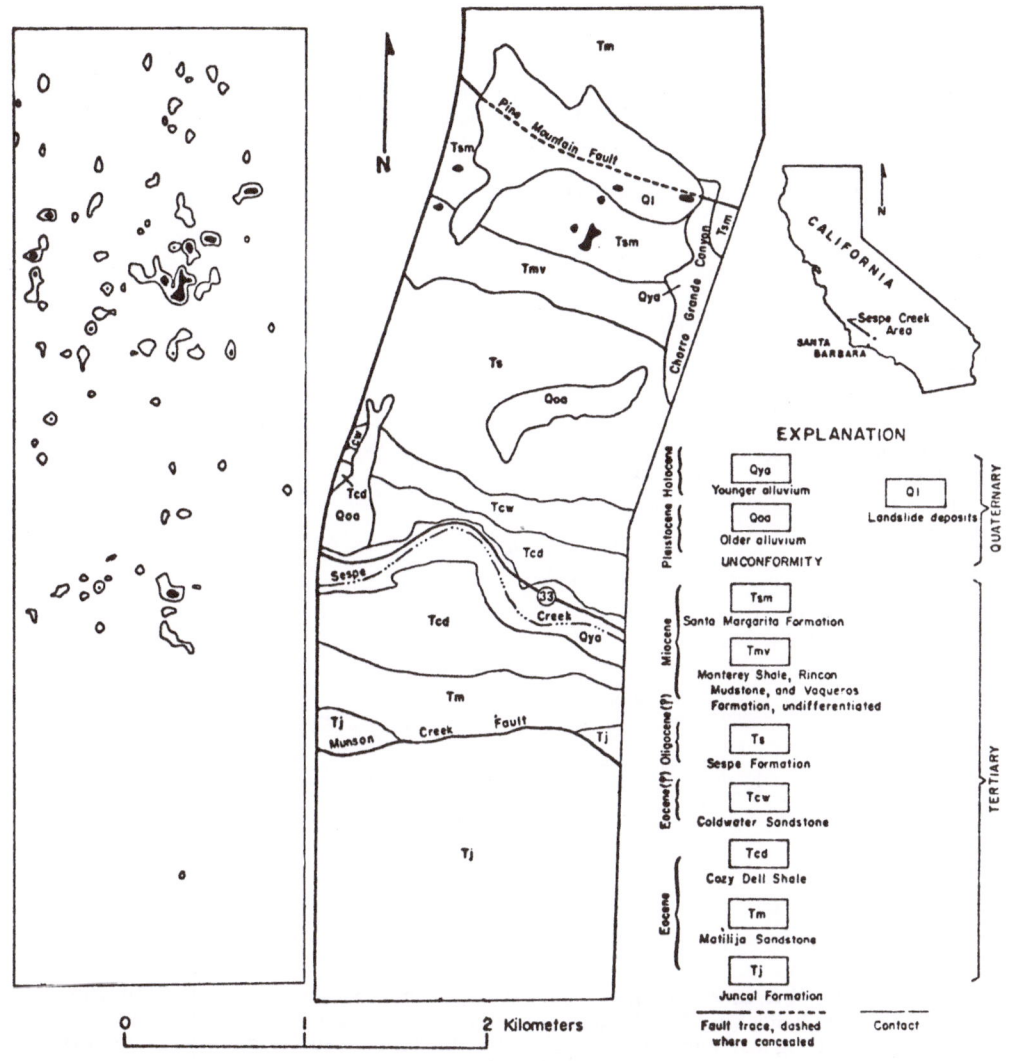

Figure 4. Luminescence image and geologic map of the Sespe Creek area, Ventura County, California. Image was acquired November 6, 1979, with the FLD imaging system operating at 486.1 nm. Black areas in both the image and the geologic map represent luminescence that exceeds 3.9 ppb rhodamine dye equivalency; luminescence highs correlate well with the occurrence of the phosphatic-gypsiferous Santa Margarita (Tsm) Formation. Lower luminescence levels (2.5 to 3.8 ppb rhodamine dye equivalency, shown by enclosed white areas) are believed to be luminescent detritus transported downslope from the Santa Margarita Formation. Outline of the geologic map is modified to compensate for distortion in the image caused by aircraft roll and drift. (Geologic map modified from Vedder and others, 1973)

FLD imagery has been acquired over oil slicks in the Santa Barbara Channel, both west of Santa Barbara and near the Dos Cuadras oil platforms. Figure 5 shows the image acquired over the Dos Cuadras platforms on March 16, 1978. Luminescence values are at background levels south of the platforms but remain fairly uniform northward, though modified by circulation patterns. At the northern end of the image the luminescence again approaches background level. Luminescence highs are attributed to natural oil slicks that could be observed visually from the aircraft at the time of overflight and were previously reported by McCulloh (1969).

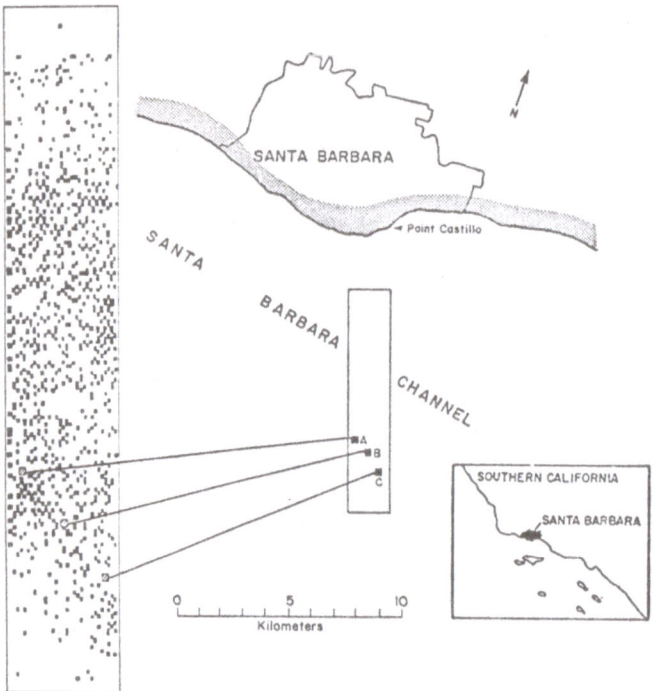

Figure 5. Luminescence image of part of the Santa Barbara Channel, California, including the Dos Cuadras oil platforms A, B, and C. Picture elements of highest luminescence, shown in black, exceed 0.5 ppb rhodamine dye equivalency and are attributed to oil seeps in the vicinity of the platforms

Other aerial images of luminescent materials include playa evaporites in Torrance County, New Mexico (Watson, 1981, p. 29 and 33); uranium-bearing outcrops in Big Indian Valley, Utah (Theisen and others, 1979; Niesen, 1981, p. 41-46); and shock-induced luminescence of rocks at Meteor Crater, Arizona (Roddy and others, 1981, p. 47-49).

AN IMPROVED IMAGING SYSTEM

Conceptual design of an improved imaging FLD has been completed. The design features newly developed image intensifiers integrated with high gain solid-state photo diodes and charge coupled devices. These components will permit operation at favorable signal-to-noise ratios, picture element size of 3 x 3 m from an altitude of 3,000 m, and simultaneous operation at the following three Fraunhofer lines: 486.1 nm, 589.0 nm, and 656.3 nm. Modification of the optical system will also permit operation of the same equipment aboard the space shuttle in circular orbit. This configuration would yield a picture element of 200 x 200 m from shuttle altitudes. It is believed that a few hours operation on a shuttle sortie would provide an opportunity to assess the feasibility of delineating the areal extent of luminescent materials on the Earth's surface from orbital altitudes.

REFERENCES

Gower, H. D., and Madsen, B. M., 1964, The occurrence of phosphate rock in California: U.S. Geological Survey Professional Paper 501-D, p. 79-85.

Hemphill, W. R., Stoertz, G. E., and Markle, D. A., 1969, Remote sensing of luminescent materials: in International Symposium on Remote Sensing of Environment, 6th, Ann Arbor, Michigan, 1969, Proceedings: Ann Arbor, University of Michigan, v. 1, p. 565-585.

Hemphill, W. R., and Watson, R. D., 1975, Ultraviolet radiation, in Manual of Remote Sensing: American Society of Photogrammetry, Falls Church, Va., p. 115-128.

Hemphill, W. R., Watson, R. D., Bigelow, R. C., and Hessin, T. D., 1975, Measurement of luminescence of geochemically stressed trees and other materials: U.S. Geological Survey Professional Paper 1015, 14 figures, 9 tables, p. 93-112.

McCulloh, T. H., 1969, Geologic characteristics of the Dos Cuadras offshore field, U.S. Geological Survey Professional Paper 679C, 8 figures, 1 table, p. 28-46.

Niesen, P. L., 1981, Aerial survey of luminescent rocks, Big Indian Valley, Utah: in Hemphill, W. R., and Settle, Mark, eds., Workshop on Applications of Luminescence Techniques to Earth Resource Studies: Lunar and Planetary Institute, Houston, Tex., LPI Technical Report 81-03, p. 41-46.

Plascyk, J. A., 1975, The MK II Fraunhofer line discriminator (FLD II) for airborne and orbital remote sensing of solar stimulated luminescence: Optical Engineering, v. 14, no. 4, p. 339-346.

Roddy, D. J., Watson, R. D., and Theisen, A., 1981, Measurements of shock-induced luminescence at Meteor Crater, Arizona from laboratory and airborne Fraunhofer line-discriminator systems: in Hemphill, W. R., and Settle, Mark, eds., Workshop on Applications of Luminescence Techniques to Earth Resource Studies: Lunar and Planetary Institute, Houston, Tex., LPI Technical Report 81-03, p. 47-49.

Theisen, A. F., Watson, R. D., and Niesen, Preston, 1979, Interpretation of luminescence imagery of mineralized areas, Big Indian Valley, Utah: U.S. Geological Survey Open-File Report 79-574, 10 p.

Vedder, J. G., Dibblee, T. W., and Brown, R. D., 1973, Geologic map of the upper Mono Creek - Pine Mountain area, California: U.S. Geological Survey Miscellaneous Geological Investigations Map I-752.

Watson, R. D., 1981, Airborne Fraunhofer line discriminator surveys in southern California, Nevada, and central New Mexico: in Hemphill, W. R., and Settle, Mark, eds., Workshop on Applications of Luminescence Techniques to Earth Resource Studies: Lunar and Planetary Institute, Houston, Tex., LPI Technical Report 81-03, p. 28-35.

Watson, R. D., and Hemphill, W. R., 1976, Use of the airborne Fraunhofer line discriminator for the detection of solar stimulated luminescence: U.S. Geological Survey Open-File Report 76202, 109 p.

Watson, R. D., Hemphill, W. R., Hessin, T. D., and Bigelow, R. C., 1974, Prediction of the Fraunhofer line detectivity of luminescent materials, in International Symposium on Remote Sensing of Environment, 9th, Ann Arbor, Michigan, 1974, Proceedings: Ann Arbor, Environmental Research Institue of Michigan, v. 3, p. 1959-1980.

Watson, R. D., and Theisen, A. F., 1977, Mapping luminescence of uranium bearing sandstone using an imaging Fraunhofer line discriminator: U.S. Geological Survey Open-File Report 77-743, 10 p.

5.5 Ozone and Water Vapor Monitoring Using a Ground-Based Lidar System

G. Mégie, J. Pelon, J. Lefrère, C. Cahen

Service d'Aéronomie du CNRS, BP 3, F-91370 Verrière-Le-Buisson, France

P.H. Flamant*

Laboratoire de Météorologie Dynamique du CNRS, Ecole Polytechnique
F-91128 Palaiseau, France

Introduction

Lidar measurements of ozone and water vapor concentrations were performed during several field experiments in 1980-1981 by means of the differential absorption laser technique. Profiles up to 30km for ozone and up to 9km for water vapor were obtained on a routine basis. Experiments were conducted at the Haute Provence Observatory which is located in southern France (44°N,5°E).

Methodology

The DIAL technique used for ozone and water vapor measurements can be summarized as follows (see Ref. 1 for a more detailed analysis): two laser pulses with different emission wavelengths (λ_{on} and λ_{off}) are transmitted into the atmosphere and due to the difference in attenuation the backscattered signals carry information about the absorbing species. In the case of discrete spectra—i.e., water vapor—one wavelength (λ_{on}) is centered on an absorption line of the species under study whereas the other one (λ_{off}) corresponds to a spectral interval free of absorption. The on and off denomination is extended to continuum spectra—i.e., ozone—where the on wavelength corresponds to the strongest absorption. Ozone measurements are made within Hartley bands ($\lambda = 0.3\mu m$) and the difference between wavelengths is 5nm. For water vapor measurements two lines of the (301) rotation-vibration band in the near infrared ($\approx 0.72\mu m$) are selected depending on the altitude range and meteorological conditions, and a spectral interval of 0.1nm between on and off wavelengths is used.

Pulsed lasers as transmitters allow one to retrieve in a direct way the spatial distribution of the molecular densities or mixing ratio by a time analysis of the backscattered signals. The local optical thickness (for a cell range determined by the laser pulsewidth and detection systems) is then proportional to the derivative with respect of the range of the logarithm of the ratio of backscattered signals at λ_{off} and λ_{on}. The molecular densities or mixing ratio are calculated knowing the absorption cross section at λ_{on} and λ_{off} which in general depends on temperature and pressure. This leads to the use of an atmospheric model for ozone to take into account the altitude dependence of the absorption cross section. For water vapor a proper choice of the lowest rotational energy level allows determination of the mixing ratio without use of a model.

---

*Present address: Jet Propulsion Laboratory, California Institute of Technology, 4800 Oak Grove Drive, Pasadena, CA 91109, USA

223

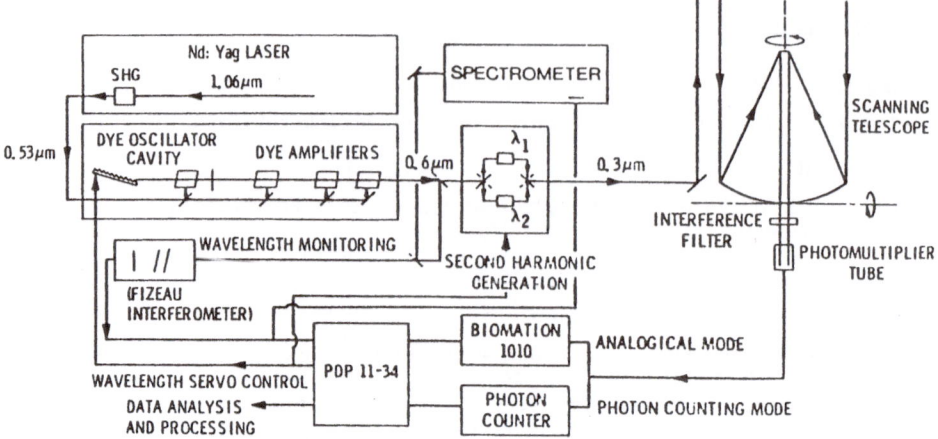

Fig..1. Synoptic diagram of the lidar system used for ozone
measurements. For water vapor the dye laser emission
is 0.72 μm and no second harmonic generation used

Experimental

The emitter of the lidar system (see Fig. 1) includes a Nd:YAG Q-switched
laser (Quantel model 480[2] and two tunable narrowline dye lasers (Jobin-Yvon
model HPHR[3]). The frequency-doubled emission of the Nd:YAG laser at 532nm is
converted into UV or IR spectral regions by means of two combinations of dye
oscillator-amplifier systems. Spectral narrowing and tuning of the dye oscillator emissions are performed by gratings near grazing incidence and results
in a 5 pm linewidth in the 570-620nm range and 1.5pm linewidth at 724nm (see
Table 1). For ozone measurements the output emissions are converted into UV
(285-310nm) using two prepositioned KDP crystals (see Table 2). This setup
allows a rapid (mechanical) switching between λ_{on} and λ_{off}. The stabilization of the output wavelengths of the dye lasers and the switching between the
on and off wavelengths are made using a servo- control device,[4] the optical
part of which is a multibeam Fizeau interferometer [5].

The backscattered signals are collected by a 36 cm diameter (O_3) or a
60 cm diameter (H_2O) telescope and detected by either a low gain or high gain
photomultiplier tubes depending on the altitude ranges and daytime or nighttime operations. Two acquisition modes and electronic processings of the
return laser signal were made at the same time by means of a transient waveform recorder (Biomation 1010) and a 256 channel photon counter. The data
provided by the two acquisition systems are then fed into a PDP 11-34 computer
and stored on floppy disk. A presummation of return signals is made to reduce
the volume of the stored data.

The PDP 11-34 computer is used both for the real time processing of the
data and the control of the full automatic sequence of the lidar measurement
procedure. The experimental parameters—sequence acquisition time, altitude
resolution, laser wavelengths, switching time between λ_{on} and λ_{off}—are typed
in to automatically start the sequence of laser firings. During the experi-
ment, the laser energy and emission wavelength are continuously monitored and
the data acquisition takes place only if all these parameters are within the
range of predetermined values.

224

TABLE I. LIDAR SYSTEM CHARACTERISTICS FOR WATER VAPOR MEASUREMENTS

Emitter		Receiver	
Transmitted energy	= 70mj	Telescope diameter	= 60 cm
Pulse duration	= 12ns	Field of view	= 3 mrd
Emission linewidth	= 1.5pm	Interference filter	
(at 724nm)		bandwidth	= 2.4nm
Repetition rate	= 10Hz		
Beam divergence	< 0.5 mrd		

TABLE 2. LIDAR SYSTEM CHARACTERISTICS FOR OZONE MEASUREMENTS

Emitter		Receiver	
Transmitted energy	= 40mj	Telescope diameter	= 36 cm
Pulse duration	= 12ns	Field of view	= 1 mrd
Emission linewidth	= 2.5pm	Interference filter	
(at 285-310nm)		bandwidth	= 70nm
Repetition rate	= 10Hz	or	= 3nm
Beam divergence	< 0.5 mrd		

Results

Ozone

Field experiments have been conducted for 40 nights spread over 1980-1981. Ozone concentration profiles up to 30km (Fig. 2) were obtained on a routine basis [6]. A profile is recorded in three steps: below 8km; between

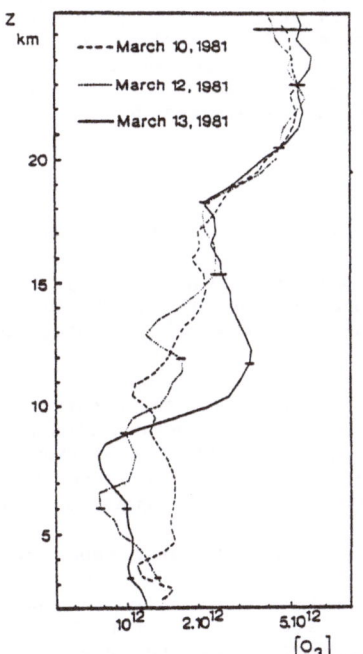

$[O_3]_{cm-3}$ Fig. 2. Vertical ozone concentration profiles

7 and 17km and above 15km. Each elementary profile requires 15 min, and
results in an overall time for measurement of 45 min. The relative accuracy (one standard deviation) on the ozone number density is better than 5%
at tropospheric levels for a vertical resolution of 0.45km and it falls to
20% at the maximum altitude levels for a vertical resolution of 1.2km.

Large variations in the ozone number density were observed on a day-to-
day basis and were correlated with a decrease of the tropopause height by
2 or 3 kilometers. Ozone variations can be related to large scale horizontal transport [7]. Occurrence of ozone bulges, with concentration up to 3.10^{12}
molecules/cm^3 at the 10km altitude level, was due to a meridional circulation. This circulation brought stratospheric air of polar origin into the
troposphere at mid latitude.

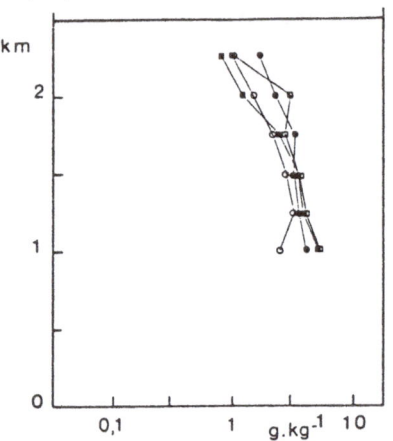

Figure 3. Short-term fluctuations of
the water vapor mixing ratio below
2.5 km

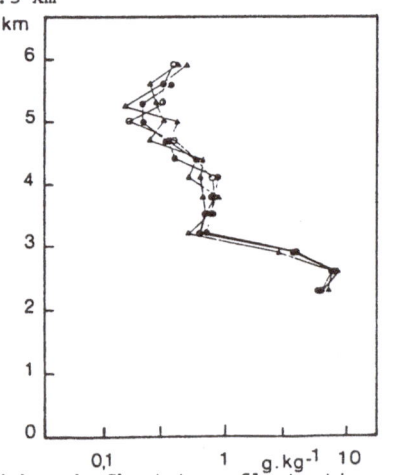

Figure 4. Short-term fluctuations of
the water vapor mixing ratio above
2.5 km

Water Vapor

Lidar measurement was evaluated by
comparison of the short-term fluctuation of the water vapor profiles and
the statistical error. Short-term
fluctuations of the water vapor mixing ratio lead us to make a distinc-
tion between measurements performed
below or above 2.5km of altitude[8].

-below 2.5km the altitude profiles recorded with an integration time of 4
min and a vertical resolution of 100m
(Fig. 3) show a variability of the
water vapor mixing ratio of 20% while
the statistical error corresponding to
one standard deviation is in this case
smaller than 10%: the observed variations are thus representative of dyna-
mical movements related to convective
processes and influenced by the local
orography.

-above 2.5km the variability of the
water vapor mixing ratio is within the
statistical error of the measurement as
shown on Fig. 4 where successive profiles of the water vapor mixing ratio
with an integration time of 8 min and
a vertical resolution of 300m are represented.

For altitude up to 9km the accuracy is
better than 15% for an acquisition time
of 50 min and a spatial resolution of
300m (the same accuracy is reached in
only 2 min in the boundary layer).

Continuous survey of water vapor profiles in the boundary layer has been
performed over a period of 24 hours.

The return signal at λ_{off} allowed us to retrieve the aerosol content. So,
we have been able to follow downward and upward motions of the water vapor
field in correlation with particulate layer motions.

The day-to-day variations of the water vapor mixing ratio in the troposphere can also be determined from the lidar measurements[9]. Figure 5 shows the average profiles representative of the water vapor content for five successive nights from July 18 to July 22, 1980. On July 20 the meteorological chart shows that a meridional circulation was established along the Atlantic coast between 60 N and 45 N. This circulation is responsible for the advection over southern France of dry air masses from higher latitudes. This behavior is confirmed up to 4km by the radio sounding data of Bordeaux (45 N, 1W), Brest (48 N, 5W) and Camborne (51 N, 5 W) which show similar altitude variations as those observed by laser sounding over the Haute Provence Observatory (44N, 5E).

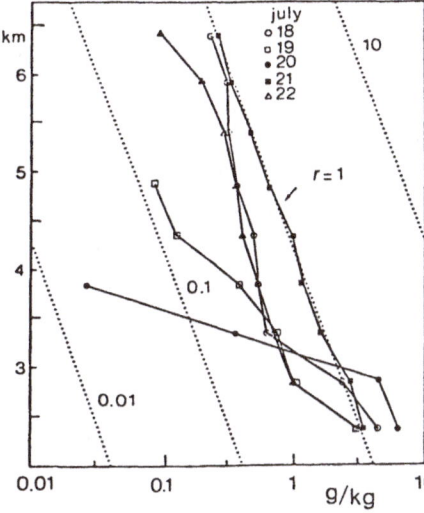

Fig. 5. Day to day variation of the water vapor tropospheric profile

Conclusion

Lidar monitoring of vertical profiles of ozone in the lower stratosphere and troposphere and of water vapor in the troposphere has been performed on an operational basis at the French lidar facility at the Haute Provence Observatory. Our measurements illustrate the potentiality of the DIAL method for atmospheric studies. In terms of accuracy, range resolution and time required for measurements, the capability of our system is already sufficient to fulfill many geophysical objectives.

References

1. E. D. Hinkley, editor, Laser Monitoring of the Atmosphere, Springer-Verlag (1976).

2. G. Brassart, G. Bret, J-M Marteau, Opt. Commun., 23, 327 (1977).

3. F. Bos, Appl. Opt., 20, 1886 (1981).

4. C. Cahen, J-P Jegou, J. Pelon, P. Gildwarg, J. Porteneuve, Rev. Phys. Appl., 71, 1255 (1981).

5. Y. H. Meyer, J. Opt. Soc. Am., 71, 1255 (1981).

6. J. Pelon, G. Mégie, to appear in J. Geophys. Res. C, 1982, accepted for publication.

7. J. Pelon, P. H. Flamant, M-L Chanin, G. Mégie, <u>C. R. Acad. Sc. Paris</u>, <u>292</u> (II), 319 (1981).

8. C. Cahen, J. Pelon, P. H. Flamant, G. Mégie, <u>C. R. Acad. Sc Paris</u>, <u>292</u> 29 (1981).

9. C. Cahen, G. Mégie, P. H. Flamant, Proc. IAMAP Symposium, Hambourg 25-28 August 1981. In Nowcasting: mesoscale observations and short-range prediction, p. 149.

5.6 The NASA/Goddard Balloon Borne Lidar System

William S. Heaps and T.J. McGee
NASA Goddard Space Flight Center, Greenbelt, MD 20771, USA

A balloon borne LIDAR system for the measurement of trace constitutents of the stratosphere has been built and operated by the Goddard Space Flight Center obtaining ozone measurements in the altitude region from 21 to 36 km and hydroxyl radical measurements in the 33 to 36 km region. These species were measured as part of NASA's program to investigate the seriousness of man-made stratospheric ozone depletion via high altitude aircraft exhaust or the release of chlorofluorcarbons into the lower atmosphere.

The experimeent was housed in the gondola approximately 7 feet on a side and weighing about 2 1/2 tons. Principal components of the experiment included a flowing liquid external radiator panel cooling system, an azimuthal and elevation pointing system, command and data telemetry communication systems, batteries ($\sim$ 1000 pounds of the total weight), and the LIDAR system itself.

The LIDAR system used two different techniques to acomplish its measurements. Hydroxyl radical was measured using remote laser induced fluorescence in which a wavelength is emitted which excites the species under investigation to fluoresce. By relating the intensity of the fluorescence wavelength following a laser pulse the species concentration can be deduced. 282 nm was the exciting wavelength and fluorescence was monitored in a band from 305 nm through 315 nm. Ozone was measured by the technique of differential absorption LIDAR (DIAL) in which two wavelengths are transmitted simultaneously and their backscattered intensities are measured. Choosing one wavelength which is absorbed by the species under study and the other wavelength in an unattenuated portion of the spectum allows one to infer the species concentration along the path of the laser beams from the relative intensities of the backscatter. Since 282 nm which was already in use for the hydroxyl radical measurement is absorbed by ozone it was chosen for the first wavelength. 355 nm was chosed for the second wavelength for reasons of convenience.

The laser source for our 282 nm was a frequency doubled tuneable dye laser pumped by a doubled Nd:Yag laser. The dye laser was an oscillator amplifier system with off axis longitudinal pumping and an etalon-diffraction grating tuning system. Energy output from the doubler was 1-2 mJ in a .002 nm wide line at a pulse repetition frequency of 10 Hz. 355 nm was produced by frequency tripling the 1.06 micron radiation from the Nd:Yag laser which remained after doubling for the dye laser pump. The 355 nm energy was a few millijoules. After generation these two wavelengths were combined using a dichroic mirror, expanded in an all reflective telescope to reduce beam divergence and transmitted along the axis of a 30 cm diameter Cassegrain telescope used as the receiver. The light returning following a laser pulse was collected by the telescope and focussed on the entrance aperture of a 1/4 meter monochromator. This monochromator had three exit apertures for the three detected wavelengths 282 nm, 305-315 nm and 355 nm. Light at each wavelength was detected by a separate photomultiplier tube. Output from the tubes was sampled at 1 microsecond time intervals and stored in high speed buffers awaiting transmission to the ground between laser shots. Data taking was controlled by an onboard microprocessor which also provided routines, callable from the ground, for dye laser tuning and system diagnostics.

Figure 1 shows some results of the fluorescence measurement of hydroxyl radical as a function of time of day. The error bars shown represent one standard deviation of statistical uncertainty in the determination of concentration. Although the signal-to-noise ratio is not outstanding in these measurements it should be recognized that these represent a concentration

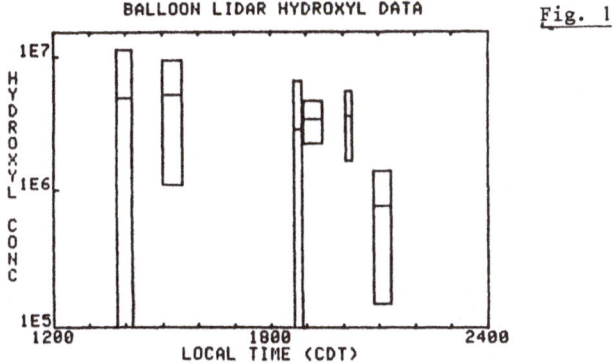

Fig. 1

determination in the 3 to 30 parts per trillion range from a distance of 300 meters. Improvements in the system are expected to improve this capability by one to two orders of magnitude. It should also be noted on this figure that substantial improvements in signal-to-noise are achieved at later times in the day since scattered solar flux is the principal interference with this measurement.

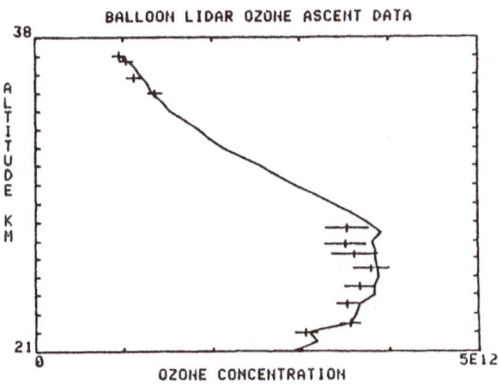

Fig. 2

Figure 2 shows a comparison between the ozone profile measured during ascent by the LIDAR and that determined by an in situ ozone analyzer carried onboard the gondola. The error bars on the LIDAR measurements represent two standard deviations of uncertainty due to counting statistics. No LIDAR measurements were taken between 27 km and 33 km because of the LIDAR was shut down for cooling during this phase of the ascent. The profile from 20 km to

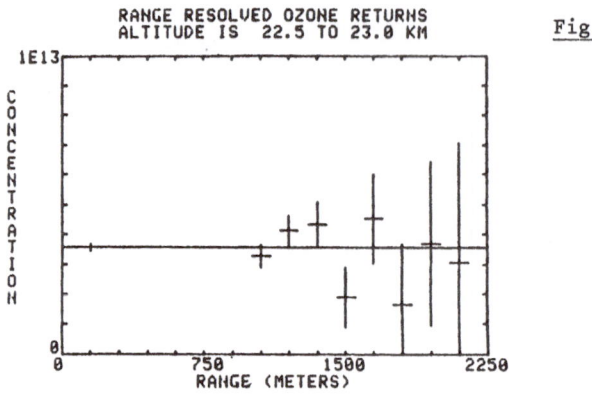

RANGE RESOLVED OZONE RETURNS
ALTITUDE IS 22.5 TO 23.0 KM

Fig. 3

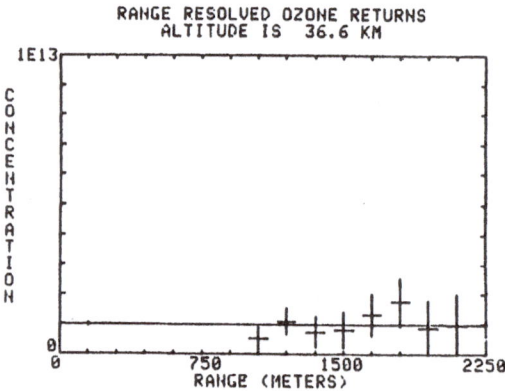

RANGE RESOLVED OZONE RETURNS
ALTITUDE IS 36.6 KM

Fig. 4

27 km represents about one hour of continuous operation. The agreement between the two sensors is rather gratifying. One should recognize that the in situ monitor reflects a measurement made at the balloon site while the LIDAR measurement represents a weighted average of the ozone over a column extending about 2 kilometers to the east of the gondola.

Figure 3 and 4 represent range resolved profiles looking east from the gondola in a horizonal direction at two different altitudes. Again the error bars represent two standard deviations. The large error bars at long range in Figure 3 point up another feature of DIAL experiments which is the need to match the strength of the expected absorption with the desired ranging capability. Best concentration determinations are achieved with an absorption of approximately (1/e). 282 nm is too strongly absorbed for use in the lower

232

stratosphere and the small amount of light returning to the detector from 2 km is reflected in the large error bars. Note that in Figure 4 at greater altitude the uncertainty in the concentration determination at 2 km range is much lower even though the actual concentration is down by about a factor of 3.

In conclusion we believe that these results have shown LIDAR to be an effective tool for stratospheric investigation. Unfortunately this sytem crashed at the end of its flight and was totally destroyed. A new system incorporating many of the improvements suggested by this flight is currently under construction and is presently expected to fly in the fall of 1982.

Part 6

Laser Sources and Detectors

6.1 Development of Compact Excimer Lasers for Remote Sensing

James B. Laudenslager, I. Stuart McDermid, and Thomas J. Pacala

Jet Propulsion Laboratory
California Institute of Technology, 4800 Oak Grove Drive,
Pasadena, CA 91109, USA

Active laser remote sensing is an emerging technique which has the unique capability of providing range-resolved measurements of gaseous and particulate species without day-night restrictions as is the case with certain passive sensing methods. The coherent output of a laser source can be used to interogate small areas, even from orbital altitudes, and pulsed lasers can be used to obtain high spatial resolution for concentration profiles by range gating the return signal. Laser measurements of chemical species can be made using differential absorption, DIAL, laser induced fluorescence, LIF, and by a new highly sensitive in-situ method resonance ionization spectroscopy, RIS. All these measurement techniques require tunable, narrow spectral bandwidth laser sources. Important atmospheric species suitable for ultraviolet or visible laser detection are: O_3, SO_2, and NO_2 by DIAL methods, OH, NO, Na, K, Li, Ca and Ca^+ by LIF, and the RIS method in conjunction with a mass spectrometer may prove to be a general point monitoring method for a wide variety of trace atmospheric molecules.

Table 1 lists different types of uv/visible laser sources. The field of laser sensing has always been constrained by the availability of efficient laser sources having spectral output in resonance with absorption features of atmospheric species of interest. Flashlamp or laser pumped dye lasers are convenient sources for tunable output in the uv/visible region and are presently being used to make LIDAR measurements. Although dye lasers are widely tunable, they are complex optical systems which require critical alignment of many optical components, have low efficiency, and have limited pulse energy in the ultraviolet. In the last six years a new class of laser, the rare gas halide excimer, has emerged as an intense source of coherent ultraviolet radiation which has great potential for use in remote sensing.

Excimer lasers are high pressure (typically 1 to 5 atmosphere) transversely excited electric discharge lasers and the devices are similar, but with subtle differences, to the TEA CO_2 laser. The typical excimer laser gas mixture contains greater than 90% of a rare gas buffer (He, Ne, or Ar) with several percent of another heavier rare gas (Kr or Xe) and a few tenths of a percent of a halogen donor (F_2, HCl, NF_3, or HBr). Volumetric preionization of the laser gas mixture using uv arc sources behind a screen electrode, a corona bar, or X-rays is necessary before the main discharge is ignited in order to initiate a stable glow discharge. The discharge electrons excite the heavy rare gas component (Ar, Kr, or Xe) to electronically excited levels and also produce positive rare gas ions as well as negative halide ions. Fast reactions of electronically excited rare gas atoms with the halogen donor and positive-negative ion recombination reactions produce the bound electronically excited rare gas halide molecule e.g. KrF, KrCl, XeF, XeCl, XeBr, and ArF. Lasing occurs from these excited bound levels to the ground states of the rare gas halide molecule which are unbound or, as the

TABLE 1

COHERENT LIGHT SOURCES
IN VISIBLE AND ULTRAVIOLET SPECTRAL REGIONS

	λ (μm)	Efficiency (%)	Energy (Joules)	Pulse Width (nsec)	Repetition Rate (Hz)	COMMENTS
Nd: YAG QUANTA RAY MODEL DCR–1A	1.06	≤1	0.7	9	22	$\lambda_1, \lambda_2, \lambda_3$ are generated by mixing in a crystal. This is a commercial laser.
	0.53	0.3	0.2	7	22	
	0.35	0.2	0.13	5	22	
	0.27	0.07	0.05	4	22	
Nd: YAG Pumped Dye Laser	>0.53	0.1	0.075	7	22	Tunable, Narrow–Bandwidth Complex System with low efficiency, but available commercially.
Frequency Doubled Dye output	>0.23	0.04	0.025	<5	22	
Flash Lamp Pumped Dye Doubled output (Phase–R model DL–1200uw)	0.96	0.5	10	300	0.3	Low efficiency, short dye life-time, long pulse widths not appropriate for lidar-complex systems with low doubling efficiency for UV.
	0.22	0.01				
Copper Pumped Dye Laser	0.55–0.8	0.5	0.001	20	6000	Good photon conversion but low pulse energy not scalable to high pulse energy–low doubling efficiency in uv.

237

Table 1 (Cont.)
COHERENT LIGHT SOURCES
IN VISIBLE AND ULTRAVIOLET SPECTRAL REGIONS

	$\lambda(\mu m)$	Efficiency (%)	Energy (Joules)	Pulse Width (nsec)	Repetition Rate (Hz)	COMMENTS
Rare Gas Halide Excimer Lasers						
XeCl	0.308					
ArCl	0.175					Many λ's each tunable over 10 to 20Å — narrow spectral width from oscillator-amplifier, 1kHz repetition rate requires gas recirculation.
ArF	0.193–0.194					
KrCl	0.221–0.223	0.1–2	0.2–5	<1–100	1–1000	
KrF	0.248–0.249					
XeCl	0.281–0.282					
XeF	0.351–0.353					
Excimer (XeCl @ 308 nm)						
Pumped dye	0.32–0.91	0.1–0.2	0.12	1–100	1–1000	A longer wavelength visible excimer would be more efficient as a dye pump.
Frequency doubled	0.22–0.32	0.1	0.8	1–50	1–1000	

Raman Shifting of XeCl Laser @ 308 nm

	Stokes Shifts			Efficiency (%)	Energy (Joules)	Pulse Width (nsec)	Repetition Rate (Hz)	COMMENTS
	1st	2nd	3rd					
H$_2$	0.353	0.414	0.499					
D$_2$	0.339	0.377	0.426					
CH$_4$	0.338	0.375	0.422	0.1–0.7	0.1–2.5	1–100	1–100	Tuning of Excimer laser produces tunable Stokes output.
N$_2$	0.337	0.359	0.392					
Ba	0.475							
Tl	0.405							
Pb	0.459							
Bi	0.475							

case with XeCl, very weakly bound. Since the terminal laser level for the rare gas halides are unstable they dissociate, depopulating the lower laser level and making these lasers very efficient. The bound-free transition also gives rise to a broad spectral gain bandwidth typically 10 to 20 Å for the excimer lasers. These lasers are scalable to large pulse energies in the ultraviolet (0.1 to 5 J) and with gas recirculation can be operated at high repetition rates (10 to 1000 Hz). The excimer lasers temporal pulsewidths can also be varied from a picosecond to several hundred nanoseconds.

Excimer lasers can be tuned directly over their 10 to 20 Å gain bandwidth using intracavity dispersive optics to obtain narrow spectral bandwidth output. However, the use of tuning elements causes severe attentuation of the laser output due to the increase in cavity losses, the need for spatial aperturing and the increase in the optical build up time. Nevertheless, efficient energy extraction from excimer lasers, while maintaining frequency and bandwidth control, can be achieved by injection locking an excimer laser amplifier with tunable, narrow bandwidth radiation from a low energy oscillator. Table 1 lists the tuning wavelength ranges and operational characteristics of the most efficient excimer lasers. Although the various developed excimer lasers do not offer total wavelength coverage in the uv, the high energy output from the excimer laser allows for efficient non Resonant Raman shifting of the tuned output of the excimer laser to other uv, visible, and vuv wavelengths. The utility of excimer lasers for Lidar has been discussed previously[1-3] but up to now only a DIAL measurement of stratospheric O_3 using an untuned XeCl excimer laser has been demonstrated.[3,4] Even though only one wavelength was used for these DIAL measurements, the increased power in the uv available from the excimer laser permited stratospheric O_3 measurements to be made easily from ground level.

The following discussion of our development at JPL of a compact tunable XeCl excimer laser for detection of atmospheric OH illustrates the unique capabilities of excimer lasers for remote sensing applications. Remote detection of atmospheric OH radicals is particularly important to support modelling of atmospheric chemistry,[5] and to serve as a possible indicator of radiation levels due to nuclear fallout. Laser-induced fluorescence (LIF) using narrow-bandwidth tunable uv lasers has been used to measure OH concentrations in the troposphere[6-8] and the stratosphere.[9] To date, atmospheric OH measurements have used frequency doubled tunable dye lasers to excite the $A^2\Sigma^+ - X^2\Pi_i$ (1,0) transition of OH near 282 nm, followed by broadband detection of fluorescence from the A-X (1,1) and (0,0) transitions centered at 314 nm and 309 nm respectively. However, it is also possible to measure atmospheric OH using laser excitation on the A-X (0,0) transition near 308 nm followed by narrowband detection at 309 nm. The advantages for LIF detection of OH using initial excitation on the (0,0) transition at 308 nm have been reported by McDermid and Laudenslager,[2] and the relative merits of (0,0) versus (1,0) excitation for shuttle-based Lidar detection of OH have been evaluated by Heaps.[10]

The fluorescence spectrum of XeCl covers the wavelength region of 307.4 to 308.7 nm. However, an untuned XeCl laser spectrum consists of two principal laser emission lines corresponding to the (0,1) and (0,2) bandheads of the B-X transition of XeCl at 307.92 and 308.17 nm,[11] respectively, with linewidths of 0.03 nm (FWHM). Since the gain coefficient determines the range of tunability of a laser, a frequency doubled nitrogen-pumped dye laser ($\Delta\lambda$ =0.01 nm) was used to measure the wavelength dependence of the net small signal gain coefficient for a discharge excited XeCl laser. Figure 1a shows the spectral profile of the gain coefficient superimposed on the XeCl fluorescence spectrum. The weakly bound ground state of XeCl gives rise to

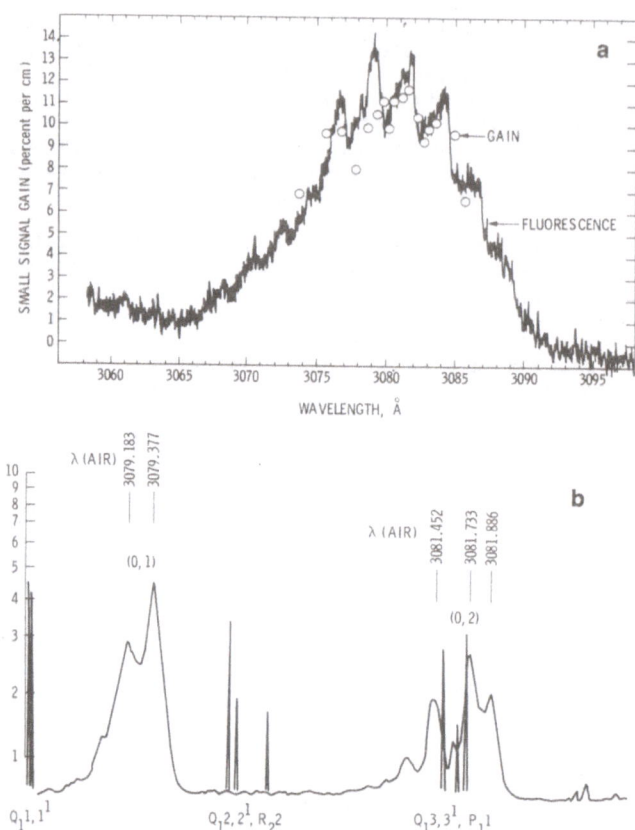

Fig. 1. a) Microdensitometer plot of the XeCℓ fluorescence spectrum. The wavelength ($\Delta\lambda \gtrsim 0.01$ nm) dependence of the small signal gain coefficient, measured in a large volume device, is shown superimposed on the spectrum.

     b) Untuned XeCℓ laser spectrum with OH absorption lines for the A-X (0,0) transition superimposed

the vibrational band structure evident in the gain and fluorescence spectra. This measured gain coefficient profile indicates that a XeCl laser should be continuously tunable from 307.5 to 308.5 nm.

    Several OH absorption lines of the A-X (0,0) transition near 308 nm lie in the wavelength region of the XeCl laser, Figure 1b. Therefore, a narrow bandwidth XeCl laser tuned to an OH absorption line would be suitable for Lidar detection of atmospheric OH radicals. Further, the high energy (5 J/pulse) and high efficiency (>1%) of the XeCl laser would facilitate Lidar detection of stratospheric OH from orbital altitudes.

    In order to determine the range of continuous narrow bandwidth tunability for XeCl, a XeCl laser was used as an amplifier for the frequency doubled output of a nitrogen laser-pumped dye laser.[12] The dye laser bandwidth in the ultraviolet was 0.01 nm with grating tuning, and 0.001 nm with an intracavity etalon. Double pass amplification, as shown in Figure 2a, was required to

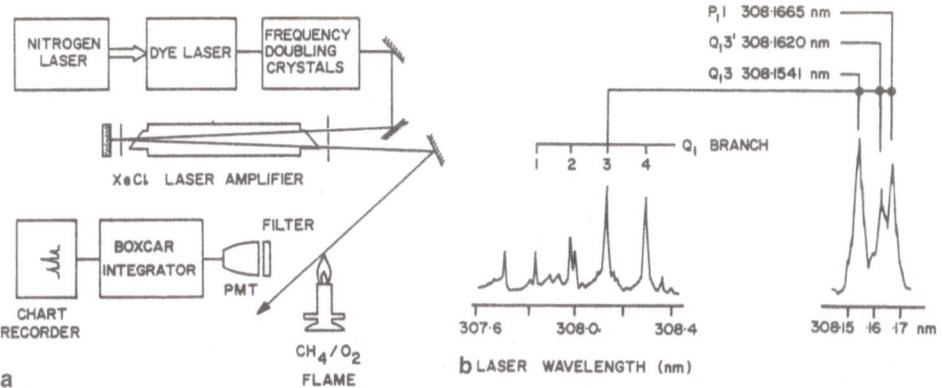

a

b LASER WAVELENGTH (nm)

Fig. 2 a) Experimental arrangement for detection of laser induced fluorescence of OH using double pass amplification of a frequency doubled dye laser in XeCℓ

b) Examples of the high and low resolution excitation spectra of OH A-X (0,0) obtained with this arrangement

approach saturation levels in the amplifier output due to the low input energy (0.5 μJ in a 3 ns pulse). Apertures of 4 mm diameter were used to suppress amplified spontaneous emission and appropriate synchronization, to within 5 ns, between the nitrogen and excimer lasers was provided by thyratron switching of each laser. The energy of the 0.01 nm spectral bandwidth amplified output was monitored with a Laser Precision pyroelectric detector to determine its wavelength dependence. Double pass amplification resulted in laser output energies across the complete XeCl fluorescence spectrum. The output varied only 30% as the dye laser was scanned from 307.6 to 308.4 nm, indicating an approach to saturation levels in the XeCl amplifier on the second pass of the dye laser beam. The peak amplified laser intensity was 1 MW/cm^2 in a 3 ns pulse. This can be compared to a laser intensity of 1.5 MW/cm^2 in a 20 ns pulse when this device was operated as an untuned oscillator.

To determine the suitability of the XeCl laser for OH detection, the tuned output of the XeCl amplifier was passed through a methane/oxygen flame and LIF from OH was detected by means of a photomultiplier and interference filter combination placed orthogonally to the intersection of the laser beam and the top of the flame, Figure 2a. The fluorescence excitation spectrum of OH using the dye laser oscillator-excimer amplifier configuration is shown in Figure 2. The low resolution spectrum of OH shows that the 0.01 nm resolution of the dye laser was maintained during amplification. The high resolution scan with the intracavity etalon in the dye laser resolved the Q_13, Q_13', and P_11 transitions and a linewidth (FWHM) of 0.0035 nm was observed. This is the Doppler width for OH at the temperatures in the flame and is approximately four times the Doppler width of OH at 298K. No OH fluorescence was observed if the two lasers were not properly synchronized or if only one of the lasers was pulsed.

As a practical and less complex alternative to a dye laser-excimer amplifier system an oscillator-amplifier excimer system was constructed to provide a simple, compact, and tunable source of narrow bandwidth, ultraviolet radiation.[12] The oscillator and amplifier were corona-preionized, transverse-discharge devices with a 1 x 1 x 40 cm^3 discharge volume. Electrical

241

excitation was provided by thyratron switching of dc charged capacitors to pulse charge ceramic capacitors close coupled to the laser bodies. A single thyratron was used and the relative timing between the oscillator and amplifier was adjusted by varying the inductance in the amplifier circuit. Both lasers used a gas mixtuure of 2.5% Xe, 0.2% HCl and 0.04% H_2 in an Ar buffer at 20 psia.

The optical configuration is shown in Figure 3. A holographic grating (3600 ℓ/mm) at grazing angle of incidence (>87°) was used to tune the oscillator. At such high angles the oscillator beam illuminated the whole length (50 mm) of the grating which served to reduce the spectral bandwidth. The zero-order reflection from the grating was used to couple the tuned laser radiation out of the oscillator cavity. The output wavelength was scanned using a stepping motor, with digital control and position readout, to turn a differential micrometer on the tuning mirror. Intracavity spatial apertures of 2.5 x 5 mm^2 were also used, with the short axis perpendicular to the grating. The oscillator output was expanded to match the amplifier cross section and a spatial filter (50-μpinhole) was used to reject the divergent amplified spontaneous emission. An optical delay was necessary so as to inject the oscillator pulse into the amplifier cavity as the gain reached threshold. The beam was injected through a quartz flat which together with two total reflectors formed the ring cavity. The ring configuration for the amplifier has the following advantages: (1) the output is unidirectional; (2) the oscillator is decoupled from the amplifier; (3) the output beam is uniform

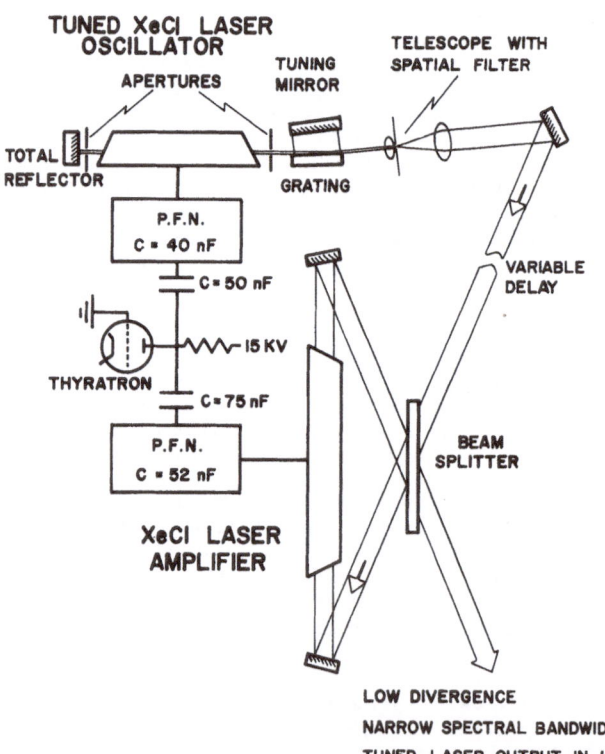

Fig. 3. Wavelength scannable XeCℓ oscillator-ring amplifier laser system

(provided the gain profile is uniform) without the hole common to unstable resonators; and (4) simple, inexpensive optical components can be used allowing variable cavity length and output coupling. An advantage of the entire system is that there is no need for special dielectric coatings; therefore, the system may be used at other wavelengths by simply changing the gas mixture and adjusting the angle of the tuning mirror. The grazing incidence grating tuning system provides narrow spectral bandwidth and can be simply tuned since the wavelength is a linear function of the tuning mirror angle over the tuning ranges typical of the excimer lasers. Alternative methods for achieving narrow bandwidth tunable radiation, e.g., with multiple elatons or etalon-grating combinations, are extremely difficult to scan continuously because of the complex interdependence of the relative angles on the wavelength.[13] Also, because of the narrow wavelength range for the reflective coatings used in the etalons additional sets of tuning optics would be required for different excimer lasers.

The spectral output from this oscillator injection-locked amplifier was narrow bandwidth, <0.005 nm, and this narrow bandwidth was obtained throughout the regions where single line injection locking was achieved.

Throughout the regions where injection locking was achieved an output energy of 12-15 mJ was consistently extracted. When the oscillator was operated as an untuned, broadband laser and its output was used to injection lock the ring, 15 mJ was also extracted. In regions where injection locking was not established 8-10 mJ was extracted. Pulse energies were measured with a Lumonics 20D detector and displayed on a Tektronix 475 oscilloscope. The amplifier output pulse width was 30 ns FWHM for both tuned and untuned injection locking. Since the output obtained from a XeCl amplified dye laser exhibited only 30% variation over the XeCl emission spectrum, an improved oscillator with single line output tunable throughout the XeCl spectrum could provide continuous, constant energy extraction from the amplifier. The injection locked output was also used as described previously to detect OH in a flame. The $Q_1 1$, $Q_1 2$, $Q_1 3$, and $P_1 1$ transitions of the A-X (0,0) band at 307.85, 308.00, 308.15, and 308.17 nm were observed. These experiments demonstrate efficient, narrow bandwidth operation of a XeCl amplifier in a ring configuration and its utility for LIF measurements of OH. This same tuned XeCl laser can also be used to measure SO_2 and O_3 by the DIAL method. The uv output from XeCl can be Raman shifted into the visible for applications to measure fluorescence from phytoplankton in the ocean and for bathymetry. Other excimer systems such as the XeBr, KrCl, ArF, and KrF lasers can all be operated in similar tuned oscillator-amplifier arrangements for remote sensing applications which require compact, tunable high energy uv/visible laser sources.

References

1. J. B. Laudenslager, R. W. Svorec, I. S. McDermid, and T. J. Pacala, "Development and Application of Excimer Lasers for Remote Sensing," Conf. Abst. Tenth Intl. Laser Radar Conf., Silver Spring, MD., p. 80, October (1980).
2. I. S. McDermid and J. B. Laudenslager, "Lifetimes and Quenching Rate Constants Relevant to Remote Sensing of Hydroxyl Radicals with 308 nm Excitation (XeCl)," Technical Digest of Spectroscopy in Support ot Atmospheric Measurements, Sarasota, FL, November 1980, Paper WP13-1. J.Chem.Phys., Scheduled for 15 Feb 1982 edition.
3. O. Uchino, M. Maeda, and M. Hirono, IEEE J. of Quantum Electron. QE-15, 1094 (1979).

4. O. Uchino, M. Maeda, J. Kohno, T. Shibata, C. Nagasawa, and M. Hirono, Appl. Phys. Lett. 33, 807 (1978).
5. M. Nicolet, "An Overview of Aeronomic Processes in the Stratosphere and Mesosphere," Can. J. of Chem. 52, 1381 (1974).
6. E. L. Baardsen and R. W. Terhune, Appl. Phys. Lett. 21, 209 (1972).
7. C. C. Wang and L. I. Davis, Phys. Rev. Lett. 32, 349 (1974).
8. D. D. Davis, W. Heaps, and T. McGee, Geophys. Res. Lett. 3, 331 (1976).
9. W. Heaps, unpublished.
10. W. S. Heaps, Appl. Opt. 19, 243 (1980).
11. T. McKee, Phys. in Can. 36, 41 (1979).
12. T. J. Pacala, I. S. McDermid, and J. B. Laudenslager, Appl. Phys. Lett. 40, 1 (1982).
13. M. A. A. Clyne and I. S. McDermid, Laser Induced Fluorescence: Electronically Excited States of Small Molecules, Advances in Chemical Physics edited by K. P. Lawley (John Wiley & Sons, England, in press).

6.2 Solid-State Laser Sources for Remote Sensing

R.L. Byer, T. Kane, J. Eggleston, and Sun Yun Long*
Stanford University, Ginzton Laboratory, Stanford, CA 94305, USA

I.   INTRODUCTION

Remote sensing with laser sources has made tremendous strides during the past decade. The progress has been paced, however, by the slow development of the high power, narrow bandwidth, tunable laser sources required as transmitters for DIAL and coherent detection methods. Only recently, with the development of Nd:YAG pumped dye and parametric oscillator sources has near infrared and ultraviolet measurement systems been demonstrated.

This paper summarizes recent progress in solid state lasers that utilize the slab geometry with a zig-zag optical path as illustrated in Fig. 1.

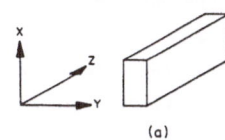

(a)

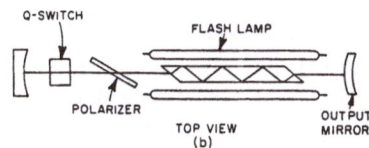

Q-SWITCH   FLASH LAMP

POLARIZER   OUTPUT
        TOP VIEW   MIRROR
           (b)

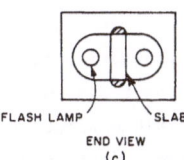

FLASH LAMP   SLAB
        END VIEW
           (c)

Figure 1.
a—Geometry showing axis of the slab.
b—Schematic of the laser resonator with Brewster angle slab and zig-zag optical path.
c—End view of the slab holder showing flashlamp position within the non-imaging flashlamp reflector structure

The slab configuration solid state laser approach first proposed and investigated at General Electric[1] has been under development in our laboratory for the past three years. Slab geometry lasers offer a number of important advantages over the conventional rod geometry lasers. These advantages include the elimination of stress induced birefringence and thermal and stress induced focusing, and higher average output power limited only by stress induced fracture of the glass or crystalline laser host material. The progress in slab Nd:Glass and Nd:YAG lasers is discussed in Section II.

Coherent detection of wind velocity by Doppler velocimetry has now been established as a viable measurement approach at the CO_2 laser infrared wavelength range.[2] We have undertaken

---

*Visiting scholar from North China Electro-Optics Institue, Peking, China.

to design a single frequency Nd:YAG laser source for coherent wind velocity measurements at 1.06 μm. The Nd:YAG source offers the potential advantages compared to CO_2 based system of more than two orders of magnitude larger backscatter coefficient, improved depth resolution, room temperature silicon diode based detection and a compact all solid state construction. The design and testing of a single frequency Nd:YAG oscillator is discussed in Section III.

II. SLAB GEOMETRY SOLID STATE LASERS

A. Introduction

Remote atmospheric measurements using the DIAL technique requires a high pulse energy tunable laser source. Only recently, with the development of the unstable resonator Nd:YAG source[3] has 1 J pulse energies at 10 Hz repetition rate become available. However, the Nd:YAG pumped 1.4 - 4.0 μm tunable parametric oscillator source[4] has been limited to 10 - 50 mJ of output energy. The doubled Nd:YAG pumped dye tunable source has output pulse energies up to 100 mJ. For SO_2 measurements in the ultraviolet, the doubled dye laser output energy is typically 10 mJ.[5,6] These energies are adequate for range resolved measurements over intermediate ranges, but are an order of magnitude below the requirements for space shuttle based LIDAR or DIAL measurements to 10 kilometer ranges. In addition, from a system viewpoint, the tunable OPO or dye laser is not as efficient or reliable as desired for long term measurement applications.

We began our slab geometry laser development effort three years ago in recognition of the need for a higher pulse energy, more reliable tunable source. Our approach was to tune the glass laser over its 200 cm^{-1} wide bandwidth and to extend the tuning range by harmonic generation and by selective Raman shifting.[7]

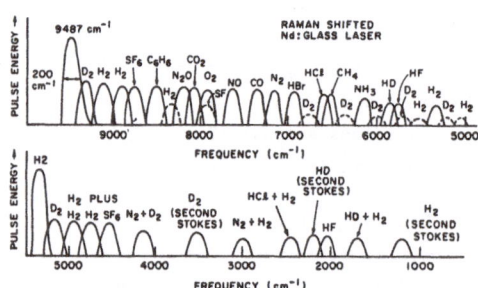

Figure 2--Tuning range of Nd:Glass laser centered at 1.055 μm followed by Raman shifting in gases. Harmonic generation in $LiNbO_3$ and KD*P can be used to extend the wavelength range into the visible and ultraviolet range

Figure 2 illustrates the tuning range available with just a few Raman active gases. Harmonic generation can be used to extend the tuning range into the ultraviolet and visible spectral regions. The tunable Nd:Glass oscillator/amplifier source, followed by nonlinear frequency extension, combines the higher pulse energies generated in Nd:Glass with solid state reliability and system simplicity. The key to successful implementation lies in the application of the slab geometry concept.

To appreciate the advantages of the

zig-zag slab teometry approach it is useful to compare the performance
limitations of rod geometry lasers to slab geometry lasers. To make the comparison
specific we consider the performance limits of a 15 cm long x 6 mm diameter
Nd:Glass rod laser and a 15 cm long x 8 mm thick three-to-one aspect ratio slab
Nd:Glass laser.

The performance limitations are summarized in Table I for LHG-8 phosphate
glass. Phosphate glass was selected because of its high gain cross-section and
negative dn/dT which helps to offset thermal and stress focusing in the rod laser.

TABLE I Rod and Slab Geometry Laser Design Limitations

Rod geometry (6 mm dia. x 15 cm length LHG-8 phosphate glass)	Average flashlamp power. (Watts)	Effect
Birefringence	100	π phase shift
Thermal focusing	1000	-115 cm focal length
Stress focusing	1000	+70 cm focal length (biaxial)
Stress fracture	1000	Surface stress fracture limit
Slab geometry (8.3 mm x 25 mm x 15 cm length, LHG-8 phosphate glass)		
Birefringence	-	Eliminated by x,y,z geometry
Thermal & stress focusing	-	Eliminated by zig-zag optical path
Stress fracture	2000	Only design limitation

In the rod geometry, the stress fracture limit occurs at 5 Hz repetition rate
at 200 J flashlamp energy or 1000 W of average input power. This stress fracture
limit was verified experimentally by fracturing rods of Nd:Glass. Stress and
thermal focal lengths are +70 cm and -115 cm at the stress fracture loading limit.
These focusing effects must be compensated in the design of the laser oscillator to
enable near diffraction limited performance. Since the stress and thermal focusing
has both radial and tangential components, the focusing is not equivalent to a lens
but is instead a biaxial focusing element that cannot be simply compensated by an
additional lens.[8]

Finally, at ½ Hz repetition rate, or 100 W of flashlamp average power, stress
induced birefringence is severe enough to cause a polarization induced phase shift
of π radians. This in turn induces a severe loss for linearly polarized resonator
designs that are required for Q-switching or frequency extension by nonlinear
processes.

The slab geometry, on the other hand, completely eliminates stress induced
birefringence by slab symmetry. If a zig-zag optical path is used, then the stress
and thermal induced focusing is also eliminated. This leaves stress fracture as

247

the only design limitation on the average power loading of the slab and thus on
the available average output power. For LHG-8 glass, the thermal loading limit set
by fracture induced breakage is 2000 W for an 8.2 mm x 25 mm x 15 cm glass slab.
At this loading, the slab can operate at up to 5 Hz repetition rate and store up to
12 J of optical energy at 3.75% storage efficiency. Clearly the slab geometry
offers a significant performance improvement over the usual rod geometry approach.

B. Nd:Glass Slab Measurements

The advantages of the zig-zag slab geometry have been known for more than a
decade.[1] Only recently, however, has the technology for fabricating the slabs
become widely available. The demands of the laser fusion research effort have
pushed the fabrication and optical coating technology to the present status where
slab fabrication is possible at a reasonable cost.

To take full advantage of the slab geometry in laser system design, we have
developed a computer program that calculates the thermal profile of the slab, the
induced stresses, the stress induced depolarization, and ray traces the zig-zag
optical path through the slab. The program displays the results and can be used
as an interactive design tool.

To be useful the computer model must be carefully calibrated and verified.
We have accomplished verification, where possible, by comparison with analytical
solutions to the thermal and stress equations. However, a test-bed Nd:Glass slab
laser source was designed and constructed to provide calibration and quantitative
verification of the model. Measurements using the test-bed laser system have
verified predictions of the model and have shown that the slab geometry approach
does, in practice, provide the projected advantages of elimination of birefringence
and thermal and stress induced focusing.

The calculated stresses in the thermally loaded slab are an example of the
model capability. Figure 3 shows the compression and tension field in the slab
assuming uniform pumping on the y faces. From the stress field, the depolarization
can be calculated and displayed as shown in Fig. 4. Note that the peak depolarization is predicted to be 32% while the average depolarization over the slab
area is near 5%. Figure 5 shows a plot of predicted depolarization (solid curve)
and measured depolarization (dots) at 1600 W average flashlamp power loading of the
slab.[10] The excellent agreement verified a number of aspects of the model including
the calculated stress field, depolarization and optical ray tracing along the
zig-zag path.

Using the slab test-bed laser we have completed measurements of gain, energy
storage efficiency, pumping uniformity, depolarization, beam divergence and
quality, thermal induced end effects and coolant effects on focusing, tuning,

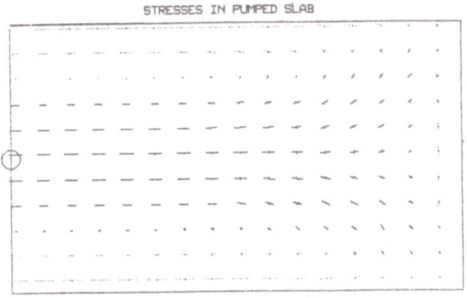

STRESSES IN PUMPED SLAB

THICK=0.860 WIDTH=2.850
CROSS-BAR WITH LENGTH EQUAL TO HORIZONTAL SEPARATION
BETWEEN POINTS IS 150.000 STRESS UNITS.
CIRCLE IS AT SLAB CENTER.
COMPRESSION   TENSION

Figure 3--Calculated tensional and compressional stresses in a pumped glass slab assuming uniform pumping on the y faces. The slab is in compression in the center and tension at the edges. The tensional stress at the surface eventually leads to fracture of the slab

Z-Z DEPOLARIZATION

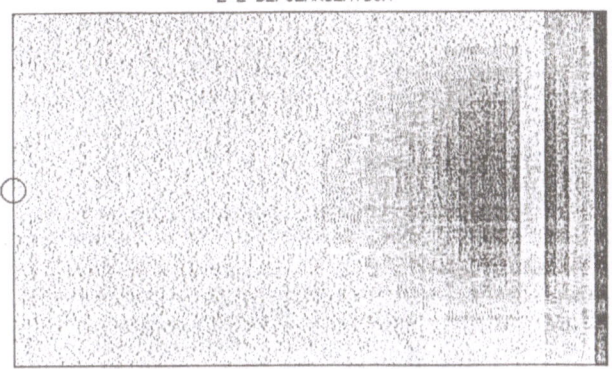

WIDTH=2.850 THICK=0.860
CIRCLE AT SLAB CENTER. ONE COLOR CHANGE= 4.010019E-02

1.427284E-10   AVG= 5.772877E-02   5.213025E-01

Figure 4--Calculated depolarization for a pumped glass slab assuming uniform pumping on the y faces. The maximum depolarization is near 32%. Note that over a large region of the slab the depolarization is insignificant

Q-switch oscillator operation, and second harmonic generation.[9]

The test-bed slab laser has operated as a long pulse oscillator at 3.75% storage efficiency, 2.0% slope efficiency and 1.6% wall plug efficiency. Typical output data for extraction from the entire slab volume is shown in Fig. 6a. We have used the test-bed slab laser to verify, by far field diffraction measurements, that the beam quality of the slab is diffraction limited. We have operated a slab glass oscillator using both an unstable resonator[3] and a radial birefringent resonator design.[11] Figure 6b shows the output pulse energy free running and Q-switched for the 6 mm diameter oscillator mode volume.

We plan to follow the oscillator by a pre-amplifier in the same slab to generate 1.2J and by an amplifier in a second identical slab to generate over 6J per

pulse at 5 Hz repetition rate. We have previously demonstrated tuning over a
± 100 cm^{-1} range with a prism beam expander and grating. We expect efficient
nonlinear frequency extension via SHG and Raman processes due to the diffraction
limited polarized output beam quality. The theoretical and experimental results
for the slab laser are being prepared for presentation and publication.[12]

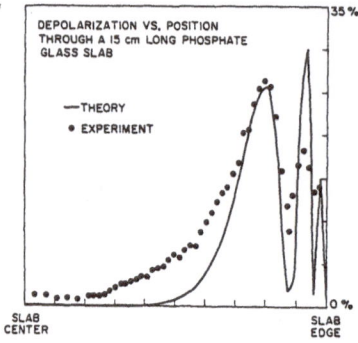

Figure 5—Predicted and measured depolarization for a pumped glass
slab. The oscillations are
caused by stress vector rotation
combined with the zig-zag
optical path through the slab

Future slab laser research will
involve the completion of a second
generation slab laser system with more
efficient flashlamp pumping, an optimized
slab thickness and improved resonator/
amplifier designs. The improved slab
glass laser system will be used in
remote sensing measurements in the
near infrared.

C. Nd:YAG Slab Laser Measurements

The slab geometry can also be
applied to crystalline laser systems.
We have carried out preliminary measurements with a cw lamp pumped Nd:YAG slab
oscillator. These measurements have
shown that thermal and stress induced

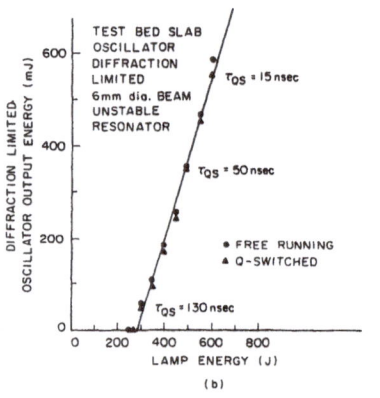

Figure 6.
a—Measured long pulse output energy and
average power from the test-bed slab
laser vs input flashlamp energy at
2.5 Hz repetition rate. The 1500 W
average flashlamp power is below the
measured 2000 W stress fracture limit
for the present 8.3 mm x 25 mm x 15 cm
long LHG-5 glass slab
b—Measured Q-switchd output energy for a
6 mm diameter beam, unstable resonator
oscillator

focusing can be eliminated with the zig-zag slab geometry in Nd:YAG at up to 4 kW of
flashlamp average power. This level of pumping is over six times that typically

250

used for Q-switched, high power, pulsed Nd:YAG unstable resonator laser systems. The present pumping limit is set by the maximum power provided by our currently available power supply and not by the thermal limitations of the Nd:YAG slab.

We have also verified that the slab geometry completely eliminates stress induced birefringence in Nd:YAG.[13] This is important for applications where Nd:YAG is to be frequency shifted by nonlinear processes. Work is continuing on the slab geometry applied to crystalline laser host materials.

In summary, the slab geometry approach offers more than an order of magnitude improvement in laser performance for both Nd:Glass and Nd:YAG laser systems. Recent model and test-bed laser experiments have verified that the expected improvements are realizable in practice.

III. SINGLE FREQUENCY Nd:YAG OSCILLATOR/AMPLIFIER SYSTEM

A. Introduction

The tremendous progress of coherent LIDAR measurements at the CO_2 wavelength range[2] has sparked an interest in long range depth resolved wind measurements using coherent LIDAR. An examination of the measurement system shows that there are advantages to be gained by using wavelengths shorter than 10 μm. These advantages include increased backscatter by more than two orders of magnitude at 1 μm as opposed to 10 μm, improved depth resolution, access to room temperature silicon diode detectors with avalanche gain, and solid state transmitters with demonstrated long operational lifetimes and small volume and weight. The principal difficulty in the approach to shorter wavelength coherent LIDAR measurements was the unknown bandwidth of the transmitter due to the unknown thermally induced chirp rate in the solid state laser host material.

During the past year we have pursued the design of a single axial mode Nd:YAG oscillator with the goal of demonstrating a frequency stability adequate for wind velocity measurements by Doppler velocimetry in the atmosphere.

B. Nd:YAG Oscillator Design

Our research efforts with single frequency Nd:YAG oscillators has clearly shown that Nd:YAG is not homogeneously saturated and that special steps such as injection locking[14] must be taken to insure single mode operation. We also learned previously that the Nd:YAG chirp rate was less than 9 MHz for pulsed flashlamp excitation. Thus there was a possibility that the chirp rate in Nd:YAG could, in fact, be much less than 9 MHz and that linewidths of less than 1 MHz required for wind measurements could be achieved.

We have designed a quasi-cw flashlamp pumped, very short cavity, Nd:YAG TEM_{oo} mode oscillator source to test the frequency linewidth limitations. The oscillator is shown schematically in Fig. 7. The cavity physical length is 7.5 cm. The

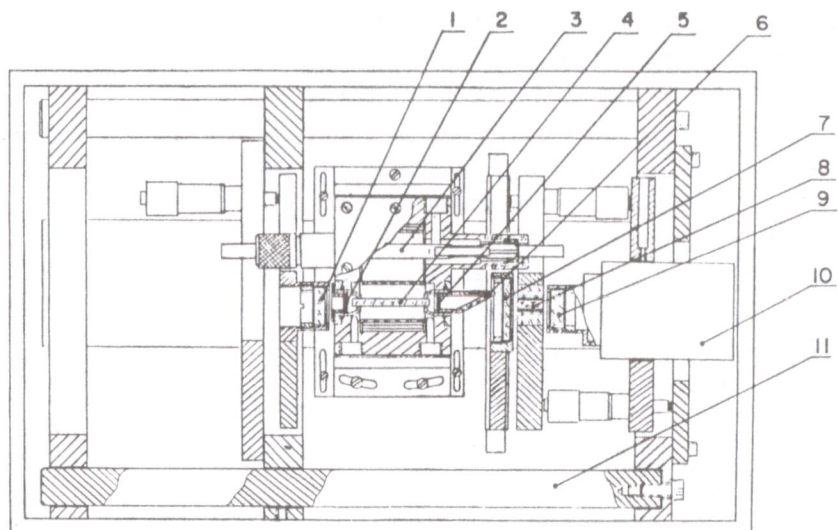

Figure 7-- Schematic of the single axial mode Nd: YAG oscillator showing resonator mirrors (1,9), YAG rod (4), quarter wave plates (2,5), flashlamp (3), Brewster polarizer (6), etalon (7), LiNbO$_3$ phase modulator (8), piezoelectric stack (10), and invar resonator spacers (11)

optical length is 10 cm due to the 2.0 cm Nd:YAG rod and thus the axial mode spacing is 1.5 GHz. A single 2 mm thick fused silica tilted etalon, with a finesse of 7 is adequate to select a single axial mode under the very broad 10 cm^{-1} Nd:YAG gain curve.

The flashlamp is simmered and pulsed for a 3 - 5 msec period at 10 Hz repetition rate. To generate constant power output during the quasi-cw pumping period, the flashlamp current must be increased during the pulse as shown in Fig. 8a. Note that the laser oscillator initially spikes when threshold is reached, but that the spiking behavior decays to steady state output in 240 µsec as expected.

The laser output frequency can be stabilized relative to the tilted etalon pass band by measuring the output power vs time as originally noted by Kuizenga.[15] The power variation is due to the chirp of the axial mode frequency relative to the etalon pass band and is one approach to determining the chirp rate. The chirp rate can also be measured by monitoring the laser output power after a high resolution confocal interferometer spectrum analyzer as shown in Fig. 8b. The position of the peak can be measured as a function of bias voltage on the piezoelectric stack of the resonator to yield an accurate measurement of the chirp rate. Both measurements showed that the present Nd:YAG oscillator has a chirp rate of 30 kHz per 1 microsecond.

This chirp rate is very constant from pulse-to-pulse and can be compensated by inducing a ramp voltage of 2 V per millisecond on the laser resonator piezo-

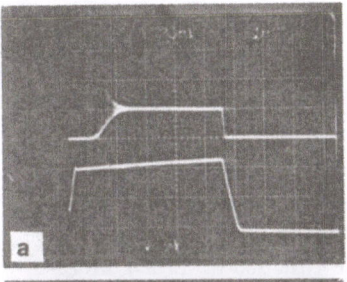

Figure 8a.
Upper trace shows the quasi-cw single axial mode Nd:YAG laser output at 0.1 W power level. The lower trace shows the pumping lamp current with ramp compensation to achieve constant laser output power

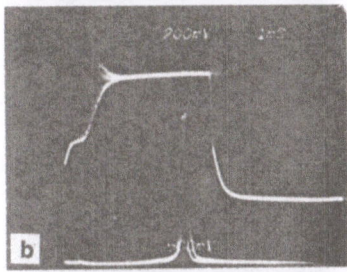

Figure 8b.
The upper trace shows the laser output power for the 3 msec pulse. The lower trace shows the output power following a 10 cm, R = 99.7% confocal interferometer spectrum analyzer. Measurement of the pulse width or delay shows that the chirp rate is 30 kHz per microsecond

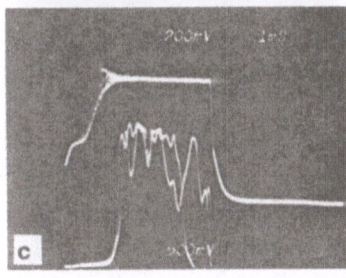

Figure 8c.
Upper trace is laser output power. The lower trace is the laser power detected after the confocal interferometer while the chirp is compensated by a 2V/msec ramp on the PZT. Acoustically induced vibrations lead to the power and frequency fluctuations which are less than ± 20 MHz on a pulse-to-pulse basis

electric stack. The result of this procedure is to stabilize the output frequency over the 3 millisecond duration as shown in Fig. 8c. The remaining oscillations in amplitude correspond to acoustic induced cavity length variations that in turn yield a ± 20 MHz pulse-to-pulse free running laser stability.

We are in the process of completing a locking loop to stabilize the Nd:YAG output frequency onto a 10 cm confocal interferometer with a 15 MHz passband. We expect to demonstrate a closed loop frequency stability of less than 1 MHz using a combination of a 1 kHz bandwidth feedback to the piezoelectric controlled mirror and a 1 MHz bandwidth feedback to a $LiNbO_3$ phase modulator within the Nd:YAG oscillator cavity.

The 3 millisecond quasi-cw operation was selected so that an early part of the output could be selected by a half-wave electro-optic switch and amplified by a Nd:YAG amplifier chain. The remaining part of the pulse is then available for use as the local oscillator for the heterodyne measurement for ranges up to 450 km. Longer pulse lengths are easily possible so that measurements from space platforms at 1000 km altitude are feasible. The quasi-cw operation significantly reduces the power consumption of the oscillator. The present system requires 30 W average power into the flashlamps while a cw lamp pumped oscillator would require close to 1000 W to achieve the same performance.

The present Nd:YAG oscillator generates 0.1 W of output power. We have demonstrated a double pass amplifier with a gain of 600 that utilized a 4 mm diameter Nd:YAG rod pumped by a pulsed flashlamp. We have verified that the laser linewidth is not broadened by the pre-amplifier. We expect, in the future, to demonstrate amplification to pulse energies of greater than 300 mJ in a 1 micro-second pulse at 10 Hz repetition rate in a rod geometry Nd:YAG system. The use of slab geometry offers an order of magnitude improvement without sacrificing beam quality or frequency stability.

In summary, we have demonstrated that a single frequency Nd:YAG oscillator operating in a 3 msec quasi-cw mode has a chirp rate of 30 kHz per microsecond and a free running stability of ± 20 MHz. The chirp can be easily compensated thus giving adequate frequency stability for wind velocity measurements by coherent LIDAR.

We plan to pursue the Nd:YAG source improvements and to apply the system to atmospheric wind studies. In the future, diode laser pumped Nd:YAG sources[16] offer the possibility of an efficient, long term operation, space qualified transmitter source for global wind velocity measurements.

IV.   CONCLUSION

We have pursued the study of advanced solid state laser concepts with the goal of meeting remote sensing transmitter requirements for future LIDAR systems. The frequency shifted, tunable glass laser source, using the slab geometry approach, offers the potential for a high pulse energy, high average power solid state transmitter without the need for a parametric oscillator or dye laser tunable source. The current Stanford test-bed slab laser source has a design goal of 6 J per pulse at 5 Hz with greater than 1% wall plug efficiency. Advanced slab glass laser designs, such as the system presently under construction at the Swiss Institute of Nuclear Research for measurements at 6.0 μm wavelength of the Lamb shift in the μp hydrogen atom,[16] call for 10 J of output energy at 50 Hz repetition rate or 500 W of average output power. Clearly these next generation laser sources will have a major impact on the tunable LIDAR measurement capability.

The demonstration of a small chirp rate in a Nd:YAG single axial mode oscillator opens the possibility for remote wind measurements by coherent LIDAR at 1.06 μm. The advantages of shorter wavelength and an all solid state transmitter offer the possibility of global wind field measurements from satellite platforms.

ACKNOWLEDGEMENTS

We want to acknowledge the support provided by the United States Army Research Office through contract number DAAG29-81-C-0038, the Department of Energy through contract number DOE 3818301 and by the National Aeronautics and Space Administration through contract number NAG1-182. We also want to acknowledge the early contributions to this work by Dr. J. Unternahrer of the Swiss Institute of Nuclear Research.

REFERENCES

1. W.S. Martin and J.P. Chernoch, "Multiple Internal Reflection Face Pumped Laser", U.S. Patent #3,633,126,(1972); see also, Y.S. Liu, W.B. Jones and J.P. Chernoch, "Recent Development of High Power Visible Laser Sources Employing Solid State Slab Lasers and Nonlinear Conversion Techniques", General Electric Report #81CRD104, May 1981.

2. R. Milton Huffaker, "Feasibility of a Global Wind Measuring Satellite System (WINDSAT)"; and also R.D. McPerson, "Data Requirements and Priorities for Operational Global Forecasting During the 1980's and 1990's" in the Proceedings of the Coherent Laser Radar for Atmospheric Sensing Conference held at Aspen Colorado, July 1980.

3. R.L. Herbst, H. Komine and R.L. Byer, "A 200 mJ Unstable Resonator Nd:YAG Oscillator", Optics Commun. 21, p.5 (1977); see also R.L. Byer, "Nd:YAG Pumped Tunable Sources and Applications", (to be published).

4. M. Endemann and R.L. Byer, "Remote Single Ended Measurements of Atmospheric Temperature and Humidity at 1.9 μm Using a Continuously Tunable Source", Optics Letters, vol. 5, October 1980; see also M. Endemann and R.L. Byer, "Remote Measurement of Trace Species in the Troposphere", presented at the A.I.A.A. 19th Aerospace Sciences Meeting, January 12-15, 1981, St, Louis, Missouri, (to be published).

5. J.G. Hawley, G.F. Wallace, L.D. Fletcher, "A Mobile Differential Absorption Lidar (DIAL) for Range Resolved Measurements of SO_2, O_3 and NO_2", presented at the A.1.A.A. 19th Aerospace Sciences Meeting, January 12-15, 1981, St. Louis, Missouri.

6. K. Fredrikson, B. Galle, K. Mystrom and S. Svanberg, "Mobile LIDAR System for Environmental Probing", Applied Optics, vol. 20, p.4181 (1981).

7. R.L. Byer, "Frequency Conversion via Stimulated Raman Scattering". Electro-Optical Systems Designs, February 1980.

8. W. Koechner and D.K. Rice, "Birefringence of Nd:YAG Laser Rods", J. Opt. Soc. Am. 61, p.758 (1971).

9. J.M. Eggleston, "Theoretical and Experimental Studies of Slab Geometry Lasers", Ph.D. Thesis, Stanford University, April 1982.

10. J.M. Eggleston, T. Kane and R.L. Byer and J. Unternahrer, "Slab Geometry Solid State Lasers", presented at the 1982 C.L.E.O. Conference, Phoenix, Arizona.

11. J.M. Eggleston, G. Giuliani and R.L. Byer, "Radial Intensity Filters Using Radial Birefringent Elements", Journ. Opt. Soc. Am. 71, p.1264 (1981).

12. J.M. Eggleston, T. Kane, J. Unternahrer and R.L. Byer, "Theoretical and Experimental Studies of Nd:Glass Slab Geometry Lasers", (to be published).

13. T.J. Kane, J.M. Eggleston and R.L. Byer, "Polarized cw Nd:YAG Laser Using a Slab Geometry", presented at the 1982 C.L.E.O. Conference, Phoenix, Arizona.

14. Y.K. Park, G. Giuliani and R.L. Byer, "Stable Single Axial Mode Operation of an Unstable Resonator Nd:YAG Oscillator by Injection Locking", Optics Letters, 5, p.96 (1980).

15. . D.J. Kuizenga, "Short Pulse Oscillator Development for the Nd:Glass Laser Fusion System", IEEE Journ. Quant. Electr. QE-17, p.1694 (1981).

16. R.J. Smith, R.R. Rice, L.B. Allen. Jr., "100 mW Laser Diode Pumped Nd:YAG Laser", S.P.I.E., vol. 247, p.144.

17. J. Unternahrer, Swiss Institute of Nuclear Research, Zurich, Switzerland, (private communication).

6.3 Progress in Laser Sources for Remote Sensing[*]

A. Mooradian, P.F. Moulton, and N. Menyuk

Lincoln Laboratory, Massachusetts Institute of Technology,
Lexington, MA 02173, USA

Substantial progress in both active and passive remote sensing using
laser sources has been demonstrated in the past several years. At present,
the development of laser remote sensing is limited by the availability of
adequate tunable laser sources. We describe here the results of recent
progress in the development of various tunable laser sources which will be
of use for both active and passive laser remote sensing. Included among
these sources are transition-metal ion–doped solid-state lasers, semiconductor diode lasers in external cavities, miniature CO_2 TEA lasers, and frequency conversion in infrared nonlinear materials.

Tunable Transition-Metal Ion-Doped Solid–State Lasers

Tunable lasers using transition-metal-doped crystals as the active
media are likely to become important sources for remote sensing applications.
The fact that such devices are continuously tunable and can be Q-switched
to generate high-peak-power output is especially useful for a variety of
experiments.

We have been studying the properties of the $Co:MgF_2$ laser,[1,2] which can
be tuned in the 1.6 to 2.3 μm wavelength region. The laser system described
here consists of a liquid-nitrogen-cooled crystal optically pumped by a
pulsed, normal-mode, 1.34-μm Nd:YALO laser. Experiments were carried out
using a 6 x 6 x 27 mm Brewster-angle crystal doped with 0.75 wt.% CoF_2.
The laser configuration was a simple two-mirror cavity; mirror radii were
50 cm and the mirror spacing was ~ 30 cm. Only the TEM_{00} mode of the $Co:MgF_2$
laser was excited by the pump and thus the laser output was diffraction-
limited. The input-output energy curves for three different sets of cavity
mirrors with coatings optimized for differing wavelengths are shown in
Fig. 1. Data was taken at a pulse repetition rate of 4 Hz; the output pulsewidth well above threshold was ~ 1 ms. Threshold pump input energy at the
1.92-μm operating wavelength was 4.5 mJ. Measurements of the transmission
losses of the pump focusing lens and cavity mirror and of the transmission
of the laser crystal showed that approximately 46% of the pump input energy
was absorbed in the crystal. When this is taken into account, the calculated output quantum efficiency (i.e., the ratio of laser output photons to
absorbed pump photons) is 79% at the 1.92-μm operating wavelength. It is
worth noting that this efficiency was obtained with a 0.9%-transmission
output coupler, an indication that the scattering losses in the laser crystal were extremely small. At a pulse rate of 50 Hz up to 7.3 W of average
output power was obtained at 1.92 μm.

---

[*]This work was sponsored by the Department of the Air Force.

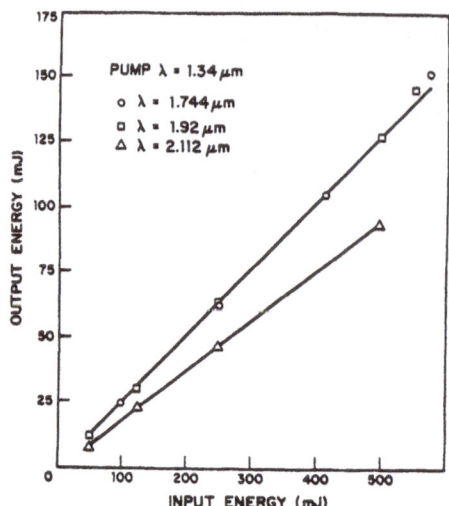

Fig. 1. Output vs input energy for Co:MgF$_2$ laser for three different operating wavelengths

Tuning of the Co:MgF$_2$ laser was accomplished by insertion of a Brewster-angle, birefringent tuning element in the laser cavity. To eliminate the need for intracavity windows and, further, to eliminate the intracavity effects of strong atmospheric water vapor absorption centered at 1.88 μm, the tuning element was contained in the vacuum section of the Dewar. Rotation of the element was done using a vacuum rotary feedthrough. The tuning curves in Fig. 2 represent results with three different sets of cavity mirrors, at a pump input energy of 500 mJ. The short-wavelength limit was determined by the finite free-spectral-range of the tuning element; as the laser was tuned to wavelengths shorter than 1.6 μm the output shifted abruptly to 1.86 μm, another peak in the tuning element transmission. With appropriate mirrors it should be possible to obtain shorter-wavelength operation, limited ultimately to 1.48 μm, the position of the zero-phonon line. The long-wavelength limit was also affected by the cavity mirrors, which started to become increasingly lossy at wavelengths beyond 2.2 μm. Thus, it may be possible to extend operation of the Co:MgF$_2$ laser beyond the demonstrated 2.3-μm limit by using more appropriate mirrors.

The long $^4T_2$-upper-state lifetime of the Co:MgF$_2$ laser, 1.3 ms at a crystal temperature of 80 K, should in principle allow almost all of the

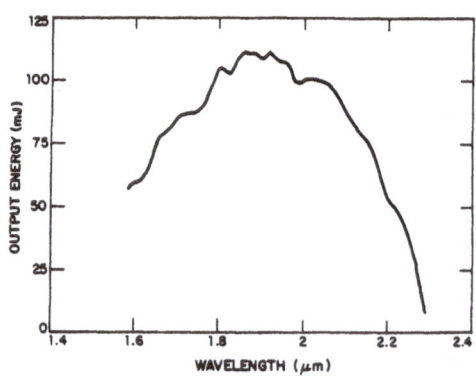

Fig. 2. Tuning range for Co:MgF$_2$ laser, a composite of results for three different sets of cavity mirrors

normal-mode output generated with a 1-ms pump pulse to be obtained in a Q-switched pulse. Because of the relatively low estimated gain cross section of the system, $\sim 10^{-21}$ cm^2, high intracavity fluences on the order of 100 J/cm^2 are present in the Q-switched mode, and increase with pump energy. At some pumping level damage to the cavity optics occurs, thus putting a practical limit on Q-switched output energy. Preliminary results with the Co:MgF$_2$ laser just described, Q-switched by a Brewster-angle, fused-silica acousto-optic Q-switch, showed that at 25 mJ of output energy, in a 200-ns-wide pulse, damage to the cavity mirrors did occur. Damage-free operation at 20 mJ of TEM$_{00}$-mode output energy in a 220-ns-wide pulse was possible; at this energy the laser, which operated at 1.92 μm with a 5% transmission output coupler, was pumped at a level 4 times over threshold. The output energies in the Q-switched and normal modes of operation were essentially the same. It is expected that by increasing the cavity mode size and by use of high-damage-resistance mirror coatings, generation of a diffraction-limited, Q-switched energy output approaching 100 mJ should be possible with the existing pump laser.

The 1.6- to 2.3-μm tuning range of the Co:MgF$_2$ laser allows detection of strong absorption bands from H$_2$O and CO$_2$ and the sensing of overtone absorption from CH$_4$ and other hydrocarbons. In addition, the output can be shifted by nonlinear techniques to longer wavelengths. For example, the first Stokes-shifted wavelengths from stimulated vibrational Raman scattering in D$_2$ gas cover the region from 3.1 to 7.4 μm, and mixing of the output of two Co:MgF$_2$ lasers in a crystal of CdSe generates a difference frequency in the range from 9.5 to 11 μm. Thus, systems based on the Co:MgF$_2$ laser can, in principle, be used to sense a wide variety of molecules with absorption in the near- and middle-infrared.

Miniature CO$_2$ TEA Lasers and Frequency Mixing

A miniaturized, line-tunable CO$_2$ TEA laser capable of operating at high-pulse-repetition frequencies has been developed at Lincoln Laboratory.[3] Two such lasers, operating in tandem, have been used as the primary radiation source in a dual-laser differential-absorption LIDAR system.[4] In addition to their use in the 10 μm wavelength region, these lasers have been used in conjunction with nonlinear CdGeAs$_2$ crystals to generate the second harmonic, and have been used in this mode for the remote sensing of molecular species with absorption spectra near 5 μm.[5,6]

The mini-TEA laser is a gas-recirculating unit capable of operating at a 500 Hz repetition rate with an average output power of up to 10 W. It is line tunable over approximately 50 CO$_2$ laser transition frequencies with an output that is almost diffraction limited. The laser has an active volume of 1.8 cm^3 (4 x 4 x 110 mm) between Rogowski-shaped main electrodes, and includes a UV-preionization system consisting of series-connected spark gaps parallel to the main electrodes. Sufficient gas circulation for the 500 Hz PRF is provided by a 3-inch fan and a small wind tunnel. A semi-closed system is used, in which a ratio of gas-replacement rate to flow rate of 1% is maintained. With this system it is possible to operate for 5×10^7 to 10^8 pulses with no reduction in output energy. After that number of pulses, significant arcing occurs, and the preionizer electrode spacing requires readjustment. The laser, including the gas-recirculating system, is shown in Fig. 3.

Pulse energy outputs of 60 mJ and a maximum average power output of slightly over 10 W have been achieved at the highest repetition rates with a high-reflectivity mirror. In a grating-tuned 3-mirror-cavity assembly, laser outputs have been obtained at over 50 transition frequencies. The

Fig. 3. High-repetition-rate gas-circulating miniaturized
CO_2 TEA laser

small active cross-sectional area (4 x 4 mm) of the laser ensures good transverse mode quality. The TEM_{10} appears to be the dominant mode at the highest output level, and about half the output is available in the TEM_{00} mode using intracavity apertures. The pulse width of a typical pulse is approximately 100 ns FWHM, with 30-40% of the total energy contained in the tail section.

When CO_2 lasers are used in conjunction with a nonlinear frequency mixing crystal, broad spectral coverage can be achieved in the infrared between 3 and 13 μm. When a CO laser and a CO_2 laser are used with such a crystal, the spectral coverage can be further expanded to encompass wavelengths greater than 16 μm.[7] In addition, a large increase in frequency diversity is obtained when these crystals are used in the sum or difference frequency mode. Since $CdGeAs_2$ is potentially one of the most efficient crystals for frequency-mixing applications in the infrared,[8] a number of these crystals were grown at Lincoln Laboratory.[9,10]

The damage threshold of these crystals has been determined to be 5-6 J/cm^2.[11] Second harmonic generation experiments carried out with these crystals have demonstrated their ability to operate efficiently and to accept irradiation at high energy and power levels without damage. As examples: (1) output pulse energies to 200 mJ at 5.3 μm have been obtained using a multi-mode CO_2 TEA laser,[10] (2) a frequency-doubled average power output of 1.9 W was obtained using a Q-switched CO_2 laser operating at 20 kHz PRF,[10] and (3) a crystal was irradiated with 16.5 W from a cw CO_2 laser without damage, yielding 73 mW frequency-doubled output power.[12]

$CdGeAs_2$ crystals have been used in conjunction with the mini-TEA laser in our DIAL system. Figure 4 shows the average power of the frequency-doubled output of a $CdGeAs_2$ crystal as a function of the average input

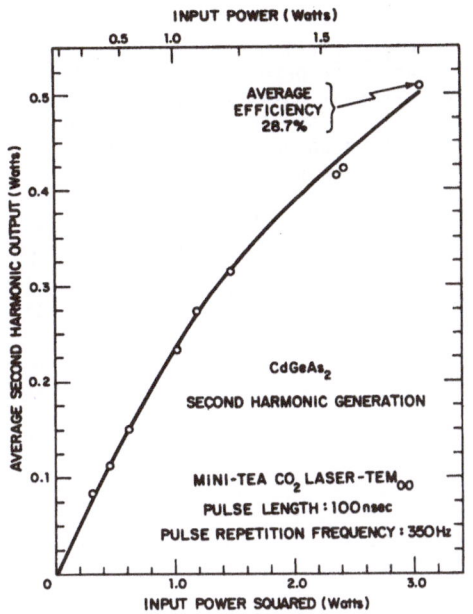

Fig. 4. Average output power at 5.3 µm versus average input power squared for second-harmonic generation in $CdGeAs_2$ with a mini-TEA CO_2 laser operating at 350 Hz PRF

power to the crystal from a mini-TEA laser operating at a 350 Hz PRF. The uppermost point corresponds to an average power second harmonic conversion efficiency of over 28.5%. This is the highest value reported to date for frequency doubling of a CO_2 laser.

Semiconductor Diode Lasers

Semiconductor diode lasers are small, efficient sources of laser radiation and have operated in the range from the near-infrared to the mid-infrared (about 35 µm). These devices are generally difficult to scale up in output energy and are usually used as local oscillators in coherent active or passive detection systems. As local oscillators, they need only put out a few milliwatts or less of power. One of the difficulties in the operation of monolithic semiconductor diode lasers has been that their spectral output usually tends to be multimode and can change with aging. This problem can be overcome by operating these devices in conjunction with an external cavity.

Extensive studies of (GaAl)As diode lasers operating in an external cavity have been carried out.[13] Figure 5 shows the results of typical measurements for a (GaAl)As diode laser operating first as a monolithic device and then in an external cavity designed to force the fundamental spatial mode to oscillate. The upper trace shows multiaxial mode operation of the monolithic device. However, when this device has been antireflection coated on both ends so that it no longer oscillates at the maximum injection current level used before coating, and subsequently operated in an external cavity, nearly all of the previous double-ended multimode output power can be extracted from one end of the external cavity in a single frequency. Typical tuning ranges are about 100 Å for these lasers. Similar results should be possible for lead salt diode lasers which operate further in the infrared. The advantages of narrow linewidth and frequency

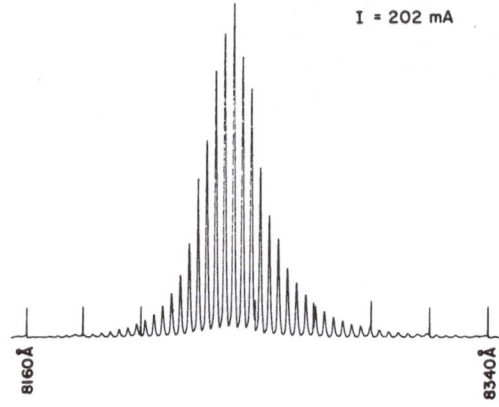

I = 202 mA

8160 Å

8340 Å

SPECTRUM OF $Ga_{1-x}Al_xAs$ LASER BEFORE OPERATION
IN EXTERNAL CAVITY

a

Fig. 5. (a) Spectrum of (GaAl) As laser before operation in external cavity. (b) Spectrum of same device antireflection coated on both facets and operated in an external cavity with a grating and a 20% output coupler. Output in single line is about 80% of total multimode double ended output of uncoated laser

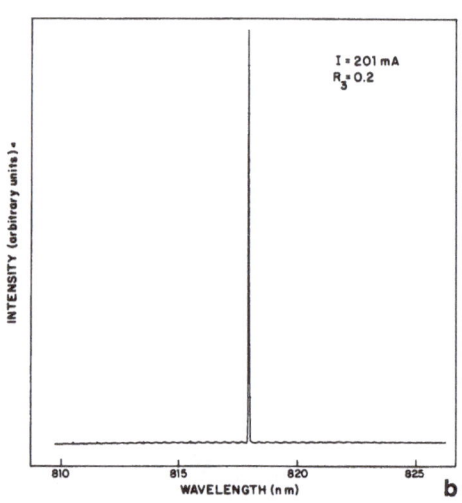

I = 201 mA
$R_3 = 0.2$

INTENSITY (arbitrary units) →

810   815   820   825
WAVELENGTH (n m)

b

control for such lasers would make them much more useful in for field applications for both active and passive coherent remote sensing systems.

REFERENCES

1. L. F. Johnson, R. E. Dietz and H. J. Guggenheim, "Spontaneous and Stimulated Emission from Co^{2+} Ions in MgF_2 and ZnF_2," Appl. Phys. Lett. **5**, 21 (1964).

2. P. F. Moulton and A. Mooradian, "Broadly Tunable CW Operation of Ni:MgF_2 and Co:MgF_2 Lasers," Appl. Phys. Lett. **35**, 838 (1979).

3. N. Menyuk and P. F. Moulton, "Development of a High-Repetition-Rate Mini-TEA Laser," Rev. Sci. Instrum. **51**, 216 (1980).

4. D. K. Killinger and N. Menyuk, "Remote Probing of the Atmosphere Using a CO_2 DIAL System," IEEE J. Quantum Electron. QE-17, 1917 (1981).

5. D. K. Killinger, N. Menyuk and W. E. DeFeo, "Remote Sensing of CO Using Frequency-Doubled CO_2 Laser Radiation," Appl. Phys. Lett. 36, 402 (1980).

6. N. Menyuk, D. K. Killinger and W. E. DeFeo, "Remote Sensing of NO Using a Differential Absorption LIDAR," Appl. Opt. 19, 3282 (1980).

7. A. Mooradian, "High Resolution Tunable Infrared Lasers," in Laser Spectroscopy, R. G. Brewer and A. Mooradian, Eds. (Plenum Press, New York, 1974), pp. 223-236.

8. R. L. Byer, "Nonlinear Optical Phenomena and Materials," in Annual Review of Materials Science, Vol. 4, R. H. Bube and R. W. Roberts, Eds. (Annual Reviews, Inc., Palo Alto, 1974), pp. 147-190.

9. H. Kildal and J. C. Mikkelsen, "Efficient Doubling and Frequency Mixing in the Infrared Using the Chalcopyrite $CdGeAs_2$," Opt. Commun. 10, 306 (1974).

10. G. W. Iseler, H. Kildal and N. Menyuk, "Ternary Semiconductor Crystals for Nonlinear Optical Applications," in Institute of Physics Conference, Series No. 35 (Institute of Physics, London, 1977), pp. 73-88.

11. H. Kildal and G. W. Iseler, "Laser-Induced Surface Damage of Infrared Nonlinear Materials," Appl. Opt. 15, 3062 (1976).

12. N. Menyuk, G. W. Iseler and A. Mooradian, "High-Efficiency High-Average-Power Second-Harmonic Generation with $CdGeAs_2$," Appl. Phys. Lett. 29, 422 (1976).

13. M. W. Fleming and A. Mooradian, "Spectral Characteristics of External-Cavity Controlled Semiconductor Lasers," IEEE J. Quantum Electron. 17, 44 (1981).

6.4 Review of NDRE Remote Sensing Program and Development of High Pressure RF Excited CO_2 Waveguide Lasers

P. Narum, T. Jaeger, and G. Wang

Norwegian Defence Research Establishment, P.O. Box 25
N-2007 Kjeller, Norway

1. INTRODUCTION

At the Norwegian Defence Research Establishment (NDRE) we have had a long and continuing interest in optical remote sensing, especially in relation to detection and identification of targets by their emission spectra. It has been recognized that this work has many applications, military as well as civilian.

Our work in this field can be divided into three different areas. Firstly, Fourier Transform Spectroscopy (FTS) for short and medium range measurements where a great number of spectral lines are of interest. Secondly, optical heterodyne techniques for high sensitivity detection of emission at a specific wavelength, and lastly, development of tunable infrared lasers with the ultimate goal of using them as local oscillators in heterodyne detection systems or as sources in differential absorption lidar. We shall in the following give a brief review of our work in these fields.

2. FOURIER TRANSFORM SPECTROSCOPY

Our FTS system consists of a slightly modified interferometer from General Dynamics, a minicomputer for real time processing of the interferograms, and a digital plotter. The system is controlled from a video display terminal. Instead of the original sandwich detector we now use two separate detectors. One InSb detector for wavelengths up to 6 μm and a MCT detector for wavelengths up to 14 μm. The detector lens is changed to a diffraction limited doublet of germanium, and we have added a collecting telescope consisting of a 15 cm diameter parabolic primary mirror and a 2.5 cm diameter secondary mirror. The field of view is 5 milliradians without and 0.85 milliradians with the telescope. The theoretical spectral resolution is 0.035 cm^{-1}, which has also been demonstrated experimentally. The processor has been developed at NDRE, and processing time for a spectrum of 4096 points is 3 seconds, which includes phase-compensation and correction for the instrument response. We have the possibility of averaging over an arbitrary number of spectra to reduce noise, generate blackbody curves, and store background spectra for background suppression. We are now building a new computer based on the MC68000 microprocessor with hardware multipliers.

Recently we have used this instrument to measure the emission from propane burners at ranges of 60 and 1400 meters; these spectra have been correlated with the spectral lines of the combustion gases. In figure 1 we show a few examples of the spectra we have obtained at a

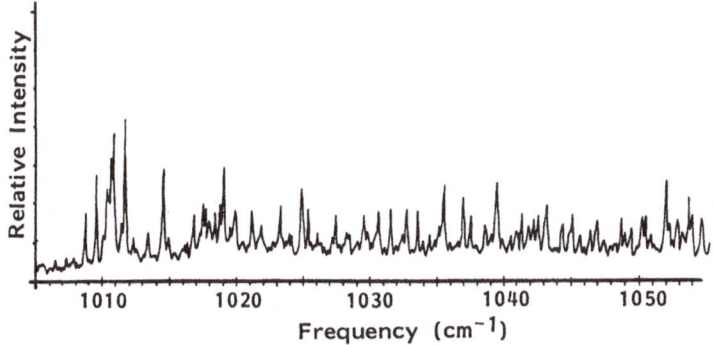

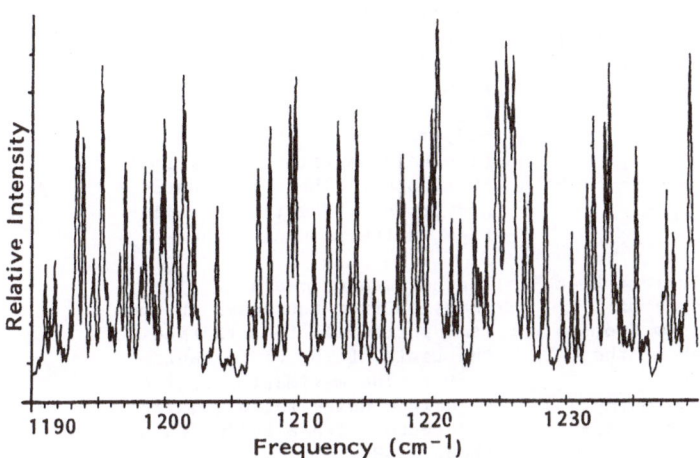

Figure 1 Emission spectra of combustion gases from a propane burner
measured at a range of 60 m with an FTS instrument.
Spectral resolution is 0.035 cm^{-1} and signal to noise ratio
is about 800

path length of 60 m. The presented spectra are the average of 128
spectra and the signal to noise ratio, calculated as the amplitude of
the strongest line divided by the RMS noise, is about 800. In these
experiments the instrument resolution is 0.035 cm^{-1}. Most of the
lines are identified as H_2O and CO_2 lines. We have repeated this
experiment several times under different weather conditions and have
observed only small changes in the intensity of the identified lines.
Some of the unidentified lines, however, show large variations. The
identification of these lines and the explanation of their large variations are currently in progress.

3. INFRARED HETERODYNE RADIOMETRY

We have built a heterodyne radiometer for laboratory measurements.
The local oscillator is a CO2 laser, and the detector is a MCT photodiode with 500 MHz bandwidth, made by Honeywell. With a 2 dB noise
figure preamplifier the effective heterodyne quantum efficiency of our

system is about 50% at low frequencies and drops almost linearly to
25% at 500 MHz.

To speed up the measurement line profiles of gases at low pressure, we
use an 8-channel synchronous detection system. The 5-500 MHz output
from the photodetector is separated into the eight frequency bands with
individual signal processing. Each channel has a 65 MHz bandwidth and
the effective integration time is 1 second. The outputs are multiplexed and fed either to an oscilloscope for direct display of the
line profile or to a digital data logger.

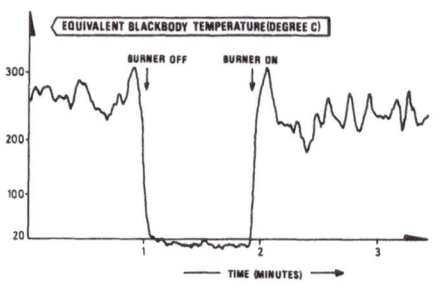

From measurements of the output
from one channel only, we have
calculated the equivalent
blackbody temperature of the
R22-line of CO_2 from the same
propane burner that we used for
obtaining the spectra in
figure 1. As shown in
figure 2, the equivalent blackbody temperature has large
fluctuations with periods up to
30 seconds. These fluctuations
are probably due to instabilities in the propane burner.

Figure 2   Measured equivalent blackbody temperature of
combustion gases from a
propane burner at the
R22 line of CO_2.
Time constant of the heterodyne
radiometer is 1 sec and rf bandwidth is 65 MHz

Further measurements of line
emission from hot gases in the
atmosphere will be performed on
the basis of the FTS measurements. The maximation of the
signal to noise ratio requires
a detector with a bandwidth of
about 3 GHz.

With such a wide bandwidth photodetector it will be possible to use
the heterodyne radiometer for identification of hot gases.

4   TUNABLE HIGH PRESSURE CO_2 LASERS

Continuous tuning between vibrational/rotational line centers of
CO_2 lasers is possible at total gas pressure of 8 - 12 atmospheres.
We are currently working with two approaches to this type of laser,
optical pumping [1], and pumping by a radio frequency electrical
discharge [2]. In the optical pumping approach, a sealed off
DF-CO_2 transfer laser is pumped with a single frequency DF laser.
The pump laser (Lumonics TEA 203) gives about 70 mJ in a 0.5 µs pulse
on the 1P7 line in DF at 3.64 µm. This radiation is strongly absorbed
by the DF and the energy is subsequently transferred to CO_2^* by a V-V
process. The output energy from the CO_2 laser is typically 3 - 4 mJ,
corresponding to a quantum efficiency of about 15%. The laser pulse
is gain switched and has a FWHM of 80 - 100 ns. Lasing between line
centers has been obtained at a DF:CO_2:He mixing ratio of 0.5:4.5:95
and with a total pressure of 10 atm, demonstrating that continuous
frequency tuning is possible with this laser. Lasing has so far been
obtained at pressures up to 18 atm.

266

In our other approach to a high pressure continuously tunable
CO_2 laser, rf excitation is used. Application of rf excitation has
several potential advantages: stable output pulses, high pulse repetition rate is easily obtained, preionization is not necessary, the
laser is compact and simple, and long lifetime is expected since the
metal electrodes are not in contact with the gas.

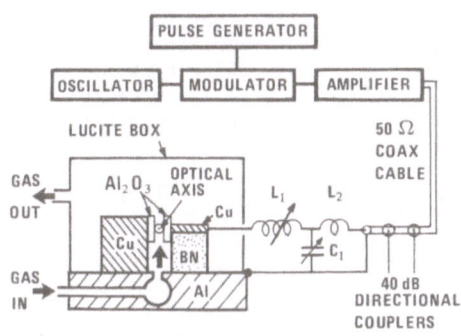

Figure 3 Cross-section of waveguide laser and the
excitation circuitry.
The optical axis is normal to
the paper plane

A cross-section of the laser
and the excitation circuitry
are shown in figure 3. The
copper electrodes are 135 mm
long. The ungrounded electrode
is 2 mm thick and is mounted to
a room temperature heath sink
with a boron nitride block as a
spacer.

The parallel plate waveguide is
formed by polished sapphire
ribbons which are 0.75 mm
thick, 150 mm long, and 12 mm
wide, and are epoxied on the
electrodes. The gas is channeled through the waveguide
transversely to the optical
axis. The rf amplifier is connected to the laser with an
impedance matching network.
The optimum gap between the
sapphire ribbons for good

plasma stability and maximum small signal gain was found to be between
1 and 2 mm. A 1.5 mm gap was used in most of the experiments. The
pulse repetition rate was varied between 0.1 and 1 kHz. The rf power
amplifier can give 8 kW of rf power at 40.68 MHz.

Arcing is a serious problem in high pressure gas discharges. The formation of arcs may determine the maximum length of the rf pump pulse,
and this will limit the maximum excitation energy obtainable with a
fixed pump excitation power. This problem has been overcome by using
low CO_2-N_2 concentrations in the gas. At 10 atm, CO_2:N_2:He mixing
ratios of 1:1:98 or 2:2:96 have been used. Almost arc free excitation

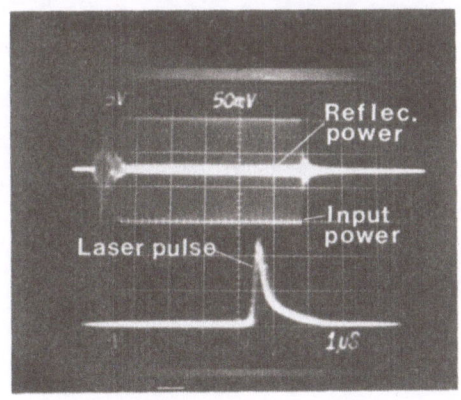

Figure 4 Oscilloscope trace of rf
excitation power, the
power reflected from the
impedance matching net-
work and the laser pulse.
The total gas pressure is 10 atm
and a 0.4% transmission output coupler is used. Excitation power is
approximately 7.5 kW and peak power
of laser pulse approximately 150 W

267

has been obtained with pulse durations of 2 to 3 times the gain decay time at 10 atm and with 7 - 8 kW pump power.

The laser output pulse consists of a gainswitched spike approximately 0.4 µs long followed by a tail with reduced power. Figure 4 shows the excitation pulse and the laser pulse at 10 atm gas pressure with 7.5 kW pump power. It is seen that perfect impedance matching is obtained for at least 6 µs pump pulse duration. As arcing would lead to impedance mismatch, this also means that we have arcfree excitation. The laser pulse traces in Figure 4 show approximately 10 consecutive pulses, illustrating the good pulse to pulse stability. Using a flat 3.5% output coupling mirror and spherical 99% dielectric reflector with 1.3 m radius of curvature, 1.5 kW of output peak power has been obtained at 10 atm with 8 kW input rf pump power.

Study of the frequency stability and chirp, the possibility of CW operation, and work towards development of an optical resonator permitting continuous tuning of this laser are in progress.

References

S. Løvold and G. Wang: "Ten atmospheres high repetition rate rf-excited veguide laser", Appl. Phys. Lett., vol. 40, p. 13, 1982.

K. Stenersen and G. Wang: "Optically pumped high-pressure DF-CO2 transfer lasers", Opt. Commun., vol. 39, p. 251, 1981.

6.5 Progress in Dye and Excimer Laser Sources for Remote Sensing*

T. Srinivasan, H. Egger, T.S. Luk, H. Pummer, and C.K. Rhodes

Department of Physics, University of Illinois at Chicago,
Chicago, IL 60680, USA

1. Introduction

In laser based remote sensing, atmospheric constituents are identified and their concentration is determined through observation of the characteristic interaction of the laser radiation with specific atmospheric constituents. Techniques such as resonance fluorescence and resonance Raman scattering involve tuning the laser to a frequency characteristic of the sample and observing the radiation from the sample at the same or a slightly different frequency. The temporal behavior of the observed signal contains the information on the spatial distribution of the sample and the intensity indicates its concentration. For general application of this method, the laser radiation must be tunable over a wide range.

TABLE I

Differential Cross-section of different laser interactions with matter, applicable to LIDAR (Ref. Laser Monitoring of the Atmosphere, E. D. Hinkley (ed.) New York, Springer Verlag, 1976)

Interaction	Differential Cross-section (cm^2/sr)
Raman Scattering, Non-resonant	10^{-26}
Raman Scattering, resonant	10^{-23}
Fluorescence, signal freq. = laser freq.	10^{-26}
Fluorescence, signal freq. $\neq$ laser freq.	10^{-24}

An estimate of the energies and repetition rates of the lasers required is represented in the following example. Table I presents the typical values of the differential cross-sections[1] for various modes of scattering. If a differential cross-section of $\simeq 10^{-25}$ cm^2/sr is assumed, the corresponding signal would be $\sim$ 0.2 photons/pulse for a laser source with an output energy of 1J at 600 nm, a sample concentration of 1 ppb at an altitude of 10 km, a range resolution $\sim$ 300 m corresponding to the laser pulse duration of 1 μ s, and an optical receiver system with an efficiency of 10^{-1} and a detection area of 1 m^2. In this estimate, the transmission losses at both laser and

---

* This work was sponsored by the Department of the Air Force.

signal frequencies are neglected. This example clearly demonstrates the need for high output energies. Furthermore, to obtain reliable information from this small signal, single photon counting techniques would have to be adopted. Therefore, laser sources with high repetition rates and high output energies are clearly desirable.

From Table I it is seen that for resonant interactions, the differential cross-section is enhanced by approximately a factor of 10^3 in comparison to that of corresponding non-resonant cases. Indeed, in order to maximize resonant coupling in processes such as resonance fluorescence, it is desirable to match the laser linewidth to the width of the transition corresponding to the sample. Optimally, therefore, the sources used for remote sensing should be capable of generating tunable high spectral brightness radiation. Dye and excimer lasers are very good candidates for producing such radiation in the visible and ultraviolet regions, respectively.

2. Dye Lasers

Dye lasers are certainly the most widely used tunable lasers. The very favorable inherent properties of dye media such as the high damage threshold, superior optical quality, controllable laser parameters and wide tunability, along with the availability of the well developed technology for accurate tuning, make dye lasers the prime choice to obtain tunable, narrow band radiation at high average power. Table II lists the laser performance characteristic of a few available dyes utilizing different sources of excitation. In the following, we will discuss both flash lamp pumped and laser pumped pulsed dye lasers because of their capability to deliver outputs of high average and peak power.

TABLE II

Comparison of dye laser energy conversion efficiency η for different pumping sources.

Dye	$\eta_{flashlamp}$ (percent)	η_{XeCl}^{+} (percent)	$\eta_{nitrogen}^{+}$ (percent	η_{YAG}^{*} (percent)	η_{KrF} (percent)
p-terphenyl	$\sim 0.03^{x}$	20.7	no lasing	not available	28 (30)
Coumarin 120	$\sim 0.25^{x}$	41.1	34.2	not available	not available
Rhodamine 6G	0.8 (31)	27.7	21.2	40	5 (30)

$^{+}$Ref. 5

*Ref. Quantel International; ^{x}Ref. Exciton

2.1 Flash Lamp Pumped Dye Lasers

Flash lamp pump dye lasers with a peak power of ~ 1.4 MW, repetition rate of 10 Hz and pulse length of ~ 1.4 μs are commercially available. These lasers can have a linewidth as low as ~ 0.001 nm and are tunable from 430 nm - 720 nm. Peak powers of ~ 30 MW with a pulse duration of ~ 10 μs have been demonstrated.[2] The beam divergence of the latter device was $\sim 3.5 \times 10^{-2}$

rad. By injection locking a narrow band continuous wave (CW) dye laser in two stages, spectrally narrow (0.7 GHz) radiation of high peak power ($\sim$ 11 MW in 2 μs pulses) has been obtained.[3] Improved beam divergence ($\sim$ 0.4 mrad) and a shorter pulse duration (800 ns) have been obtained using prepulse-preionized linear flash lamps.[4] The peak power of the radiation obtained by this method is $\sim$ 250 kW at repetition rate of $\sim$ 10 Hz with an efficiency for energy conversion of 0.3%.

A major disadvantage of flash lamp excited dye systems is the small conversion efficiency, since only a small percentage of the lamp spectrum can be usefully coupled to the dye pump band. The infrared as well as most of the ultraviolet part of the pumping radiation is transformed into heat. This causes thermally induced schlieren, and hence, a significantly degraded beam quality. Unfortunately, degradation of the dye is also enhanced by heating. The limited lifetime associated with flash lamps is another serious drawback of such systems. Nevertheless, by optimizing the dye flow and improving the properties of flash lamps, it is expected that both the beam quality and the average power of these systems can be substantially improved.

2.2 Laser Pumped Dye Lasers

The main advantage of laser excitation for dye systems is the high conversion efficiency that can be achieved with the use of spectrally narrow, low divergence radiation. Indeed, conversion efficiencies as large as 40% have been observed[5] for some coumarin derivatives. In sharp contrast to systems involving flash lamps, effects due to excessive heating such as thermal schlieren and dye degradation are minimal. Since both Nd:YAG and excimer lasers represent the principal high power sources for dye laser excitation, we will confine our discussion to these two systems.

2.2.1 Nd:YAG PUMPED DYE LASERS

Frequency doubled (530 nm) or tripled (353 nm) radiation derived from Nd:YAG sources represent the most commonly used pump sources. Dye lasers with output energy up to 100 mJ, pulse duration $\sim$ 10 ns, bandwidth $\sim$ 0.02 nm, tunability from 320 nm - 1000 nm, and repetition rate up to 50 Hz are commercially available. Systems with higher output energy and narrower bandwidth are also commercially available. But in the latter cases, the repetition rate is reduced to a value < 20 Hz. Since the YAG lasers are generally excited by flash lamps, the essential and intrinsic drawbacks of flash lamp technology, such as the limited lifetime of the flash lamp, are present here. Hence, improvements in the flash lamps for YAG lasers will immediately translate favorably into corresponding improvements in the overall properties of dye lasers.

2.2.2 Excimer Laser Pumped Dye Lasers

Commercial dye lasers pumped by excimer lasers generate essentially transform limited radiation with an output energy of $\sim$ 50 mJ, pulse duration $\sim$ 10 ns and a repetition rate of 100 Hz. Repetition rates up to 300 Hz can be obtained at the expense of the output energy (< 10 mJ). Detailed information on excimer pumped dye lasers can be obtained from a recent review article[6] on this subject. The scaling of these dye lasers depends principally on the scalability of the excimer laser technology, an issue that is discussed in the next section.

3. Excimer Lasers

Excimer lasers are the best sources available for coherent ultraviolet radiation. These systems can deliver both high peak and average powers of ultraviolet radiation at several frequencies throughout a wide range of the ultraviolet spectrum. Furthermore due to the breadth of the molecular resonances involved, a relative tuning range of $\sim$ 1% of the quantum energy for each excimer system can be obtained. In addition, frequencies not directly available with excimer lasers can be generated with high efficiency by Raman shifting.

TABLE III

Output Characteristics of Commercially Available Excimer Lasers

Laser medium	ArF	KrCl	KrF	XeCl	XeF
Wavelength (nm)	193	222	248	308	351
Output energy (J)	0.5	0.05	>1.0	0.5	0.4
Average power (W)	10	1	>20	10	6
Bandwidth	$\sim$ 100 cm^{-1}			Multi-Line	
Pulse duration	typically 10 - 20 ns				
Beam divergence	$\sim$ 10 mrad				

Table III lists the commercially available rare gas halide lasers along with their output characteristics. As Table III illustrates, when operated as simple free-running oscillators, these systems have extremely poor spectral brightness. Furthermore, spectroscopic remote sensing techniques, such as resonance fluorescence and resonance scattering, would require the introduction of wavelength selectors to narrow the bandwidth in addition to the need to tune accurately the frequency of the output. Standard commercial instruments normally are not equipped to meet these requirements. These conditions, however, for precise optical control can be established by various methods such as the introduction of combinations of different wavelength selectors and filters[7-14] and by injection of a small, tunable, narrow band and low divergence seed radiation into either an unstable oscillator[14,15] or a simple linear chain of amplifiers[16,17]. In the first technique, with the introduction of intracavity prisms or gratings, tunable radiation with a bandwidth[13] of $\sim$ 0.6 cm^{-1} and an output energy $\sim$ 80% of free running system[7] has been obtained. Although this approach is simple, it requires synchronous tuning and alignment of all the wavelength dependent elements, a procedure that may be overly cumbersome. With longer pulses, and hence, a larger number of round trips in the cavity, intracavity filtering directly in the oscillator is more efficient. Therefore, this technique is expected to be more effective for systems that operate with excimer pulse lengths > 100 ns. Indeed, pulse durations $\gtrsim$ 100 ns have been obtained in microwave discharge pumped XeCl lasers.[18]

For the injection locking technique, a small tunable pulse of radiation having well defined spectral properties is used to control the extraction of energy from a larger oscillator. For an oscillator with good discrimination between its lowest order spatial mode and higher oder modes, the amplitude of the seed radiation can be minimized by matching the spatial mode of the seed pulse as closely as possible with the lowest order mode of the oscillator. For optimum performance, the timing of the master oscillator and injection locked slave oscillator has to be accurately controlled to within a few nanoseconds.[14] Peak powers of $\sim$ 10 MW at 248 nm (KrF*) have been obtained[14] by injection locking with as little as 100 mW in the seed pulse. The near-diffraction limited beam obtained was tunable over $\sim$ 10Å while the bandwidth was $\sim$ 0.1 Å. This scheme has the advantage that with relatively small amounts of precisely controlled seed radiation, the entire energy stored in the medium can be extracted with a single slave oscillator. However, in experiments requiring tuning, the cavity length must be tuned simultaneously with the wavelength of the input signal and the careful match of the spatial modes must be maintained to provide effective discrimination against higher order modes.

The ultimate results involving tunability, high spectral brightness and accurate wavelength control have been simultaneously achieved with systems utilizing pulse-amplified continuous wave dye lasers, nonlinear frequency conversion to the desired ultraviolet (UV) wavelength, and subsequent amplification of the UV radiation.[16,17] Fig. 1 shows the schematic of an ArF* (193 nm) system of this general nature. To produce spatially and spectrally transform limited pulse, frequency upconversion of the dye laser radiation is preferred over the external filtration of the output of a free-running excimer laser. It is easily shown that for the latter scheme, the external filters will act in such a way so as to reduce the output power available to a level which is very close to the theoretical minimum necessary for subsequent amplification with good discrimination. Hence, adoption of this technique generally provides marginal reliability of operation for the system. Furthermore, accurate tuning and wavelength calibration techniques have yet to be fully developed in the ultraviolet. Contrarily, the commercially available sophisticated tuning and spatial beam control systems characteristic of high quality dye lasers make them the ideal choice for the front end of a high brightness system. In this scheme the wavelength of the output can be readily determined to an accuracy of 1 part in 10^6 using a commercial wavemeter. Radiation at any available excimer laser wavelength can, thereby, be obtained by proper choice of the dye and nonlinear medium.

In the ArF* system shown in Fig. 1 the output of a frequency-stabilized, single mode, continuous wave dye laser (Coherent 599-21, $\Delta\nu \sim$ 5 MHz, power $\sim$ 50 mW at 580 nm) is amplified in a multi-stage, XeF* pumped amplifier chain producing a visible pulse $\sim$ 10 ns with an energy > 20 mJ at a repetition rate up to 10 Hz. The linewidth and divergence of these pulses represent the corresponding transform limits for the pulse duration and aperture, respectively. In cases not requiring absolute wavelength calibration, the wavemeter can be eliminated and the CW dye laser can be replaced by a commercially available pulsed dye laser which generates transform limited outputs. We note, however, that pulsed wavemeters capable of measuring the wavelength to an accuracy of 1 part in 10^6 can be constructed.[19]

In order to generate the desired ultraviolet wavelength (193 nm), the output of the dye amplifier chain is focussed with a lens (f = 50 cm) into a strontium heat pipe. The frequency of the third harmonic at 193 nm generated by the strontium vapor, which serves as the high quality seed radiation, can be readily tuned over the ArF* gain profile by tuning the dye laser. With this method, narrow band ultraviolet radiation of peak power $\sim$ 200 mW and pulse duration $\sim$ 5 ns was obtained for subsequent amplification in one double pass and two single pass ArF* amplifiers (Lambda Physik EMG 200). The final output of this system has the following measured characteristics: energy $\sim$ 450 mJ, pulse duration $\sim$ 7 ns, and wavelength $\sim$ 193 nm. The beam divergence

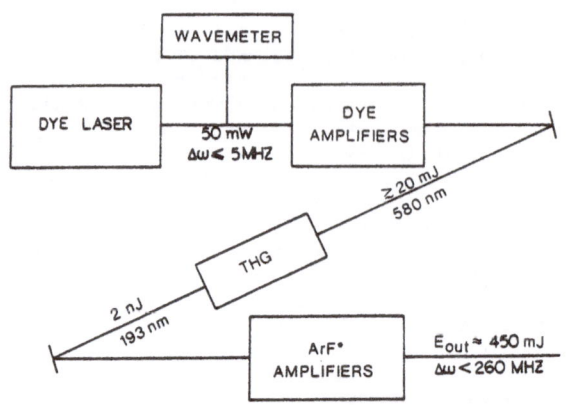

Fig. 1. Schematic of high spectral brightness ArF[*] system

($\sim 5 \times 15$ μrad) and the bandwidth (< 260 MHZ) are maintained at their fundamental limits by introducing spatial and spectral filters in both the visible and ultraviolet beam paths. The pulse synchronization of the lasers and amplifiers is achieved by monitoring the actual firing time of the lasers and correcting for the time drift before the next firing, using a PDP11 computer and CAMAC.[20]

In principle, with this general technique, radiation at all available excimer laser wavelengths can be produced by changing the dye and using the appropriate nonlinear medium and amplifier. Wavelengths outside the gain profile of the available excimer systems can be obtained by nonlinear scattering, principally Raman shifting and frequency sum and difference mixing. In fact, with KrF* as the pump radiation and H_2 as the scattering medium, Raman shifted radiation from 2059 Å, corresponding to second anti-stokes, to 6501 Å, corresponding to sixth stokes, has been generated.[21] Atomic vapors have also proved to be efficient media for electronic Raman scattering[22,23]. Indeed, with the use of good spatial quality and narrow bandwidth excimer radiation, electronic Raman scattering in atomic Ba vapor with photon conversion efficiency as high as 90% has been demonstrated.[24] Moreover, the divergence of this Stokes radiation was comparable to that of the pump beam and the linewidth measurement was limited to the width of the spectrograph used ($\lesssim$ cm^{-1}). Raman scattering experiments in the infrared[25] indicate that bandwidth may be broadened by Raman shifting, probably due to the ac stark effect or the presence of electrons and ions produced by ionization. These frequency broadening effects, however, will normally be insignificant in comparison with the linewidths of the transitions of the materials being detected under common atmospheric conditions.

It must be mentioned that current excimer systems are operating far from the limits of their scalability. Output energies of $\sim$ 10 J in discharge-excited amplifiers and 1 kJ in electron-beam pumped systems with average powers of $\sim$ 1 kW and 10 kW, respectively, at an overall efficiency that could exceed one percent, appear feasible. With such systems, the average power of tunable, high quality ultraviolet radiation can be increased by two or three orders of magnitude from currently used values.

4. Short Pulse Radiation

Recently, tunable, high energy picosecond pulses with bandwidth and divergence corresponding closely to their fundamental limits have been generated in the visible and ultraviolet regions at $\sim$ 580 nm and 193 nm, respectively.[27]

In order to implement such a system, only relatively simple modifications of the instrument illustrated in Fig. 1 are required. The CW dye laser at the front end of the system shown in Fig. 1 is replaced by a synchronously pumped, mode-locked dye laser (Coherent 599-04, $\lambda \sim 580$ nm, pulse duration < 10 ps). In this case, frequency is measured with a commercial wavemeter with an accuracy of ~ 1 cm^{-1} and the corresponding pulse duration is determined with a standard autocorrelator. A single pulse from the pulse train is amplified in a three stage, XeF* pumped amplifier chain. The saturable absorber cells, with DQOCI as saturable absorber, positioned between successive amplifiers serve to suppress the amplified spontaneous emission. In addition, a grating-pinhole combination between the second and third amplifier stage provides additional isolation and eliminates the amplified spontaneous emission during the time when the saturable absorber is bleached. The pulse duration of this radiation, as measured with a streak camera, is ~ 6 ps with an output energy of ~ 1 mJ and a repetition rate up to 10 Hz. Given the availability of suitable dyes and saturable absorbers, it should be possible to generate picosecond pulses in this manner at any wavelength within the visible spectral range.

This short visible pulse can be used to generate short ultraviolet pulses at 193 nm in the same way that the 10 ns visible pulses were used to generate 10 ns ultraviolet pulses. The 1 mJ, picosecond visible pulse is focussed into the strontium heat pipe with a 50 cm focal length lens to generate ~ 2 nJ at the third harmonic (~ 193 nm) of the visible radiation. This radiation is amplified in a double pass ArF* amplifier, passed through a grating-pinhole combination of bandpass ~ 1 Å to suppress the amplified spontaneous emission, and is subsequently amplified in two single pass ArF* amplifiers. The output energy is typically ~ 40 mJ with a pulse duration ~ 10 ps and a repetition rate up to 10 Hz. By direct measurement, the bandwidth and beam divergence of the picosecond pulses are found to be close to the transform limit. With this radiation, ignoring atmospheric distortion but allowing for normal dispersion, intensities of ~ 1 GW/cm^2 can be maintained over a distance up to 1 km. At such intensities, very substantial nonlinear effects have been observed[28,29] in laboratory experiments. Hence, with these sources, the observation of nonlinear effects may allow alternative or supplementary techniques useful for remote sensing.

5. Conclusions

In conclusion, tunable radiation of high average power and narrow bandwidth can be obtained at any wavelength in the visible and ultraviolet region from ~ 800 nm to ~ 193 nm. Depending on the specific requirements of the measurements, different techniques can be used for optimum performance in either region. With the enhanced output of excimer lasers, the output characteristics of the dye and modified excimer lasers, and in particular the energy and average power, are expected to improve substantially, greatly facilitating a wide range of LIDAR measurements. In addition, the new higher range of intensities that are now possible with the picosecond systems may open up further techniques, such as observation of nonlinear processes, for remote sensing.

REFERENCES

1. Laser Monitoring of the Atomosphere, E. D. Hinkley (ed.), New York, Springer Verlag, 1976.

2. F. N. Balkatov, B. A. Barikhin and L. V. Sukhanov, JETP Lett. 19, 174 (1974).

3. T. Okada, M. Maeda and Y. Miyazoe, IEEE J. Quantum. Electron. QE-15, 616 (1978).

4. A. Hirth, Th. Lasser, R. Meyer and K. Schetter, Opt. Commun. 34, 223 (1980).

5. O. Uchino, M. Maeda and M. Hirono, IEEE J. Quantum. Electron. QE-15, 1094 (1979).

6. K. Hohla, Laser Focus 18, 67 (1982).

7. T. R. Loree, K. B. Butterfield, and D. L. Banker, Appl. Phys. Lett. 32, 171 (1978).

8. J. Bokor, J. Zavelovich, and C. K. Rhodes, Phys. Rev. A21, 1453 (1980).

9. J. Liegel, F. K. Tittel, W. L. Wilson, Jr., and G. Marowsky, Appl. Phys. Lett. 39, 369 (1981).

10. A. J. Andrews, A. J. Kearsely, M. C. Gowen, and C. E. Webb. Topical Meeting on Excimer Lasers (1979). Digest of Technical Papers Optical Soc. of America IEEE Cat. Number 79CH1470-4QEA.

11. R. S. Hargrove, J. A. Paisner, Topical Meeting on Excimer Lasers (1979). Digest of Technical Papers Optical Soc. of America IEEE Cat. Number 79CH1470-4QEA.

12. J. Goldhar and J. R. Murray, Opt. Lett. 1, 199 (1977).

13. J. R. Murray, J. Goldhar, and A. Szöke, Appl. Phys. Lett. 32, 551 (1978).

14. I. J. Bigio, and M. Slatkine, Opt. Lett. 6, 336 (1981).

15. J. Goldhar, W. R. Rapaport, and J. R. Murray, IEEE J. Quantum. Electron. QE-16, 235 (1980).

16. H. Egger, T. Srinivasan, K. Hohla, H. Scheingraber, C. R. Vidal, H. Pummer, and C. K. Rhodes, Appl. Phys. Lett. 39, 37 (1981).

17. R. T. Hawkins, H. Egger, J. Bokor, and C. K. Rhodes, Appl. Phys. Lett. 36, 391 (1980).

18. A. J. Mendelsohn, R. Normandin, S. E. Harris and J. F. Young, Appl. Phys. Lett. 38, 603 (1981).

19. J. J. Snyder, Laser Focus 18, 55 (1982), Laser Spectroscopy, J. L. Hall and J. L. Carlsten (ed.), Berlin, Springer Verlag, 1977; L. S. Lee and A. L. Schawlow, Opt. Lett. 6, 610 (1981).

20. D. F. Muller, H. Egger, and B. Yost, Rev. Sci. Instrum. 52, 1575 (1981).

21. T. R. Loree, R. C. Sze and D. L. Barker, Appl. Phys. Lett. 31, 37 (1977).

22. J. C. White and D. Henderson, IEEE J. Quantum Electron. QE-18, 941 (1982).

23. R. Burnham and N. Djeu, Opt. Lett. 3, 215 (1978).

24. N. Djeu and R. Burnham, Appl. Phys. Lett. 30, 473 (1977).

25. D. Cotter, D. C. Hanna, P. A. Kärkkäinen and R. Wyatt, Opt. Commun. 15, 143 (1975).

26. H. Egger, H. Pummer and C. K. Rhodes, Laser Focus 18, 59 (1982); High Power Lasers and Applications, K. L. Kompa and H. Walther (ed.), Berlin, Springer Verlag, 1978.

27. H. Egger, T. S. Luk, K. Boyer, D. F. Muller, H. Pummer, T. Srinivasan, and C. K. Rhodes, to be published.

28. J. Reintjes, C. She and R. C. Eckardt, IEEE J. Quantum Electron. QE-14, 581 (1978); H. Puell, K. Spanner, W. Falkenstein, and W. Kaiser, Phys. Rev. A14, 2240 (1976); H. Scheingraber, H. Puell, and C. R. Vidal, Phys. Rev. A18, 2585 (1978).

29. H. Egger, R. T. Hawkins, J. Bokor, H. Pummer, M. Rothschild, and C. K. Rhodes, Opt. Lett. 5, 282 (1980); H. Pummer, T. Srinivasan, H. Egger, K. Boyer, T. S. Luk, and C. K. Rhodes, Opt. Lett 7, 93 (1982).

30. V. I. Tomin, A. J. Alcock, W. J. Sarjeant and K. E. Leopold, Opt. Commun. 26, 396 (1978).

31. J. Y. Allain, Appl. Optics 18, 287 (1979).

6.6 IR Detectors: Heterodyne and Direct[*]

D.L. Spears

Lincoln Laboratory, Massachusetts Institute of Technology,
Lexington, MA 02173, USA

Introduction

In recent years there have been considerable advances in the performance of wide-bandwidth infrared detectors, both for direct and heterodyne detection. In this paper I will discuss briefly the major areas of current detector development, compare various aspects of heterodyne and direct detection, discuss amplifier considerations for wide-bandwidth detection operation, and describe state-of-the-art performance in direct and heterodyne detection. In addition, differences and trade-offs between photodiodes and photoconductors will be discussed, in particular the operating temperature dependence of direct and heterodyne detectors at 10 μm. The content of this paper will be limited to photodiodes and photoconductors in the 1- to 20-μm region with bandwidths in the range of 1 to 100 MHz. This encompasses most of the present detector needs of LIDAR systems for wavelengths beyond the range of efficient photomultipliers.

Major areas of detector development

Most current infrared detector research and development is directed towards the needs of fiber optic communications and thermal imaging systems. Fiber optic systems operate in the 0.9- to 1.7-μm region (where fibers have very low loss) and require receiver bandwidths of 1 to 1000 MHz. Devices actively investigated for these systems are PIN photodiodes and avalanche photodiodes (APD) made from silicon, germanium, and sophisticated heterostructures of the alloy semiconductors GaInAsP, GaAlAsSb and HgCdTe.[1-4] With alloy semiconductors the detector response can be peaked at the desired wavelength, thereby minimizing thermally-generated dark current.

Thermal imaging systems generally operate in the 3- to 5- and 8- to 14-μm atmospheric windows. Large focal-plane arrays with thousands of elements are being developed for systems with bandwidths in the range of 0.1 to 1 MHz.[5] Although a great deal of progress has been made in developing high-performance arrays of HgCdTe photoconductors,[6,7] current efforts are on photovoltaic detectors (i.e. zero-biased photodiodes) and charge injection devices because of power dissipation considerations in these high-density arrays. Many different materials have been used to make high-performance infrared photovoltaic detectors, including InSb, PbSnTe, PbSnSe, InAsSb, and HgCdTe.[8,9] However, HgCdTe has become the preferred material because, by

---

[*]This work was supported by the U.S. Department of the Air Force, the Defense Advanced Research Projects Agency and the National Aeronautics and Space Administration.

using the appropriate alloy of HgTe and CdTe, the detector response can be peaked at any wavelength from 0.9 to over 40 μm; and HgCdTe has a dielectric constant about 20 times smaller than that of the Pb-salts.

Theoretical photodiode sensitivities

The major applications of IR detectors use the devices in the direct detection mode, rather than the heterodyne mode where an optical local oscillator is needed. IR photomixers have been used primarily for passive heterodyne radiometry and CO_2 laser radar applications. In direct detection the detector is generally sensitive to any wavelength shorter than the detector cutoff wavelength, whereas in heterodyne detection the detector responds only to wavelengths very close to the local oscillator wavelength (typically $\lambda_{LO} \pm 0.003$ μm). However, much lower power levels can be detected in the heterodyne mode than in the direct detection mode. The ratio of the signal-to-noise power for heterodyne and direct detection with a photodiode is shown below.[10]

HETERODYNE DETECTION

$$S/N = \frac{2(\eta e/h\nu)^2 P_{LO} P_S}{2eB[(\eta e/h\nu)P_{LO} + I_D + I_E]}$$

$$I_D \sim \frac{2kT}{e\,R_o}$$

$$I_E = \frac{2kT_{eff}}{e\,R_L}$$

$$NEP_H^{SNL} = \frac{h\nu B}{\eta}$$

DIRECT DETECTION

$$S/N = \frac{(\eta e/h\nu)^2 P_S^2}{2eB[(\eta e/h\nu)(P_S + P_B) + I_D + I_E]}$$

$$NEP_D^{BLIP} = \left(\frac{2h\nu B}{\eta}\,P_B\right)^{1/2} \quad (P_B \propto A)$$

$$NEP_D^{AL} = \frac{h\nu}{\eta}\left(\frac{2I_E B}{e}\right)^{1/2}$$

$$NEP_D^{DL} = \frac{2h\nu}{\eta}\left(\frac{kTB}{e^2\,R_o}\right)^{1/2} \quad (1/R_o \propto A)$$

$$NEP_D^{SL} = \frac{2h\nu B}{\eta}$$

Here, η is the detector quantum efficiency, P_S is the optical signal power, P_{LO} is the optical local oscillator power, P_B is the incident background radiation power, I_D is the dark current in the photodiode, and I_E is the equivalent noise current of the preamplifier. For a diffusion-limited photodiode, $I_D \sim 2kT/eR_o$, where R_o (the parameter generally reported for thermal imaging devices) is the zero-biased impedance of the photodiode. Note, S/N is proportional to P_S for heterodyne detection and P_S^2 for direction detection (except for very large values of P_S).

Direct detection

In direct detection the sensitivity can be limited by several different noise sources: shot noise from the background flux, dark current noise, or amplifier noise. A background-limited infrared photodetector (BLIP) is a high-performance device, which usually requires cryogenic cooling. The direct detection noise equivalent power NEP_D is proportional to the square root of the detector area in the background-limited and dark-current-limited situations. The area does not enter in the case of amplifier-noise-limited performance. However, in general the detector area must be small in order for I_D to be less than I_E for state-of-the-art amplifiers. The signal-noise limit, $NEP_D^{SL} = 2h\nu B/\eta$, is not achieved in photodiodes, but is in the case of photomultipliers, where virtually noiseless gain occurs.

Amplifier noise is a primary consideration for a high-quality photodiode operated under low background conditions.[1] Three basic amplifier configurations frequently used in systems are shown in Fig. 1. The simplest and most common is the voltage amplifier, where the effective noise is the Johnson noise of the load resistor R_L (typically 50 ohms). Increasing R_L improves the noise performance. In the equalized-voltage configuration a very large load resistor giving $2\pi R_L C_T B > 1$ is used to reduce noise, where C_T is the total capacitance (photodiode, amplifier input, lead, etc.).

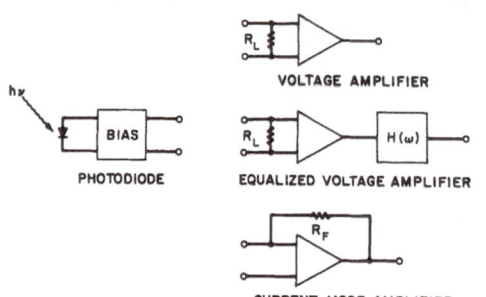

VOLTAGE AMPLIFIER

EQUALIZED VOLTAGE AMPLIFIER

CURRENT-MODE AMPLIFIER

Fig. 1 Amplifier configurations used with IR detectors. (Ref. 1)

The resulting degraded frequency response is compensated for by a post-amplification equalization network. In the current-mode amplifier the noise contribution is the Johnson noise of the feedback resistor R_F and the effective input resistance is R_F/A, where A is the open-loop gain of the amplifier. In Fig. 2 we compare noise equivalent current I_E as a function of bandwidth for a GHz-bandwidth 50-ohm voltage amplifier and that of an optimized amplifier, which is based upon the measured and projected performance of state-of-the-art current-mode or equalized voltage amplifiers.[1] To utilize the optimized amplifier, C_T must be less than 1 pF, requiring careful hybrid integration of the amplifier with the photodiode. Note that over the 1- to 100-MHz range, I_E is proportional to B. Thus with an optimized amplifier NEP_D^{AL} is proportional to B, similar to the heterodyne case.

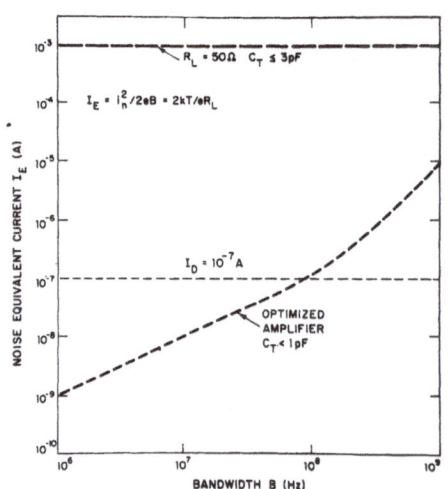

Fig. 2 Noise equivalent current as a function of bandwidth for a typical GHz-bandwidth 50-ohm amplifier and that of a state-of-the-art optimized amplifier. (Ref. 1)

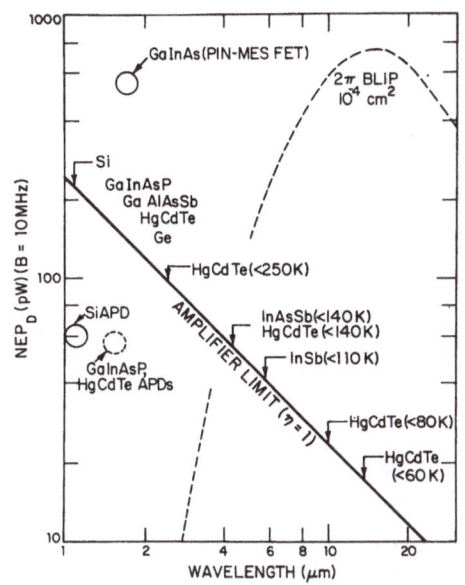

Fig. 3 Direct detection $\overline{NEP}_D$ at 10 MHz as a function of wavelength, showing the 2π-300K BLIP limit, the amplifier noise limit ($\eta = 1$) and various photodiodes with the operating temperatures required to reach the state-of-the-art amplifier limit. (Refs. 1-4, 9, 11-13)

In Fig. 3 is plotted the direct detection NEP_D for B=10 MHz as a function of wavelength, showing the amplifier noise limit described above. Also indicated in Fig. 3 are various photodiodes with A = 10^{-4} cm^2 and the temperatures at which they have demonstrated either dark currents or zero-bias impedances R_0 necessary to reach this amplifier-noise-limited performance. At short wavelengths this condition is achieved at or near room temperature. However, the required operating temperature falls rapidly with increasing wavelength, down to a temperature of less than 60K for a 12-µm device. At 10 µm the amplifier limit is considerably below the 2π-300K BLIP limit. Many commercial devices in the 3- to 14-µm region achieve BLIP performance, but generally at much lower frequencies. Considerable improvement in performance at 10 MHz could be obtained by careful integration with a preamplifier. The required low device capacitance could be achieved with application of a moderate bias (e.g. about 1 volt for a carrier concentration in the mid 10^{14} cm^{-3} range). GaInAs PIN photodiodes[2] matched to GaAs MESFETs have achieved sensitivities close to this limit at 1.6 µm.

Avalanche gain within the photodiode enables one to obtain sensitivities better than this amplifier limit. A commercial Si APD has demonstrated this at 1.06 µm,[13] and GaInAsP[1] and HgCdTe[4] APDs should also be capable of this performance. Although avalanche multiplications well in excess of 1000 have been achieved in many APDs, dark current noise is also multiplied so that useful gains tend to be in the range of 10 to 100.[1,2] Unfortunately, avalanche gain is only attainable at short wavelengths, $\lambda < 3$ µm, because of the large dark currents produced in long-wavelength photodiodes by interband tunneling.

Heterodyne detection

In heterodyne detection, if P_{LO} is sufficiently large the LO-induced noise can become the dominant noise source, and the noise-equivalent signal power is simply $h\nu B/\eta$. This sensitivity is indeed realized with heterodyne receivers,[14] as demonstrated in Fig. 4, which shows a spectrum-analyzer display of a 40-MHz IF beat signal between two CO_2 laser beams detected with

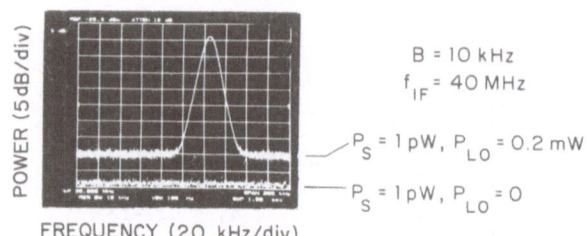

B = 10 kHz
f_{IF} = 40 MHz

P_S = 1 pW, P_{LO} = 0.2 mW

P_S = 1 pW, P_{LO} = 0

Fig. 4 CO_2 laser beat signal detected by a 77K HgCdTe photodiode

a 77K HgCdTe photodiode. From the measured S/N = 36 dB, the noise bandwidth B = 10 kHz, and the signal power P_S = 1 pW, we get NEP_H = 2.5 x 10^{-20} W/Hz, which is only 1 dB from the theoretical limit. Note, when the local oscillator is removed the noise level drops by about 9 dB, indicating shot-noise limited operation. Here, the base noise level is due to the 50-ohm preamplifier (T_{eff} ≈ 100K).

In Fig. 5 are shown the best values of NEP_H at 10.6 μm obtained to date as a function of frequency.[14-17] At 1 GHz, NEP_H is only a factor of 2 above that of an ideal device, and average sensitivities of 12-element arrays[14] (4.3 x 10^{-20} W/Hz) have been very close to this value. Quite good sensitivities (1.1 x 10^{-19} W/Hz) have been achieved as far out as 4 GHz.

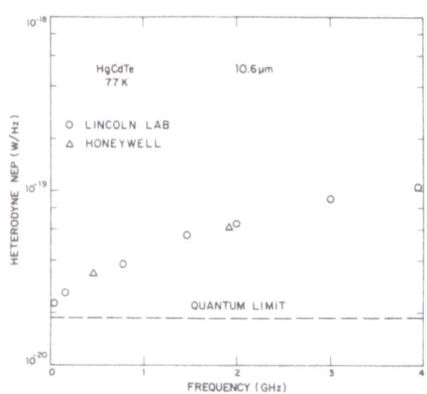

Fig. 5 State-of-the-art heterodyne NEP_H as a function of IF frequency

Comparison of heterodyne and direct detection

Let us now compare heterodyne and direct detection sensitivities. In Fig. 6 we show the signal-to-noise ratio as a function of signal power for heterodyne and direct detection at 10 μm. The dashed curve shows the direct-detection performance which could be obtained with the use of a cold filter (to reduce the background radiation) and an optimized amplifier. Even with an optimized amplifier the direct-detection NEP_D is about 100 times higher than the heterodyne NEP_H, and this factor is independent of λ and B. With a BLIP-limited 10-μm detector there is a factor of 3000 difference between direct and heterodyne sensitivities at 10 MHz, and a 10,000 difference at 1 MHz. At high S/N, as required in digital communications systems,[1] the difference between heterodyne and direct detection is much less. Note that with an optimized amplifier S/N becomes signal-noise limited at a signal power level about 30 times NEP_D. At high signal power levels heterodyne and direct detection have about equivalent S/N.

282

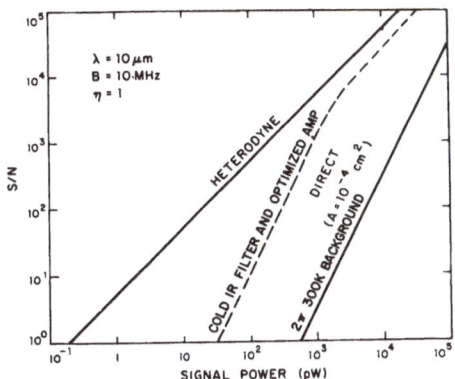

Fig. 6 Signal to noise vs signal power for heterodyne and direct detection at 10 μm. Note: the heterodyne and amplifier-noise-limited curves can be applied to other wavelengths and bandwidths by simply multiplying signal power by $(10 \ \mu m/\lambda)(B/10 \ MHz)$

Elevated-temperature 10-μm detectors

Up to this point we have discussed only photodiodes, although photocon-ductors have long been an important type of IR detector[7] and are particularly useful for high-temperature operation using thermoelectric coolers. The basic expressions for the direct- and heterodyne-detection NEPs for photo-diodes and photoconductors are shown in the table below, where n is the carrier concentration, τ is the carrier lifetime, and t is the photoconductor thickness. In direct detection with a photoconductor the factor ntA/τ is equivalent to the dark current in a photodiode. Since the response time of a photoconductor is determined by the minority carrier lifetime, there exists a bandwidth-sensitivity trade-off. This is not the situation in a photodiode where the response is determined by the RC time constant and carrier dif-fusion. Slow carrier diffusion can be particularly detrimental to wide-bandwidth heterodyne performance, as it results in excess noise as well as loss in signal.[14] This has been minimized, however, in well-designed photo-diodes (see Fig. 5). In heterodyne operation the ideal photoconductor has a factor of two higher NEP_H than a photodiode due to generation-recombination noise. In the table, P_2^D is the optical local oscillator power required for the shot noise to equal all other sources of noise in the photodiode, and P_2^C is the LO power required for the LO-induced g-r noise to equal all other noise in a photoconductor. P_2^D and P_2^C are functions of temperature-sensitive material parameters, such as the intrinsic carrier concentration, the minority carrier lifetime, and the energy gap.

COMPARISON OF PHOTOCONDUCTORS AND PHOTODIODES

	NEP_D	NEP_H	RESPONSE TIME
PHOTOCONDUCTOR	$\dfrac{h\nu}{\eta}\left(\dfrac{2ntAB}{\tau}\right)^{1/2}$	$\dfrac{2h\nu B}{\eta}\left(1 + \dfrac{P_2^C}{P_{LO}}\right)$	CARRIER LIFETIME (τ)
PHOTODIODE	$\dfrac{h\nu}{\eta}\left(\dfrac{2I_D^B}{e}\right)^{1/2}$	$\dfrac{h\nu B}{\eta}\left(1 + \dfrac{P_2^D}{P_{LO}}\right)$	RC CARRIER DIFFUSION

In Fig. 7 we show the calculated g-r-noise limited direct-detection NEP_D for an optimized 10-μm HgCdTe photoconductor as a function of temper-ature, along with the heterodyne LO-power requirement P_2^C for $A = 50 \times 50 \ \mu m^2$,

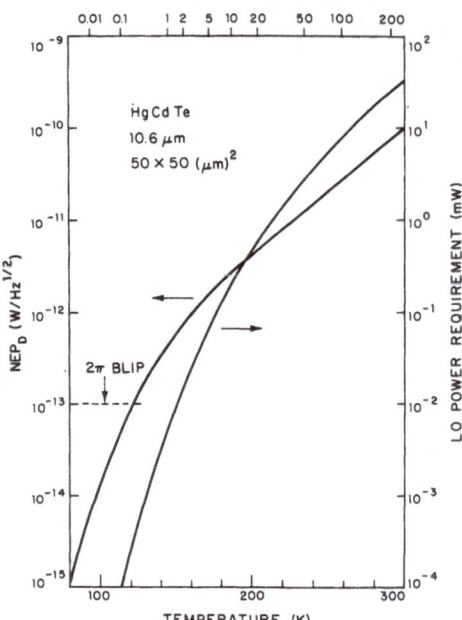

PHOTOCONDUCTOR BANDWIDTH (MHz)

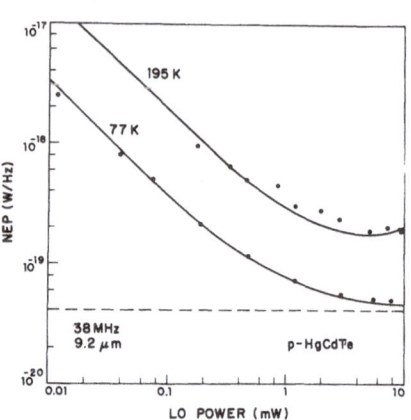

Fig. 7 Direct-detection NEP_D and CO_2 local-oscillator-power requirement as a function of temperature for 10.6-µm HgCdTe photoconductors with the HgCdTe alloy composition optimized for each temperature. The increase in bandwidth with temperature is shown at the top of the figure

Fig. 8 Measured and calculated heterodyne NEP_H as a function of CO_2 local oscillator power for a 100 x 100 µm² p-type HgCdTe photoconductor at 77K and 195K, where the device bandwidths were 140 MHz and 30 MHz, respectively

an area typical of imaging array devices. We have carried out similar calculations for HgCdTe photodiodes and have found that over the 80 to 200K range NEP_D and P_2^D for a photodiode are very close to that shown in Fig. 7 for a photoconductor. Both 10-µm photodiodes and 10-µm photoconductors have demonstrated 2π-300K BLIP performance at about 110K, in agreement with these calculations. At temperatures above 100K, 10-µm HgCdTe detectors are either dark-current-noise limited (photodiodes) or g-r-noise limited (photoconductors) so $NEP_D \propto \sqrt{A}$. At room temperature $NEP_D = 3 \times 10^{-9}$ W/Hz$^{1/2}$ has been reported[18] for a 300 x 300 µm² photoconductor with a 500-MHz bandwidth. The loss in sensitivity with increasing temperature is a result of increased intrinsic carrier concentration and low minority carrier lifetime (corresponding to increased bandwidth in the case of photoconductor). Indeed, there is between three and four orders of magnitude loss in direct detection sensitivity from 77K to 300K; nevertheless, the HgCdTe photoconductor is one of the most sensitive room-temperature 10-µm detectors at bandwidths over about 1 MHz. (Pyroelectrics, thermocouples, etc. are superior below 10 kHz.)

Turning now to heterodyne operation of the photoconductor we note from Fig. 7 that the LO power required to overcome g-r noise (which is proportional to the volume of the photoconductor) rises very rapidly with temperature, reaching a value of 30 mW at 300K. This power level creates severe heating problems and loss in sensitivity from that of an ideal photomixer.

In Fig. 8 we show the measured and calculated heterodyne NEP_H for a p-type HgCdTe photoconductor at 38 MHz as a function of CO_2 LO power at 77K and 195K.[15] At 77K the NEP_H asymptotically approaches the quantum limit with increasing LO power. At this temperature the LO-power requirement was dictated by amplifier noise. At 195K the NEP_H no longer has the simple 'asymptotic LO-power dependence, but reaches a minimum at about 5 mW. This turnaround occurs as a result of heating; the ratio P_2^C/P_{LO} actually increases with LO power. Nevertheless, at 195K a sensitivity of 1.8×10^{-19} W/Hz was realized with less than 10 mW of total power dissipation (LO plus bias power) making these devices quite compatible with simple thermoelectric coolers.

In Fig. 9 we show the heterodyne NEP_H at a 10-MHz bandwidth as a function of temperature for HgCdTe photoconductors[15] and photodiodes[14,19] along with the calculated theoretical limits for 50×50 μm^2 devices. Although relatively little work has been done at temperatures above 77K, these early results are very encouraging. The loss in sensitivity with increasing temperature is much less than in the case of direct detection. A CO_2 laser photomixer using a HgCdTe photoconductor could be operated at about 250K with only a 10-dB loss in sensitivity from that of an ideal photomixer, and at this temperature the short lifetime in HgCdTe gives this device a bandwidth of about 100 MHz (see Fig. 7). One of our p-type HgCdTe photoconductors (designed for 200K operation) is being used at 300K to monitor the 15-MHz frequency offset between the CO_2 LO and transmitter lasers in Lincoln Laboratory's LIDAR system. An NEP_H of 1×10^{-17} W/Hz was measured at 300K; much better performance should be possible with a device designed for 300K operation.

<u>Summary</u>

The theoretical and measured sensitivities of wide-bandwidth (1 to 100 MHz) infrared (1 to 14 μm) photodiodes in the heterodyne and direct

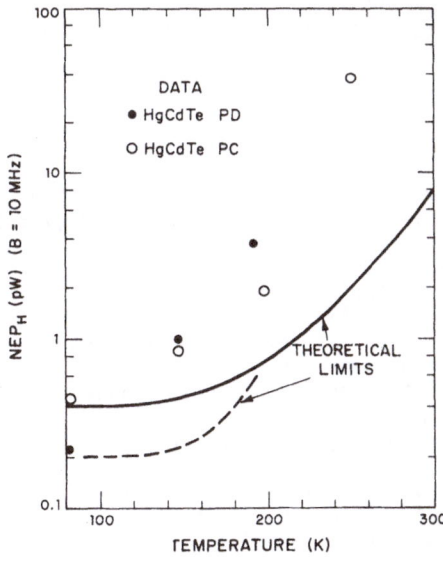

Fig. 9 Heterodyne NEP_H at a 10-MHz bandwidth as a function of temperature for CO_2 laser photomixers. The curves correspond to theoretical limits for 50×50 μm^2 HgCdTe photoconductors and photodiodes, with the HgCdTe alloy composition optimized for each particular temperature

285

detection modes of operation have been reviewed. In the direct-detection mode where amplifier noise is an important consideration the state-of-the-art NEP_D is approximated by 500 (B/10 MHz) pW. At short wavelengths ($\lambda < 1.7~\mu m$) avalanche photodiodes offer superior performance $NEP_D \sim 50$ (B/10 MHz) pW. With careful integration of photodiodes (non-avalanche) with low-noise amplifiers direct detection sensitivities could be increased to ~ 25 (B/10 MHz) (10 $\mu m/\lambda$) pW, representing a considerable improvement at long wavelengths.

In heterodyne operation near-ideal sensitivities ~ 0.3 (B/10 MHz) (10 $\mu m/\lambda$) pW have been achieved with 77K photodiodes using a few hundred μWs of LO power. At 10 μm HgCdTe photoconductors offer very good heterodyne performance at elevated temperatures; 2 (B/10 MHz) pW at 195K, and potentially 3 (B/10 MHz) pW at 250K.

1. C. E. Hurwitz, "Detectors for the 1.1 to 1.6 μm Wavelength Region," Optical Engr. 20, 658 (1981).
2. S. R. Forest, et al., "A High Gain $In_{0.53}Ga_{0.47}As/InP$ Avalanche Photodiode With No Tunneling Leakage Current," IEEE Trans. Elect. Devices ED-28, 1212 (1981).
3. G. H. Olsen, "Low-Leakage, High Efficiency, Reliable VPE InGaAs 1.0 - 1.7 μm Photodiodes," IEEE Electron Device Letters EDL-2, 217 (1981).
4. J. Meslage, et al., "Fast, High-Gain 1.3 μm HgCdTe Photodiode," presented at the 7th European Conf. on Optical Communications, Copenhagen, 8-11 September 1981.
5. K. Chow, et al., "Hybrid Infrared Focal-Plane Arrays," IEEE Trans. Electron Devices ED-29, 3 (1982).
6. See articles in IEEE Trans. on Electron Devices ED-27, No. 1 (1980). (Special Issue on Infrared Materials, Devices and Applications).
7. R. M. Broudy and V. J. Mazurczyck, "(Hg,Cd)Te Photoconductive Detectors," in Semiconductors and Semimetals vol. 18 (R.K. Willardson and A.K. Beer, eds) Academic New York (1981).
8. M. B. Reine, A. K. Sood, and T. J. Tredwell, "Photovoltaic Infrared Detectors," in Semiconductors and Semimetals vol. 18 (R.K. Willardson. and A.K. Beer, eds.) Academic, New York (1981).
9. See Proc. IRIS Specialty Group on Infrared Detectors, Vol. I (1976-1980) (Infrared Information and Analysis Center, ERIM, Ann Arbor, Michigan).
10. R. H. Kingston, Detection of Optical and Infrared Radiation, Vol. 10, Springer Series in Optical Sciences, Springer-Verlag, New York, 1978.
11. M. Lanir and K. J. Riley, "Performance of PV HgCdTe Arrays for 1-14-μm Applications," IEEE Trans. Electron Devices ED-29, 274 (1982).
12. M. Chu, A.H.B. Vanderwyck and D.T. Cheung, "High-Performance Backside-Illuminated $Hg_{0.78}Cd_{0.22}Te$ (λ_{co} = 10 μm) Planar Diodes," Appl. Phys. Lett. 37, 486 (1980).
13. A. V. Lightstone, R. J. McIntyre and P. P. Webb, "Avalanche Photodiode for Single Photon Detection," IEEE Trans. Elect. Devices ED-28, 1210 (1981).
14. D. L. Spears, "Wide-Bandwidth CO_2 Laser Photomixers," SPIE Vol. 227 (CO_2 Laser Devices and Applications (1980)), pp. 108-116, 1980.
15. D. L. Spears, "Theory and Status of High Performance Heterodyne Detectors," SPIE Vol. 300 (to be published November 1981).
16. D. L. Spears, "Extending the Operating Temperature, Wavelength, and Frequency Response of HgCdTe Heterodyne Detectors," Proc. of the International Conf. on Heterodyne Systems and Technology, Williamsburg, March 25-27, 1980, pp. 309-326 (NASA CP-2138).
17. J. F. Shanley and C. T. Flanagan, "n-p (Hg,Cd)Te Photodiodes for 8-14 Micrometer Heterodyne Applications," ibid, pp. 263-280.
18. E. Igras, J. Piotrowski and T. Piotrowski, "Ultimate Detectivity of (Cd,Hg)TGe Infrared Photoconductors," Infrared Phys. 19, 143 (1979).
19. J. F. Shanley, et al., paper presented at the Lasers '81 conference, New Orleans, December 1981.

Part 7

Advanced Optical Techniques

7.1 Optical Remote Sensing of Environmental Pollution and Danger by Molecular Species Using Low-Loss Optical Fiber Network System

Humio Inaba

Research Institute of Electrical Communication, Tohoku University, Katahira 2-1-1, Sendai 980, Japan

1. Introduction

The recent progress in fabrication and cabling techniques for extremely low-loss optical fibers in the near infrared region has stimulated considerable interest in the practical implications of the technology. The present excitement is due to the tremendous possibility that optical fiber technology could eventually find application not only to long distance optical communication and image transmission but also to remote measurement and control as those required in various scientific and industrial fields.

Following the first proposal by the present author [1], sensitive and low-cost optical network systems utilizing low-loss optical fibers have been analyzed and studied experimentally for remote monitoring of air pollution as well as specific concentrations of gas/vapor for mine and manufacturing safety, and also spilled dispersals of LNG and LPG in various industrial complexes, by the differential absorption method [2, 3, 4, 5, 6]. This method of fiber remote detection has various advantages over the previous laser remote monitoring schemes in the open atmosphere [7] in that a low-power laser or even a nonlaser source can be used in conjunction with low-loss low-cost optical fibers. Therefore, a purely optical, economical, real-time and nonhazardous, e. g., eye-safe, monitoring technique can be realized for various remote environments and severe and/or extreme conditions.

This paper is devoted to describe the fundamental configuration and operational principle of a low-loss optical fiber network system for this purpose and to summarize some experimental results which have been obtained recently for remote monitoring of NO_2 molecules in the visible region near 0.5 μm and also CH_4 molecules in the near infrared spectrum around 1.33 μm.

2. Basic Configuration and Principle of Operation

Figure 1 shows schematically the basic configuration of the system based on the differential absorption technique incorporated with a low-loss optical fiber link [1, 2]. The measurement is basically implemented by employing a laser source located at the control center, from where an optical fiber is connected to a sample cell, such as with a multiple-pass geometry, e. g., multireflection White cell [8], placed at a remote point to be monitored. If an extremely low-loss optical fiber is available in the frequency range concerned, a conventional nonlaser source such as a light emitting diode and a multiple-wavelength discharge tube could also be utilized in place of the laser source which can be tuned to or coincide with two frequencies, f_A and f_W, within and outside of an appropriate absorption line of the species, as depicted in the insert of Fig. 1. In such a case, the only requirement of the conventional source is, of course, to have at least one frequency in absorptive region and another with little attenuation.

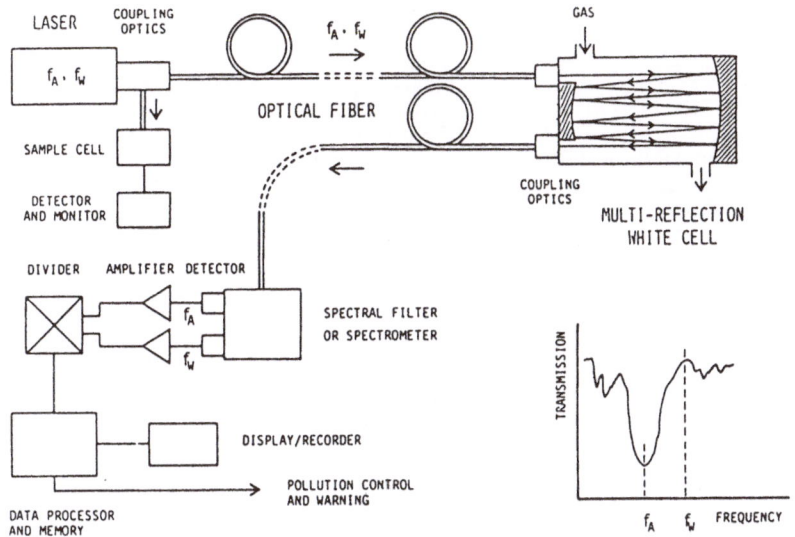

LASER
COUPLING OPTICS
$f_A \cdot f_W$
GAS
$f_A \cdot f_W$
OPTICAL FIBER
SAMPLE CELL
DETECTOR AND MONITOR
COUPLING OPTICS
MULTI-REFLECTION WHITE CELL
DIVIDER   AMPLIFIER   DETECTOR
f_A
f_W
SPECTRAL FILTER OR SPECTROMETER
TRANSMISSION
DISPLAY/RECORDER
POLLUTION CONTROL AND WARNING
DATA PROCESSOR AND MEMORY
f_A f_W FREQUENCY

Fig. 1   Schematic diagram of basic configuration for optical monitoring
of gaseous species involved in environmental pollution and danger
based on the differential absorption method employing a set of
low-loss optical fibers connected to a sample cell placed at a
remote location [1,2]

The resulting signals are then compared to isolate the absorption from
other attenuation and interference components at the control center by using
the optical fiber to return both the optical signals after passing through the
sample cell, so that the species concentration is monitored in real time.

Hence this method is fully based on the optical version, with the round-
trip transmission of the optical beam through low-loss optical fibers instead
of the open atmosphere, analogous to the laser radar technique [7]. Although
it has no range or depth resolution, parallel or sequential measurements employing a number of pairs of optical fibers connected to individual sample cells
appropriately distributed at remote locations, as illustrated in Fig. 2 as an
example of air pollution monitoring network system in the urban area, should
allow one to obtain useful information on the spatial and temporal distributions of species concentrations as well.

Figure 3 shows the transmission loss of the presently available lowest-
loss silica fiber [9, 10] as a function of the wavelength, together with typical absorption wavelengths of various molecular species present in the atmosphere and environment for comparison. The loss of a hollow quartz fiber filled with tetrachloroethylene (C_2Cl_4) at 3.39 μm [11], and that of a fluoride
glass fiber made of GdF_3-BaF_2-ZrF_4-AlF_3 core with Teflon FEP cladding [12] in
the range of 1.8 - 3.6 μm are also depicted.

Because the optical energy can be concentrated and transmitted only in a
low-loss optical fiber at a low cost even for long distances, especially with
the silica based optical fibers, this method associated with extended optical
fiber network systems would be most feasible and economical as well as completely safe for the eye, in real-time monitoring of a large but specified area
of an urban region and industrial or mining complex. Furthermore, it will be
sufficient to use low-power lasers which are well within the present state of

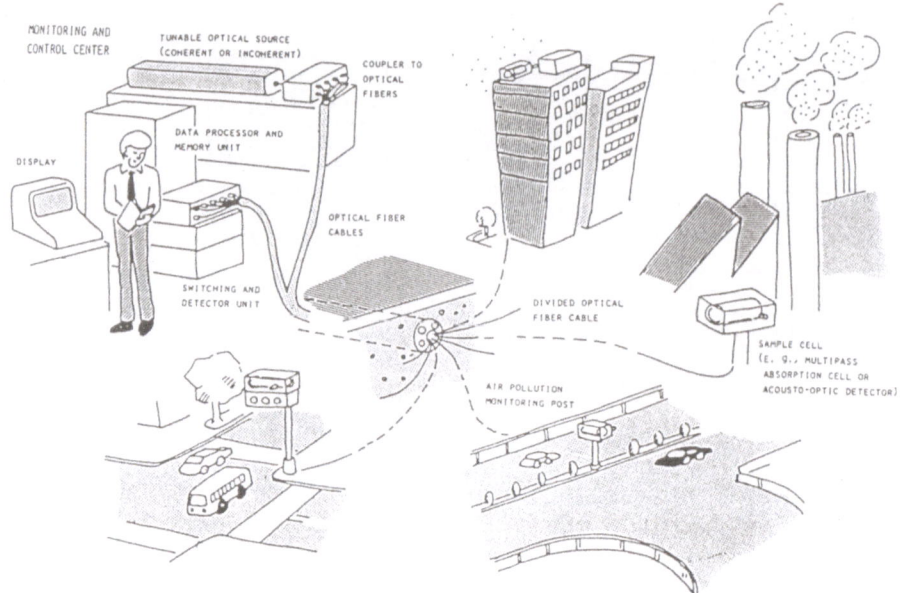

Fig. 2   Schematic illustration of a low-loss optical fiber network system
          for air pollution monitoring over a wide area by the differential
          absorption method [1]

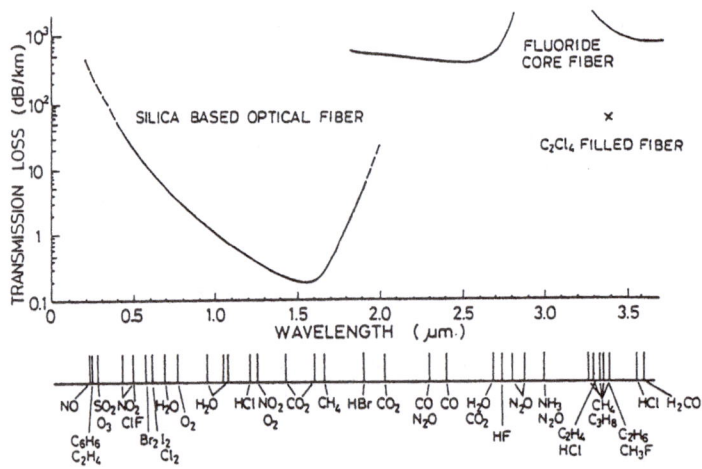

Fig. 3   Transmission loss of optical fibers in the visible and near infra-
          red regions against wavelength and typical absorption wavelengths
          of various molecular species present in the atmosphere and envi-
          ronment

the art of tunable lasers and nonlinear optical devices [13]. A considerable
simplification in the system operation can be provided if one performs a time-
sequential measurement by programmed computer control, with point-wavelength
tunable c. w. or pulsed lasers tuned to several kinds of specific species to
be monitored at the control center.

Besides the main advantages of this new method — it is feasible, inexpensive, nonhazardous, fully optical and two or three dimensional — other advantages are that it is capable of excellent signal-to-noise ratio (SNR) with little optical interference and of continuous surveillance with easy calibration and no electrical induction. Moreover, there could be various modifications to this monitoring system for incorporation with optical fiber networks. For instance, it is possible to replace a multiple-pass sample cell with an open optical path in the atmosphere after the fiber transmission to the remote location. Also, an acousto-optic detector operating with a sensitive microphone[14] can be used if higher sensitivity is demanded. In such cases, the detected signal is sent back to the control center by means of a telephone link or an optical fiber, by converting it again to the optical frequency falling in the lowest loss band of that optical fiber, using, for example, a semiconductor laser diode or a light emitting diode.

3. Remote Monitoring of NO_2 Molecules in Visible Region around 0.5 μm

It is known that in air pollutants only the NO_2 molecule exhibits significant absorption in the visible spectrum. Then we have performed an experiment to demonstrate the feasibility of an optical fiber system for differential absorption measurement of NO_2 molecules using an argon ion laser [3]. Long-path differential absorption [15], differential absorption laser radar [16], and the laser-induced fluorescence technique [17] have already been reported as optical methods for ambient NO_2 monitoring.

The block diagram of the experimental system is schematically shown in Fig. 4. An argon ion laser with multiline oscillation at 476.5, 488.0, 496.5, 501.7 and 514.5 nm was used, and the laser output beam was coupled by a 20 x microscope objective into a transmitting multimode silica fiber, 39 μm diameter core and 125 μm diameter clad. The laser beam was transmitted to a remotely located multireflection sample (White) cell of 1 m mirror spacing with optical fiber coupling structure. The output beam was then returned to the transmitter /receiver location by the multimode silica fiber with large diameter core 150 μm and clad 350 μm. These optical fibers were 20 m and 500 m long, as examples.

After measuring the absorption coefficients of NO_2 molecules for the five laser lines, we selected the maximum absorption line at λ_1 = 496.5 nm and the minimum absorption line at λ_2 = 514.5 nm for the differential absorption mea-

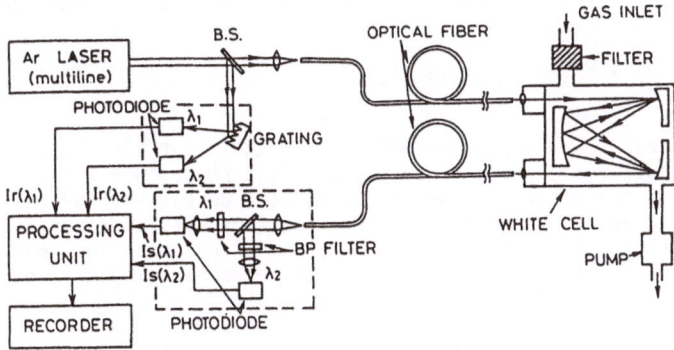

Fig. 4   Block diagram of an experimental system employing optical fiber link for differential absorption measurement of NO_2 concentrations in a sample cell at a remote location [3]

surement. The differential absorption coefficient between these two lines was $\Delta\alpha = 5.9 \times 10^{-4}/\text{ppm} \cdot \text{m}$, which agrees well with the value measured by O'Shea and Dodge [15]. These two lines at λ_1 and λ_2 in the returned signals were selected by two interference filters and detected by two silicon photodiode detectors. Also the two reference lines from the laser were delivered via beam splitter and the diffraction grating to the two photodiodes. These four detector outputs were processed by the analog processing unit to derive the concentration of NO_2 molecules in the sample cell by compensating and calibrating the fiber transmission loss, the transmitting laser power instability, and the system efficiencies. Then the result was plotted on the recorder.

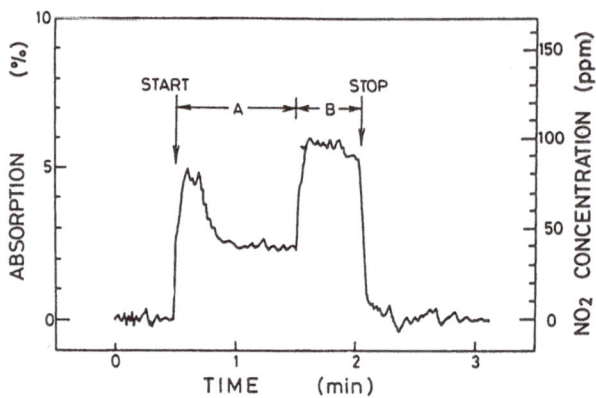

Fig. 5   Typical example of real-time remote monitoring of the NO_2 concentration using optical fibers in the exhaust of a two-stroke motorcycle engine emitted into the atmosphere [3]

A typical example of the results of real-time measurement of NO_2 differential absorption and concentration is illustrated in Fig. 5. Exhaust gas emitted into the atmosphere from a two-stroke motorcycle engine was introduced into the sample cell with optical path length L_c = 1 m and fiber length L_f = 20 m. Engine speed varied from idling ($\sim$500 rpm in region A) to high speed ($\sim$3000 rpm in region B) conditions.

Detection sensitivity of this system is determined by the voltage SNR, and the minimum detectable concentration is given by

$$N_{min} = 1 \,/\, \Delta\alpha L_c (SNR) = [\ (SNR)_p^{-2} + (SNR)_f^{-2}\ ]^{1/2} \,/\, \Delta\alpha L_c. \tag{1}$$

Here $(SNR)_p$ is the received-power dependent SNR, i.e., $(SNR)_p = P_s T^{1/2}/NEP$ in the case of photodiode detectors, where P_s is the received power, NEP is the noise equivalent power of the detectors, and T denotes the detection time constant. $(SNR)_f$ is the SNR associated with the modal noise or optical fiber transmission instability originating from mode conversion and radiation loss due to the change in fiber physical parameters, such as temperature, pressure, and curvature in the multimode fiber [18].

Figure 6 shows the relation of these two SNRs to fiber length. In our system the maximum experimental values of $(SNR)_f$ were 100 for L_f = 20 m and 67 for L_f = 500 m, whereas $(SNR)_p$ could be easily increased to more than these values by using two-line laser power of about 10 mW even with fibers of rather

292

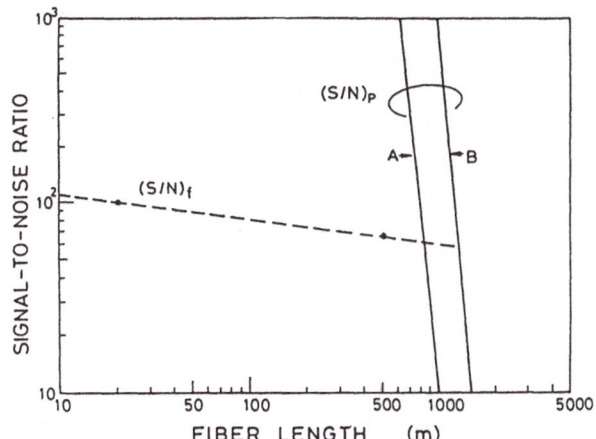

Fig. 6   Relation of two voltage signal-to-noise ratios (SNRs) to optical
         fiber length. Two points indicate experimental values for $(SNR)_f$
         and the lines for $(SNR)_p$ are estimated using the system parame-
         ters. Curve A corresponds to a fiber loss of 50 dB/km and an opti-
         cal path length of L_c = 1 m for the sample cell; curve B corres-
         ponds to a fiber loss of 20 dB/km and L_c = 1 km, with 10 mW laser
         power, NEP = 1 x 10^{-15} W/Hz$^{1/2}$ and detection time constant of T
         = 1 sec [3]

high transmission loss of 50 dB/km as shown by the estimated curve A. Hence,
it should be noted that the detection sensitivity is severely limited by $(SNR)_f$,
fiber modal noise, in the analog or AM operation mode in the optical fiber transmission system.

     In Fig. 5, with L_f = 20 m and L_c = 1 m, $(SNR) \simeq (SNR)_f$ was 100, and N_{min}
was estimated from Eq. (1) to be 17 ppm. Also the value of N_{min} = 1.3 ppm
was obtained for L_f = 500 m and L_c = 20 m with (SNR) = 60. In this experiment,
L_c was limited by the low reflectivity of the aluminum-coated mirrors of the
White cell, R = 0.67 at the laser wavelengths.

     From a system analysis based on the above results we estimate that N_{min} will
be lowered to 0.03 ppm which is enough for ambient NO_2 level monitoring of the
urban atmosphere by utilizing the White cell of L_c = 1 km, with the high reflectivity mirrors such as R = 0.99, and the lower-loss 20 dB/km optical fibers,
up to about 1.3 km long (curve B in Fig. 6). Moreover, we mention that since
this technique does not depend upon intermediate chemical processes as in conventional chemical methods, there is no chance of error owing to concentration
effects on calibration.

4. Remote Monitoring of CH_4 Molecules in Near Infrared Region near 1.33 μm

     Although the transmission loss of existing silica fiber is certainly a li-
mitation to the applicability of the method presented in this paper to measurements of species concentration using visible and ultraviolet absorption, the
system sensitivity could be considerably improved utilizing the near infrared
wavelengths, where the silica optical fiber exhibits a minimum transmission
loss as low as 1 dB/km or less [9, 10] as shown in Fig. 3. In fact, some specific species such as CH_4, which is dispersed in the atmosphere in a spill of
LNG and LPG, and HCl, HBr and CO_2, which pollute some industrial areas, will be

293

detected with high sensitivity over a wide area with tunable InGaAsP or InGaSb
P semiconductor laser diodes (LD) or even with these light emitting diodes (L
ED) operated in suitable wavelength ranges.

It is known that the bandgap of InGaAsP quaternary compound, grown by LPE
method on InP substrates, can be compositionally tuned from 0.92 to 1.65 μm
[19]. InGaAsP LDs and LEDs with emission peaks spanning approximately from 1.1
to 1.6 μm are primarily prepared as practical sources for optical communications in conjunction with silica optical fibers which possess extremely low transmission loss as well as zero dispersion regions in this range. However, their
emission wavelengths and output powers, along with their compactness and low-
power drive, make them also quite suitable for spectroscopic absorption studies
of a variety of species and their remote sensing incorporated with extremely
low-loss optical fiber networks.

From this point of view, we are carrying out precise measurements and analyzes of absorption spectra for various molecules in this range of the near infrared. For instance, the combination and overtone absorption bands of $\nu_2 + 2\nu_3$
and $2\nu_3$ of the CH_4 molecule ($\nu_2 = 1533$ cm^{-1} and $\nu_3 = 3019$ cm^{-1} [20]) exist
around 1.33 μm and 1.66 μm, respectively [5, 6], although their detailed study
has not been reported yet. For light sources at these spectral regions, InGa
AsP LDs were operated under the threshold current level as LEDs, whose central
wavelengths are located at 1.34 μm and 1.61 μm. Figure 7 shows a part of the
measured absorption spectra of CH_4 molecules in 1.31 - 1.34 μm band with the
resolution of 0.2 nm [5, 6], where the assignment of Q, P and R branches has
also been provided [21]. CH_4 gas was contained in a 50 cm long sample cell with
pressure of 400 Torr. FWHM and output power of the LED emission were about 100
nm and 0.1 mW, respectively. Absorption near 1.33 μm was found to be relatively
strong, whose estimated absorption coefficient is approximately 5.5×10^{-6}/ppm·m
at the spectral resolution of 0.2 nm.

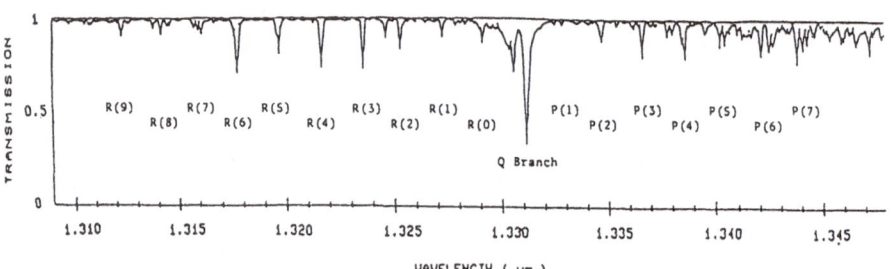

Fig. 7   Typical example of measured absorption spectra of the CH_4 molecule
at 1.33 μm band utilizing an InGaAsP LED [5, 6].
Spectral resolution : 0.2 nm, CH_4 pressure : 400 Torr

Based on these results, an experiment has been performed for the first time
to demonstrate the feasibility of remote monitoring of low-level CH_4 gas using
long distance, low-loss optical fiber link and a compact absorption cell in conjunction with an InGaAsP LED around 1.33 μm. Figure 8 depicts schematically the
block diagram of the experimental setup. Transmitting and receiving multimode
silica fibers, 50 μm diameter core and 125 μm diameter clad, were both 1 km long
with a transmission loss of nearly 1 dB/km at 1.3 μm. Partial pressure of CH_4
gas in a remotely located sample cell of 50 cm long was appropriately changed
by keeping the total pressure of CH_4 - N_2 mixture at 1 atm. The Ge detector was

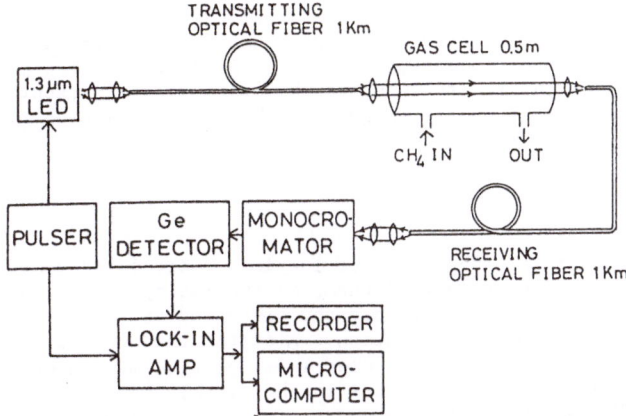

<u>Fig. 8</u>   Block diagram of an experimental system for remote absorption measurement of low-level CH$_4$ gas in a sample cell using extremely low-loss optical fiber link [5, 6]

cooled by dry ice-methanol mixture to increase the gain and suppress noises. A synchronous detection with about 110 Hz modulation of the LED output was performed by a lock-in amplifier and the signal was processed by a microcomputer.

A typical measured result in the case of 6.5 % CH$_4$ gas content in 1 atm. CH$_4$ - N$_2$ mixture in the remotely located sample cell is shown in Fig. 9. It was confirmed that the detection limit of CH$_4$ molecules in air is lower than 1000 ppm, i.e., about 1/50 of its lower explosion limit, based on the present preliminary experiment.

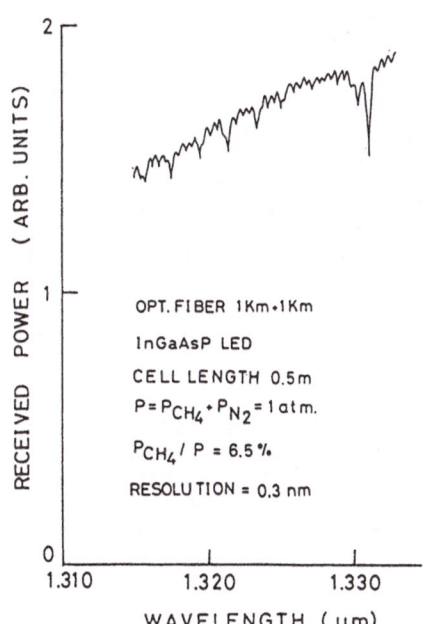

OPT. FIBER 1Km+1Km

InGaAsP LED

CELL LENGTH 0.5m

P = P$_{CH_4}$+P$_{N_2}$ = 1 atm.

P$_{CH_4}$/ P = 6.5 %

RESOLUTION = 0.3 nm

<u>Fig. 9.</u> Measured result of absorption spectra in 1.33 μm band in the case of 6.5% CH$_4$ gas in 1 atm. CH$_4$ - N$_2$ mixture contained in a remotely located sample cell employing a 2 km long low-loss optical fiber link and an InGaAsP LED

This sensitivity for remotely monitoring of low-level CH_4 gas will be much improved, if necessary, by using an InGaAsP LD tuned to the Q-branch of CH_4 absorption spectrum near 1.33 µm. Furthermore, the 1.66 µm band of CH_4 absorption appears to be more adequate for the remote detection by means of the present method, since the absorption coefficients are larger and the spectral widths are broader than those in 1.33 µm band [6].

5. Summary and Conclusion

In this paper an outline of the advantages of low-loss optical fiber network system for remote sensing of a variety of chemical species by the use of differential absorption scheme has been described and discussed as well as the basic configuration and the principle of operation. Also the experimental results for demonstrating the potentiality and usefulness of this method were presented on remote measurements of NO_2 molecules in the visible region using an argon ion laser and of CH_4 molecules in the near infrared region employing an InGaAsP LED.

This technology utilizing extremely low-loss optical fibers incorporated with low-power c.w. or pulsed tunable laser sources or even with nonlaser sources, especially semiconductor LDs or LEDs, and multiple-pass sample cells or their alternatives distributed over a wide area could offer a quite feasible and nonhazardous method for remote monitoring of specific dangerous chemicals for mine and industrial safety, of spilled dispersals of LNG and LPG in chemical and electrical industry complexes, and of air pollution in general, as well as of various sorts of species constituents even in liquids and solids in a variety of environments. Because longer wavelengths of the infrared distributed in the so-called "fingerprint" region [7], which are extensively used for chemical analysis and spectroscopic detection of gases, should be practically useful in the differential absorption measurement of various kinds of molecular species in different situations, the realization of very low-loss optical fibers in this wavelength region ranging approximately from 2 µm to 20 µm is much desirable and to be explored.

However, the presently available lowest-loss silica fiber exists in the near infrared region, primarily developed for long distance, high bit-rate optical communications. Hence it should be quite valuable and necessary to pursue extensive studies for a great wealth of spectroscopic information of a number of substances in the range of about 1.0 - 1.8 µm, that is, spanned by InGaAsP and InGaSbP LDs and LEDs as well. Also system level demonstrations are to be planned which will secure the capability of a major markets for this method. Thus the viability of this technology will be established and practical applications are expected during the next few years.

Acknowledgment

The author is much indebted to Dr. K. Inada of Fujikura Cable Works Ltd., Dr. N. Uchida of Ibaraki Electrical Communication Laboratory, Nippon Telegraph and Telephone Public Corporation, Dr. M. Nakamura of Central Research Laboratory, Hitachi Ltd., and Mr. K. Goto of Research and Development Center, Toshiba Corporation for their continuing attention and kind supports for the experimental works. He is also grateful to members of his Laboratory who have been involved in the experimental and analytical studies described in this paper.

References

[1] H. Inaba, "Laser radar studies and applications in Japan", Sec. III — Optical fiber network system for air-pollution monitoring by differential absorption method, Conference Abstracts, 9th International Laser Radar Conference, Invited paper 2-2, pp. 61 - 67, Munich, July 1979.

[2] H. Inaba, T. Kobayasi, M. Hirama and M. Hamza, "Optical-fibre network system for air-pollution monitoring over a wide area by optical absorption method", Electron. Lett., 15, pp. 749 - 751, 1979.

[3] T. Kobayasi, M. Hirama and H. Inaba, "Remote monitoring of NO_2 molecules by differential absorption using optical fiber link", Appl. Opt., 20, pp. 3279 - 3280, 1981.

[4] M. Hirama, T. Kobayasi and H. Inaba, "High-sensitive optical fiber remote measurement system of air pollution using derivative spectroscopy", Technical Digest of 7th National Laser Radar Symposium in Japan, Paper 42, pp. 85 - 86, Hamanako, February 1981 (in Japanese).

[5] K. Chan, H. Ito., T. Kobayasi and H. Inaba, "Remote absorption measurement of CH_4 molecules in 1.3 μm band using low transmission-loss optical fibers", Technical Digest of 8th National Laser Radar Symposium in Japan, Paper 7, pp. 13 - 14, Nagano, July 1982 (in Japanese).

[6] K. Chan, H. Ito, T. Kobayasi and H. Inaba, "Near infrared absorption spectroscopy of methane molecules and its application to remote measurement using optical fibers", Technical Research Report of the Institute of Electronics and Communication Engineers of Japan, 82, No. 101, OQE 82-49, pp. 43 - 48, July 1982 (in Japanese).

[7] E. D. Hinkley, Ed., "Laser Monitoring of the Atmosphere", Springer, Berlin, 1976.

[8] J. U. White, "Long optical paths of large aperture", J. Opt. Soc. Am., 32, pp. 285 - 288, 1942, and "Very long optical paths in air", J. Opt. Soc. Am., 66, pp. 411 - 416, 1976.

[9] T. Moriyama, O. Fukuda, K. Sanada, K. Inada, T. Edahiro and K. Chida, "Ultimately low OH content V. A. D. optical fibres", Electron. Lett., 16, pp. 698 - 699, 1980.

[10] F. Hanawa, S. Sudo, M. Kawachi and M. Nakahara, "Fabrication of completely OH-free V. A. D. fibre", Electron. Lett., 16, pp. 699 - 700, 1980.

[11] A. K. Majumdar, E. D. Hinkley and R. T. Menzies, "Infrared transmission at the 3.39 μm helium-neon laser wavelength in liquid-core quartz fibers", IEEE J. Quantum Electron., QE-15, pp. 408 - 410, 1979.

[12] K. Jinguji, M. Horiguchi, S. Mitachi, T. Kanamori and T. Manabe, "Infrared power delivery in the 2.7 μm band in fluoride glass fiber", Jpn. J. App. Phys., 20, pp. L392 - L394, 1981.

[13] P. G. Harper and B. S. Wherrett, Ed., "Nonlinear Optics", Academic Press, London, 1977, and references therein.

[14] E.g., Y.-H. Pao, Ed., "Optoacoustic spectroscopy and detection", Chapts. 1 - 3, pp. 1 - 77, Academic Press, New York, 1977.

[15] D. C. O'Shea and L. G. Dodge, "NO_2 concentration measurements in an urban atmosphere using differential absorption techniques", Appl. Opt., 13, pp. 1481 - 1486, 1974.

[16] K. W. Rothe, U. Brinkmann and H. Walther, "Applications of tunable dye lasers to air pollution detection : measurement of the atmospheric NO_2 concentrations by differential absorption technique", Appl. Phys., 3, pp. 115 - 119, 1974.

[17] A. W. Tucker, M. Birnbaum and C. L. Fincher, "Atmospheric NO_2 determination by 442-nm laser induced fluorescence", Appl. Opt., 14, pp. 1418 - 1422, 1975.

[18] W. B. Gardner, "Microbending loss in optical fibers", Bell Syst. Tech. J., 54, pp. 457 - 465, 1975.

[19] M. A. Pollack, R. E. Nahory, J. C. Dewinter and A. A. Ballman, ".Liquid phase epitaxial $In_{1-x}Ga_xAs_yP_{1-y}$ lattice-matched to <100> InP over the complete wavelength range $0.92 \le \lambda \le 1.65$ µm", Appl. Phys. Lett., 33, pp. 314 - 316, 1978.

[20] W. V. Norris and H. J. Unger, "Infrared absorption bands of methane", Phys. Rev., 43, pp. 467 - 472, 1933.

[21] K. Chan, H. Ito and H. Inaba, to be published.

7.2 In situ Ultratrace Gas Detection by Photothermal Spectroscopy: An Overview

Nabil M. Amer

Applied Physics and Laser Spectroscopy Group, Lawrence Berkeley Laboratory, University of California, Berkeley, CA 94720, USA

Two photothermal detection schemes, photoacoustic and photothermal deflection, are reviewed in the context of *in situ* ultrance detection of atomspheric constituents.

OVERVIEW

In recent years, optical heating has been employed in novel ways to measure minute absorption coefficients of gases, liquids, and solids. The physical principle underlying these measurements is that when a beam of electromagnetic radiation is absorbed by a given medium, heating will ensue. The conversion of light into heat is what is used to measure optical absorption coefficients as low as approximately $10^{-9} - 10^{-10}$ cm^{-1}.

In the case of gases, photothermal detection can be accomplished in one of two ways:

a. The optical heating, when modulated, will cause a time-dependent pressure fluctuation which can be detected with a suitable transducer, typically a microphone. This type of detection is known as photoacustics[1-8], and can be performed in either acoustically resonant or acoustically non-resonant regimes.

b. The modulated optically-induced heating will cause a corresponding modulation of the index of refraction of the absorbing material. The gradient of the modulated index of refraction is used to periodically deflect a weak laser probe beam propagating through the material[9,10]. The deflection, which can be as small as 10^{-10} radian, is readily detected with a position sensor.

In both approaches, the amplitude and the phase of the output signal are related quantitatively to the absorption coefficient.

IN SITU DETECTION SCHEMES

The following two schemes illustrate the high sensitivity and the *in situ* capability of photothermal detection.

a. Windowless Resonant Spectrophone [2]

In conventional acoustically resonant photoacoustic detection[11], the optically exciting beam passes through the windows at normal incidence to the cylindrical cavity. This implies that the beam enters and leaves the spectrophone at points of high pressure amplitude; thus window absorption will contribute an unwanted background signal.

This problem can be alleviated altogether by placing the windows at nodes of the mode being excited. Since the amplitude $A_j(\omega)$ of the j'th acoustical mode excited at frequency ω by the absorption of light by the gas is given by:

$$A_j(\omega) = f(\omega)(\alpha/V_c) \int P_j^*(\vec{r}) \, I(\vec{r}) \, dV$$

here $f(\omega)$ is the resonance lineshape, α is the absorption coefficient, V_c is the call volume, $I(\vec{r})$ is the laser intensity, and $p_j^*(\vec{r})$ is the j'th mode normalized eigenfunction. Hence, for most efficient excitation of the j'th mode, it is desirable that the integrand have the same sign throughout the rather small region of the cell volume in which $I(\vec{r})$ differs significantly from zero.

A significant consequence of placing the windows at pressure nodes is that one can then eliminate such windows completely, i.e., have uncovered openings in the structure of the spectrophone where the windows would have been mounted. Since these openings are located at pressure nodes, one would expect that the quality factor Q of the resonant cavity should not be degraded significantly. Such a spectrophone is shown in Fig. (1).

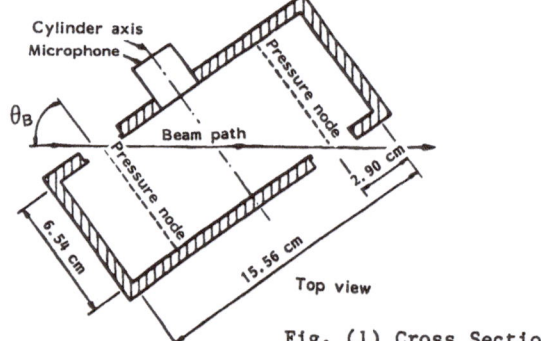

Fig. (1) Cross Section of Windowless Spectrophone

We have demonstrated that this windowless spectrophone has a high Q of 509 (as opposed to 560 when operated with windows), and achieved a detection sensitivity of about 20 parts per trillion for ethylene in nitrogen.

Clearly then, windowless operation permits the *in situ* continuous monitoring of atmospheric constituents in the field; thus obviating the need for sampling.

b. <u>Photothermal</u> <u>Deflection</u> <u>Detection</u> [9,10]

For low modulation frequency, we have shown that the amplitude of the deflection ϕ of the probe beam is given by

$$\phi \sim (dn/dT) \; (P/\kappa \; \pi^2 \; x_o) \; [1 - \exp(1-\alpha\ell)] \; [1 - \exp(-x_o^2/a^2)]$$

where dn/dT is the temperature coefficient of the refractive index, P is the incident-laser power, κ is the gas thermal conductivity, x_o is the distance between the intensity maxima of the pump and probe beams, α is the optical absorption coefficient, ℓ is the interaction length of the pump and probe beams, and a is the pump beam radius at the 1/e intensity. Thus, for small $\alpha\ell$ ($\lesssim 2$), the amplitude of the deflection is proportional to $\alpha\ell$ and to the power. Furthemore, ϕ exhibits a maximum near $x_o/a \sim 1$; this then defines the optimal separation between pump and probe beams.

A typical photothermal deflection detection scheme is shown in Fig.2).

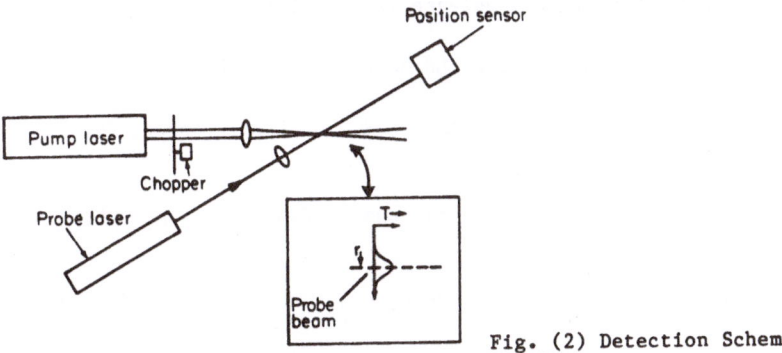

Fig. (2) Detection Scheme

To maximize the signal, the angle between the pump and probe beams can be minimized. However, it should be noted that collinearity is not required.

We have demonstrated the feasibility of this scheme for performing *in situ* measurements in the absence of sample cells or containers, hence eliminating the drawbacks associated with sampling. Typical minimum detectivity is $\sim 10^{-9} - 10^{-10}$.

CONCLUSION

Photothermal deflection and photoacoustic detections have been shown to be highly sensitive, relatively simple, and readily amenable for *in situ* detection and measurements. This is primarily due to the well developed theoretical understanding of the physics of signal generation in both techniques.

ACKNOWLEDGEMENT

This work was supported by the Director, Office of Energy Research, Pollutant Characterization and Measurements Division of the U.S. Department of Energy under Contract No. W-7405-ENG-48.

REFERENCES

(1) For an overview of photoacoustics, see Optoacoustic Spectroscopy and Detection, Y.-H. Pao, Ed. (Academic press, New York, 1977).

(2) R. Gerlach and N.M. Amer, Appl. Phys. 23, 319 (1980).

(3) R. Gerlach and N.M. Amer, Appl. Phys. Lett. 32, 228 (1978).

(4) D.R. Wake and N.M. Amer, Appl. Phys. Lett. 34, 379 (1979).

(5) L.J. Thomas, M.J. Kelly and N.M. Amer, Appl. Phys. Lett. 32, 736 (1978).

(6) C.K.N. Patel and R.J. Kerl, Appl. Phys. Lett. 30, 578 (1977).

(7) L.B. Kreuzer and C.K.N. Patel, Science 173, 45 (1971).

(8) C.K.N. Patel, E.G. Burkhardt, C.A. Lambert, Science 184, (1974).

(9) D. Fournier, A.C. Boccara, N.M. Amer, R. Gerlach, Appl. Phys. Lett. 37, 519 (1980).

(10) W.B. Jackson, N.M. Amer, A.C. Boccara, and D. Fournier, Appl. Opt. 20, 1333 (1981).

(11) C.F. Dewey, R.D. Kamm, and C.E. Hackett, Appl. Phys. Lett. 23, 633 (1973).

7.3 Laser-Induced Breakdown Spectroscopy (LIBS): A New Spectrochemical Technique*

Leon J. Radziemski, Thomas R. Loree, and David A. Cremers

Los Alamos National Laboratory, University of California, Los Alamos, NM 87545, USA

ABSTRACT

We have used the breakdown spark from a focused laser beam to generate analytically useful emission spectra of minor constituents in air and other carrier gases. The medium was sampled directly. It was not necessary to reduce the sample to solution nor to introduce electrodes. The apparatus is particularly simple; a pulsed laser, spectrometer, and some method for time resolution. The latter is essential in laser-induced-breakdown spectroscopy (LIBS) because of the strong early continuum. High temperatures in the spark result in vaporization of small particles, dissociation of molecules, and excitation of atomic and ionic spectra, including species which are normally difficult to detect. In one application, we have monitored beryllium in air at concentrations below 1 μg/m^3, which is below 1 ppb (w/w). In another we have monitored chlorine and fluorine atoms in real time. LIBS has the potential for real-time direct sampling of contaminants in situ.

INTRODUCTION

In atomic emission spectrochemistry, the light from an excited sample is spectrally analysed to yield qualitative and quantitative information about the elemental constituents. The more traditional emission techniques employ arc or spark excitation. Recently atomic flame fluorescence and the inductively coupled argon plasma (ICP) have become useful analytical tools. None of these techniques is particularly portable or usable outside of the analytical laboratory. We have created a field-deployable version of spark spectroscopy by using a pulsed laser to generate a free-standing spark by dielectric breakdown (1,2). We use the acronym LIBS for laser-induced breakdown spectroscopy; its time resolved version is sometimes called TRELIBS. Except where noted, we discuss only the time-resolved version.

LIBS has been used to analyze airborne samples (both aerosols and particles) (3), to detect beryllium (3), chlorine, and fluorine (2,3), and to detect species in the product stream of a coal gasifier (4). For details we refer the reader to the references. Here we summarize the important features of the technique and some of the results.

EXPERIMENTAL

The basic apparatus for observing LIBS with time resolution can be quite simple. A version we have used extensively is shown in Fig. 1. A 5- to 20-cm focal length lens focused the beam to fluence levels which were sufficient to break down ambient air. Light from the plasma was collected and imaged on the slit of a scanning spectrometer. A variety of photomultiplier tubes with extended uv or ir response were used as detectors. The photomultiplier output was processed by a boxcar averager for time resolution and signal averaging. Observations with the photomultiplier and spectrometer gave the maximum sensitivity over the range 200 to 900 nm. By contrast, the diode array detection described

---

*Sponsored by the US Department of Energy.

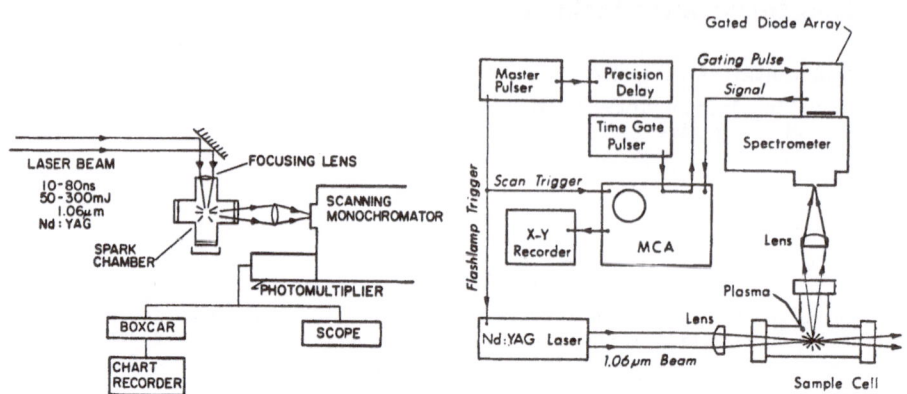

Fig. 1. Diagram of the photomultiplier tube and boxcar apparatus for time-resolved LIBS

Fig. 2. Diagram of apparatus used in the time-gated diode-array version of LIBS

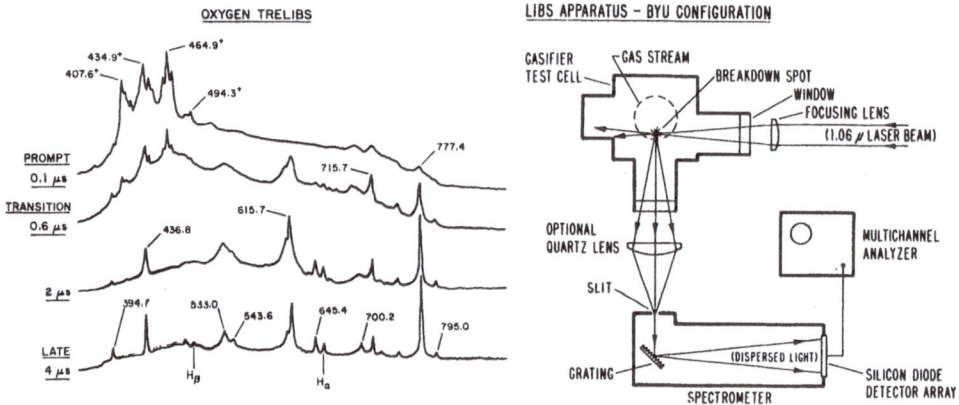

Fig. 3. Spectra of once ionized and neutral oxygen atoms at various times following plasma formation

Fig. 4. Apparatus used to monitor atomic and molecular species contained in the effluent of an experimental coal gasifier

below monitored a wide spectral range, useful for survey work not requiring maximum sensitivity.

In the time-gated diode-array version of LIBS, we coupled the output of the diodes to a multichannel analyzer. Figure 2 shows the schematic of this arrangement. Often the light impinging on the array went through a microchannel plate image-intensifier, which can have a gain of 25,000. This system was sensitive to wavelengths between 350 and 800 nm. Its utility is illustrated in Fig. 3, which contains the LIBS spectrum of oxygen at various times t_d after spark initiation. Figure 3 also illustrates the necessity for time resolution. At early times the continuum and ionic lines dominate, but at late times the spectrum becomes quieter and neutral lines which have analytical utility are prominent.

A simple, time integrated version of diode-array LIBS is shown schematically in Fig. 4. This apparatus was used to observe species in the effluent stream of an experimental coal gasifier at Brigham Young University. Species detected in this experiment are listed in Table I.

TABLE I

BYU TEST DATA: DETECTED SPECIES
    (Preliminary Results)

Atomic
Na°
K°
H°
O°, O^+
O°, C^+, C^{2+}
Ca°, Ca^+
Si°
Mg^+

Molecular
CN
N_2
CO
O_2

RESULTS

LIBS was first applied to the real-time monitoring of airborne beryllium particles. These particles were generated in the laboratory by ablation from a solid beryllium block, or by a nebulizer/heat-pipe arrangement. The high temperature of the spark (20,000 K at t_d = 0.5 µs; 10,000 K at t_d = 10 µs) vaporized the particles and excited neutral and once ionized beryllium. The strongest beryllium feature was the Be II doublet at 313.1 nm. Figure 5 shows the results of monitoring that feature while beryllium particulate was admitted to the sampling cell. Concentrations in the spark chamber were between 1 and 10 µg/m³.

MONITORING BERYLLIUM WITH Be II 313.1 nm

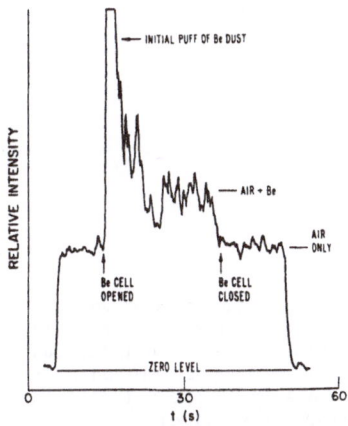

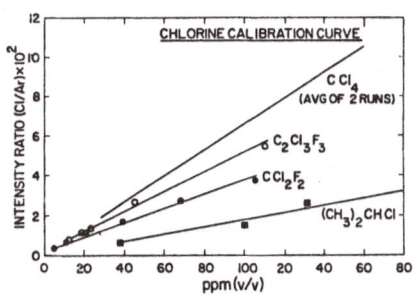

Fig. 5. Variation of Be II emission signal as beryllium particulate was admitted to the sample chamber

Fig. 6. Demonstration of the feasibility of atom counting with LIBS. The slopes of the working curves indicate the number of chlorine atoms in the parent molecule

305

In a second application, LIBS was used to detect fluorine and chlorine in air. As sources of these atoms we used a variety of freons, SF_6, and CCl_4. The high spark temperature dissociated the molecules completely and excited the upper energy levels of Cl and F. Their concentration was then monitored by infrared transitions. Working curves relating signal to concentration for chlorine are shown in Fig. 6. The slopes of these curves were proportional to the number of chlorine or fluorine atoms in the source molecule. These slopes were normalized to see how well the LIBS technique could identify the number of atoms in the molecule. The results are contained in Table II, and are promising.

In addition to demonstrating real-time detection and atom counting we measured the limit of detection for many atoms. We calculated the limit of detection (C_L) from

$$C_L = \frac{2\sigma}{S}$$

where σ is the rms noise on either the background or the signal, and S is the slope of the working curve. Results are shown in Table III. The limits of detection are sufficiently low to allow several applications which we are considering.

TABLE II

COMPARISON OF RATIOS OF SENSITIVITY TO THE NUMBER OF
CHLORINE OR FLUORINE ATOMS IN A MOLECULE

Parent Molecule	Number of Cl Atoms	Relative Sensitivity
$(CH_3)_2CHCl$	1	1.02
CCl_2F_2	2	1.92
$C_2Cl_3F_3$	3	2.81
CCl_4	4	4.00 (reference)

Parent Molecule	Number of F Atoms	Relative Sensitivity
CCl_2F_2	2	2.08
$C_2Cl_3F_3$	3	3.06
SF_6	6	6.00 (reference)

TABLE III

DETECTION LIMITS OF VARIOUS ELEMENTS IN AIR USING LIBS WITH TIME RESOLUTION

Element	Analytical Line	C_L ppm (w/w)	Delay Time (μs)
Cl	837.5 nm	20	2
F	685.6 nm	40	2
P	253.3 nm	1.2	2
S	921.2 nm	200	2
Be	313.1 nm	0.0006 (0.8 μg/m^3)	4
As	228.8 nm	0.5	2

SUMMARY AND CONCLUSIONS

We have demonstrated that analytically useful atomic spectra can be generated by LIBS and analyzed in real time. The detection apparatus is relatively simple; even the current laboratory apparatus is movable. With the utilization of small lasers and electronic miniaturization, a very small portable unit is feasible. We are working on a prototype model of such a unit, one which would be capable of making in situ spectrochemical measurements in the field.

LIBS has a number of distinct features which are spectrochemically advantageous. These are summarized in Table IV. Table V contains a number of projected applications.

TABLE IV

ADVANTAGES OF LIBS AND THE SPECTROCHEMICAL IMPLICATION

LIBS Advantages	Consequences
Atomic emission spectroscopy	* multielement capability
High temperature	* vaporizes particulates * excites species which are normally difficult to detect, such as chlorine and fluorine
Absence of electrodes	* no electrode wear * reduces interferences and eliminates plasma perturbation by electrodes
Direct sampling	* removes some analytic flicker noise * eliminates sample handling, offers real-time detection
Conventional sampling	* all the advantages of being electrodeless, but pulsed operation may be disadvantage.
Simple apparatus	* field operation

TABLE V

APPLICATIONS

* Direct air sampling
* Point detection
* Monitoring of secure-room air
* Monitoring of surfaces and clothing

ACKNOWLEDGMENTS

We gratefully acknowledge the use of the Brigham Young experimental coal gasifier, and the assistance of Dr. Paul Hedman of that University. The able technical assistance of Ron Martinez and Dean Barker was greatly appreciated. Dr. Thomas Niemczyk (University of New Mexico, Department of Chemistry), and Dr. Robert Schmeider (Sandia National Laboratory, Livermore) made valuable suggestions during portions of this work. The encouragement and support of Dr. Allan Hartford is appreciated.

REFERENCES

1. T. R. Loree and L. J. Radziemski, "Laser-Induced Breakdown Spectroscopy: Time Integrated Applications," J. Plasma Chem. and Plasma Proc. 1, 271 (1981).

2. L. J. Radziemski and T. R. Loree, "Laser-Induced Breakdown Spectroscopy: Time-Resolved Spectrochemical Applications," J. Plasma Chem. and Plasma Proc. 1, 281 (1981).

3. L. J. Radziemski, "Time-Resolved Spectroscopy of Laser-Induced Air Plasmas Seeded with Phosphorus, Chlorine, and Beryllium," J. Opt. Soc. Am. 71, 1594 (1981).

4. T. R. Loree and L. J. Radziemski, "Laser-Induced Breakdown Spectroscopy: Detecting Sodium and Potassium in Coal Gasifiers," 1981 Symp. Instr. and Control for Fossil Energy Prog., Argonne Nat. Lab. Press ANL 81 - 62/ CONS - 810607, 768 (1982).

7.4 The High Spectral Resolution Lidar

E.W. Eloranta, F.L. Roesler, J.T. Sroga
University of Wisconsin, Madison, WI 53706, USA

The High Spectral Resolution Lidar (HSRL) system was developed at the University of Wisconsin[*] to remotely measure optical properties of the atmosphere. The extinction cross section, the aerosol to molecular backscatter ratio and the backscatter to extinction ratio are measured without assumptions regarding aerosol scattering characteristics. The HSRL is both eyesafe (range >200 m) and capable of daylight operation. It has operated in both aircraft and ground based studies.

Principles of operation

The HSRL measurements are obtained by separating the backscattered signal into aerosol and molecular channels. This separation is based on the Doppler broadening of the backscattered spectrum. The thermal motion of molecules is much faster than the Brownian motion of aerosol particles and therefore molecular scattering produces more spectral broadening than aerosol scattering. The separation is achieved using a high spectral resolution Fabry-Perot optical interferometer. By tuning the Fabry-Perot etalon transmission peak to the wavelength of the narrow band laser transmitter ($\Delta\lambda \sim 2 \times 10^{-4}$nm), lidar backscatter contributions near the laser wavelength (primarily aerosol contributions) are transmitted through the etalon to an "aerosol" channel detector. Light not transmitted is reflected to a "molecular" channel detector which measures light that has been Doppler shifted.

The spectral distribution of laser light backscattered from the atmosphere depends upon the laser spectral distribution, the scattering cross section of particulates, the molecular density and the Doppler broadening of laser light backscattered from the gaseous constituents. The Doppler spectral distribution of monochromatic light backscattered from gas molecules moving randomly due to thermal motions is given by (Fiocco and DeWolf, 1968)

$$\frac{1}{N_m} \frac{dN_m}{d\sigma} = \frac{1}{\sqrt{\pi}} \frac{\sqrt{\bar{m}} c}{\sqrt{8\sigma_o^2 kT}} \exp\left[-\frac{\bar{m} c^2}{8\sigma_o^2 kT} (\sigma-\sigma_o)^2\right] \tag{1}$$

where

N_m = total number of photons backscattered from molecules
m = mean molecular mass
c = speed of light
k = Boltzman's constant
T = absolute temperature
σ = wavenumber of scattered light = 1/wavelength
σ_o = wavenumber incident light.

The spectral distribution of the molecular backscattering is obtained from the convolution of Eq. 1. with the spectral distribution of the laser transmitter. The aerosol backscatter spectral distribution is essentially that of the laser since the aerosol Doppler broadening is two orders of magnitude smaller than the molecular Doppler broadening.

[*] NASA Langley Contract NAS1-14136

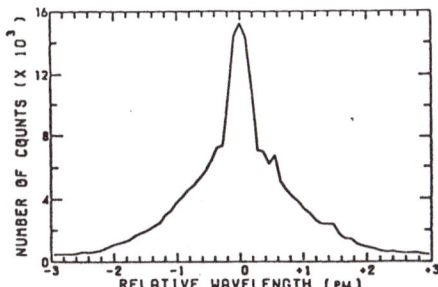

Figure 1. The spectral distribution of backscattered light measured by the HSRL during ground based testing. Wavelengths are in picometers (10^{-12}m). The aerosol "spike" in the backscatter spectrum is clearly visible over the broader molecular signal

The backscatter spectral distribution from the atmosphere is the super-position of the aerosol and molecular backscatter spectra. Fig. 1 shows an example of the spectral distribution of atmospheric backscattering obtained from HSRL measurements.

HSRL data is analyzed by writing separate lidar equations for the rate at which photons are received from molecular scattering, N_m, and the rate at which they are received from aerosol scattering, N_a

$$R^2 N_m(R) = \gamma_m(R) \, \beta_m(R) \, \frac{I\!P_R(180°)}{4\pi} \, \exp[-2 \int_o^R \beta_\varepsilon(R')dR'] \tag{2}$$

$$R^2 N_a(R) = \gamma_a(R) \, \beta_a(R) \, \frac{I\!P_a(R,180°)}{4\pi} \, \exp[-2 \int_o^R \beta_\varepsilon(R')dR'] \tag{3}$$

where

$R = c\,t/2$ = range

$\beta_a(R)$ = volume cross sections for aerosol

$\beta_m(R)$ = molecular scattering

$\beta_\varepsilon(R)$ = volume extinction coefficient

$\dfrac{I\!P_a(R,180)}{4\pi}$ = aerosol backscatter phase function

$\dfrac{I\!P_r(180)}{4\pi} = 3/8\pi$ = Rayleigh backscatter phase function

$\gamma_a(R)$, $\gamma_m(R)$ = calibration constants for "aerosol" and "molecular" channels. γ_a and γ_m are constant at ranges greater than the range where the laser beam divergence and receiver field of view completely overlap.

At ranges past beam overlap (i.e., where γ_a and γ_m are constant) dividing equation 3 by equation 2 provides the ratio of the aerosol to the molecular volume backscatter cross section

$$S(R) = \frac{\gamma_m R^2 N_a(R)}{\gamma_a R^2 N_m(R)} = \frac{\beta_a(R) \, I\!P_a(R,180°)/(4\pi)}{\beta_m(R) \, 3/(8\pi)} \quad . \tag{4}$$

The molecular backscatter cross section, $\beta_m(R)$ $3/(3\pi)$, can be computed from the atmospheric density profile using a temperature profile.

A direct measurement of the aerosol backscatter cross section $\beta_a(R) \dfrac{\mathbb{P}_a(R,180°)}{4\pi}$ can be calculated from the scattering ratio, and the molecular backscatter cross section computed from the temperature profile

$$\beta_a(R) \frac{\mathbb{P}_a(R,180°)}{4\pi} = S(R) \quad \beta_m(R) \; 3/(8\pi) \qquad . \qquad (5)$$

When temperature profile is known, the only unknown in Eq. 2 is the integrated optical depth, $\tau(R)$

$$\tau(R) = \int_0^R \beta_\varepsilon(R')dR'$$
$$= \ln[\beta_m(R)] \; - \ln[\; \frac{8\pi R^2 \; N_m(R)}{3 \; \gamma(R)} \;] \qquad . \qquad (6)$$

The volume extinction coefficient, $\beta_{\varepsilon xt}(R)$, can then be determined from the slope of the optical depth

$$\beta_{\varepsilon xt}(R) = \frac{d}{dR} \tau(R) \qquad . \qquad (7)$$

If the molecular extinction cross section is known at the operating wavelength, it can be subtracted from $\beta_{ext}(R)$ to provide the aerosol extinction cross section. The aerosol extinction cross section and equation 5 allow calculation of the aerosol backscatter phase function $\mathbb{P}_a(R, 180°)/4\pi$.

The HSRL Transmitter

The pulsed dye laser is optically pumped by a 337.1 nm N_2 laser radiation: It produces visible radiation at 467.88 nm, with a bandwidth of less than 0.3 pm (3 mÅ). A schematic diagram of the dye laser configuration is given in Figure 2.

The dye laser design is of the oscillator-amplifier type described by Lawler, Fitzsimmons, and Anderson (1976). A spherical mirror images the 25 by 8 mm rectangular exit aperture of the N_2 laser onto the dye cells with a magnification of about 0.5. A quartz cylindrical lens of 25 mm focal length in front of each cell produces a line focus about 0.3 mm wide in the dye, immediately behind the quartz UV window. An uncoated fused silica beam splitter directs 9% of the UV into the oscillator dye cell; the remainder goes to the output amplifier through a variable delay line consisting of a 1/4 meter White cell.

In operation, diffuse super-radiant emission from the early excitation of the dye is fed back to the dye cell by one uncoated surface of a glass wedge, which serves as the output mirror. This fed-back beam is greatly amplified after passing through the cell and proceeds through a beam expanding telescope and into a pressure chamber containing the dispersive elements. An echelle grating with 316 lines per mm, blazed at 63.4° and operating in 13th order, produces by itself a linewidth (FWHM) of about 13 pm (130 mÅ). With care, we have obtained linewidths as narrow as 4.8 pm at 464 nm. An off-axis air spaced Fabry Perot etalon

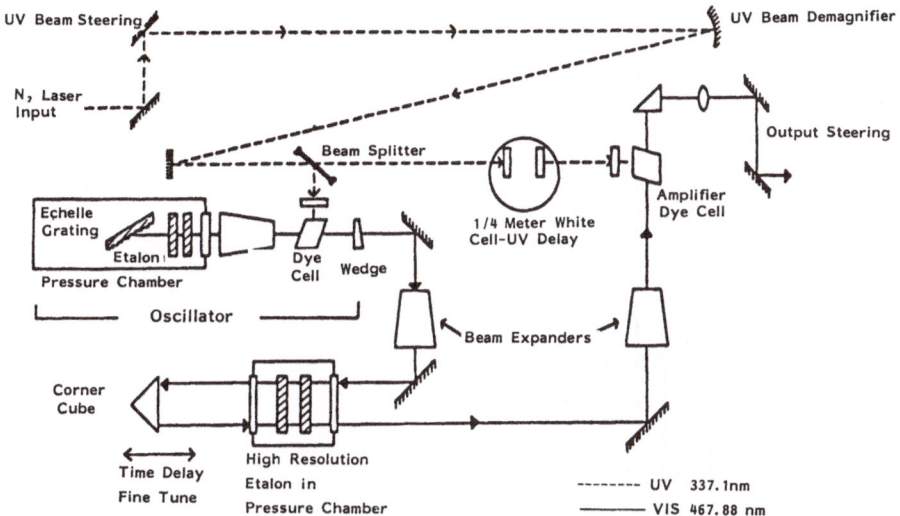

------- UV 337.1nm
——— VIS 467.88 nm

Figure 2. Dye laser configuration

with a 3 mm spacer further narrows the linewidth to 0.9 pm. The spectrally selected light is now focussed back into the dye cell where it is amplified and forms the output beam from the oscillator. The gain of the dye is such that multiple traversals of the cell are not required to generate laser action.

Before amplifying the oscillator beam, it is necessary to further reduce the bandwidth to 0.3 pm or less. This is accomplished by passing the beam through a 13 mm air spaced Fabry-Perot etalon. Because the amplifier is operated in a gain saturated mode, the bandwidth tends to increase upon amplification unless the wings of the input spectral profile are strongly suppressed. Wing suppression is achieved in this design by double-passing a high resolution etalon.

The oscillator output is considerably delayed and temporally stretched by passing twice through the high resolution etalon. Accordingly, provision is made for delaying the UV pumping radiation by up to 4 meters by means of the White cell. Fine tuning of the timing between UV and oscillator signals, to ensure their simultaneous arrival at the amplifier cell, is accomplished by translating the corner cube retroreflector which double passes the passive etalon.

The HSRL Receiver

Both the laser transmitter and receiver spectrometer share a common 355 mm diameter f/11 Schmidt-Cassegrain telescope. The secondary mirror obscures the central 115 mm of the telescope aperture. In addition, a small area of the aperture near the periphery, 5 to 10 mm in diameter, is dedicated to the outgoing laser beam.

To minimize the scattering of laser light into the receiver during the laser pulse, the laser and receiver do not share the same field stop, but rather two conjugate stops. These, together with a laser injection mirror, are mounted rigidly in a coupler assembly, independent of the main telescope body. Once the mirror is adjusted to make the two stops conjugate, the relative alignment of transmitter and receiver beams is assured, independent of any moderate motions of the telescope body with respect to the field stops.

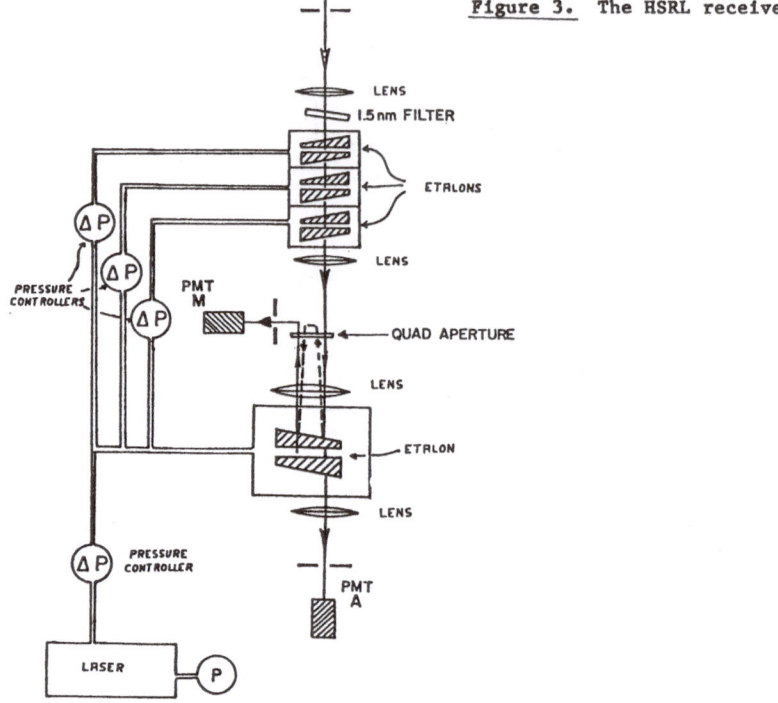

Figure 3. The HSRL receiver

Coupling of the laser amplifier to the telescope is achieved by means of a single lens of 200 mm focal length which simultaneously images the laser beam onto the laser field stop of the telescope, and images the dye cell itself approximately onto the telescope primary.

A diagram of the HSRL double reflection receiver system is given in Figure 3. A three Fabry-Perot etalon train and interference filter are used to suppress background radiation. The telescope forms an image of the scattering field at the entrance aperture to the spectrometer. After passing through this aperture, the beam is expanded to a diameter of approximately 24 mm before reaching the collimator lens. Collimated light then passes through the PEPSIOS section of the spectrometer, consisting of an interference filter and three 60 mm diameter etalons in series, see Mack et al. (1963). The collimated beam is then brought to a second focus at aperture 1 of the quad aperture assembly by a lens with a focal length of 130 mm. At this point, a spectral element 3.5 pm in width centered on the laser wavelength has been isolated, and the beam has a divergence of f/5.5, which matches the opics surrounding the final high resolution etalon.

Light which passes through aperture 1 of the quad assembly diverges to fill the 152 mm diameter f/5 lens, which in turn collimates the light reaching the high resolution etalon. This etalon separates the light into a high resolution trans-mitted beam ("spike") and a spectrally broad reflected beam ("notch"). The trans-mitted beam is brought to a focus at the aperture in front of PMT A, a lens which completes the optical path for the high resolution aerosol channel. The reflected beam is directed into a double passing arrangement to improve light rejection in the notch. This light passes through aperture 2 of the quad aperture assembly, to be retroreflected through aperture 3 by a field lens and roof prism behind the focal plane. The retroreflected beam is then recollimated, reflected from the high etalon resolution, and focussed a second time before it finally exits through aper-ture 4 of the quad aperture assembly. This doubly-passed light is then collected by PMT M, which completes the optical path of the channel which measures the Doppler broadened scattering from molecules.

HSRL DATA SYSTEM

The HSRL system uses a computer controlled photon counting data acquisition system. Counting individual photons allows the HSRL to make stable measurements at very low power levels; accurate measurements are easily achieved at signal levels corresponding to one photon returned into a 200 nsec range gate for each 20 laser pulses. By using high laser pulse repetition rates (100 Hz, presently being increased to 10^4 Hz) the system also achieves excellent dynamic range. Photon counting is also used to provide precise measurements of background skylight, which can be subtracted from observed signals.

HSRL data is accumulated into "aerosol" and "molecular" channels by a dual channel counting system. Photomultiplier tubes with fast pulse response are coupled to a dual channel pulse height discriminator. The photomultiplier output pulses are amplified by X10 DC coupled amplifiers and transmitted to the discriminators with 50 Ω coaxial cable. Individual photoelectron generated pulses have a measured time duration of 3.5 nsec at the discriminator input. The discriminator generates 3 nsec wide output pulses with a maximum dead time of 6 nsec. Practical counting rates > 30 mHz can be achieved with this system. For each data channel a pair of 8-bit ECL counters alternately accumulate photoelectron counts for 200 nsec intervals. While one counter is active, the sum accumulated in the other counter is added to the contents of a 16-bit memory location corresponding to the time interval measured from the emission of the last laser pulse. This memory accumulates photocounts over a predetermined number of laser shots; it is then transmitted to the data system computer and the memory is cleared to begin another data cycle.

HSRL Operation

Tuning and calibration of the HSRL are performed under the control of a LSI-11 computer. The receiver is tuned by adjusting gas pressures, inside the etalon chambers with computer operated pressure difference controllers. To begin the tuning the receiver, a computer activated mirror directs a white light source into the receiver. The computer then performs an algorithm which maximizes the white light transmission to the "aerosol channel" by adjusting the pressure differences between the PEPSIOS etalon chambers. In the second phase of the tuning process light from the laser transmitter is heavily attenuated, diffused and directed into the receiver. The computer then adjusts the pressure difference between the high resolution etalon and laser in order to tune the receiver to the laser wavelength. Light from an iron hollow cathode lamp is then directed

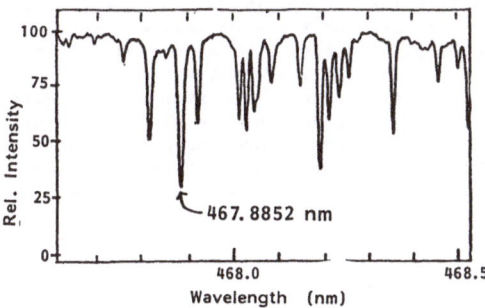

Figure 4. Solar spectrum showing the iron Fraunhofer line at 467.8852 nm. (Uttrecht Solar Atlas)

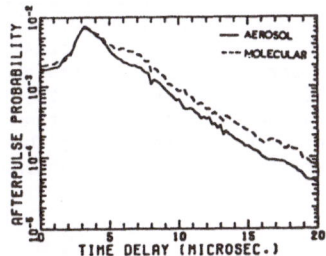

Figure 5. Probability/(.2 μsec) of an afterpulse for a single photoelectron pulse measured for HSRL phototubes (RCA C31024)

313

into the receiver and the absolute pressure in the laser etalon and grating chambers is adjusted. The pressure is selected such that transmission of light from the 467.8852 nm iron line is maximized. Because the pressure differences between all etalon chambers are held constant while the laser pressure is changed, the receiver maintains white light and laser tune at the new wavelength. This iron line was selected as a tuning reference because iron produces a deep Fraunhofer line (Fig. 4) in the solar spectrum; this decreases sky brightness by approximately a factor of 3 and therefore decreases background light in HSRL measurements.

HSRL DATA

Figure 6 shows data collected with the HSRL system operated from an Electra aricraft. Analysis of this data has required detailed modeling of the HSRL system. It has been found necessary to correct lidar profiles for the effects of ion feedback in the photomultiplier tubes. Ion feedback generates "afterpulses" which appear with a small probility several microseconds after a photoelectron generated pulse (Fig. 5). It has also been found that the effective spectral bandpass of the HSRL receiver changes slightly with range to the backscatter volume. This effect which is due to changes in the mean angle at which photons are incident on the spectrometer plates must be modeled in order to recover aerosol scattering parameters. It appears that small changes in the HSRL geometry will substantially reduce this sensitivity.

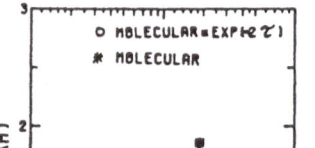
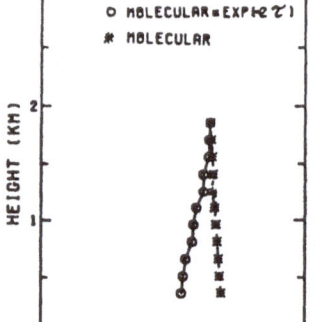
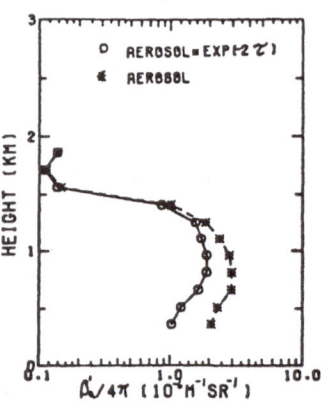

Figure 6. Height profiles of molecular and aerosol backscatter cross sections obtained with the HSRL between 18:38 and 18:51 EDT on 7 Aug. 1980. The dashed line on the left is the molecular backscatter cross section for a summer, midlatitude atmospheric model by McClatchey, et al. (1970). The open circles on the left show the observed molecular return normalized to the McClatchey model at a height of 2.2 km. A direct measure of atmospheric extinction is provided from the decrease of the HSRL molecular measurements with increasing range from the aircraft. The open circles on the right graph show the aerosol return before correction for attenuation. The dashed line on the right is the aerosol backscatter cross section corrected for the two way attenuation measured from the molecular signals on the right

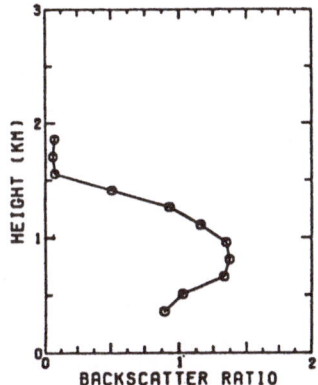 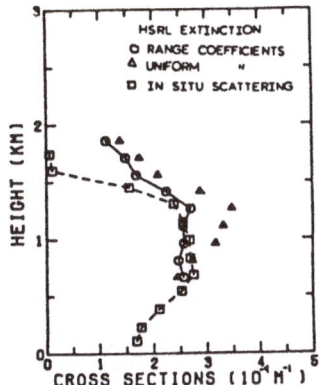

Figure 7. The ratio of aerosol to molecular backscattering measured with the HSRL is shown on the left. On the right a height profile of the extinction cross section measured with the HSRL, O, is compared with simultaneous in situ nephelometer measurements, □. Also shown is the HSRL derived extinction profile, ▲, calculated without correction for the range dependence of the spectrometer passbands

Refrences

1) Fiocco, G., and J.B. DeWolf, J. Atm. Sci., 25, 488-496, (1968).
2) Mack, J.E., D.P. McNutt, F.L. Roesler, and R. Chabbal, App. Opt., 2, 873-885, (1963).
3) Lawler, J.E., Fitzsimmons, W.A. and L.W. Anderson, App. Opt., 15, 1083-1090, (1976).
4) Shipley, S.T., D.H. Tracy, J.T. Trauger, E.W. Eloranta, M. Green, R.J. Parent, F.L. Roeseler, J.A. Weinman, "Interim progress report laser radar for meteorological measurements" NASA contract NAS1-14136 (1977).

Part 8

Lidar Technology

8.1 Lidar Measurements of Clouds

Allan I. Carswell

Department of Physics and CRESS, York University, 4700 Keele St.
Downsview, Ontario, Canada

Introduction

Clouds present one of the most difficult challenges in the quantitative description of radiative transfer in the atmosphere. They occur widely in the atmosphere and have a great diversity of physical characteristics such as their position, extent, water content, droplet size, ice content, etc. Optically they present a broad range of properties which are generally described in terms of parameters such as, extinction, reflectance and emissivity. One of the dominant characteristics of all clouds is their large variability in both space and time. It is this aspect which makes an accurate modelling of their optical properties in any radiative transfer calculations very difficult - particularly since cloud optical data with good spatial and temporal resolution are still quite limited.

With the development of lidar systems, however, we have available a remote sensing method with excellent space and time resolution so that the optical properties of clouds can now be probed in a manner not previously possible. In general, lidar returns from clouds are very large because of the large optical scattering cross-section of cloud droplets (typical droplet radii are in the 1 to 10μm range) at the lidar wavelength employed (usually between 0.3 and 10.6μm). Lidar can be used to measure not only the denser tropospheric water clouds but also the very high altitude subvisual cirrus clouds whose importance has only been recognized in recent years. In dense clouds, multiple scattering is important and this tends to degrade the lidar information by making the return more difficult to interpret. However, even in this case the lidar offers new possibilities for investigating the importance of multiple scattering to radiative transfer in clouds. The ability to control the source and receiver characteristics in lidars offers new capabilities for the quantitative investigation of cloud properties. In our group we have been utilizing such measurements for a number of years and in this paper some of the recent findings are summarized.

A special interest in our work has been the investigation of the polarization information of lidar returns. It is well known that an arbitrarily polarized light beam can be described in terms of four parameters, which can be written as the components of a four-vector, the Stokes vector of the wave.[1] The lidar return can thus be represented as a Stokes vector which is related to the Stokes vector of the transmitted wave by a linear transformation. This transformation can be described by a 4 x 4 matrix, the scattering matrix of the atmosphere, which contains a complete description of the scattering process in terms of atmospheric variables. By employing a number of polarization states for the transmitted wave and by measuring the four Stokes components of the backscattered signal, it is possible to determine all elements of this scattering matrix. As yet such a complete polarization analysis of lidar returns has not been undertaken.

In our work we have made extensive lidar polarization measurements using linearly polarized lidar beams. One feature of great use in lidar applications is the fact that the backscattering of linearly polarized radiation from perfectly spherical particles retains the incident polarization. Thus the single scattering by spherical droplets in the backward direction is linearly polarized. Any asphericity in the scatterers introduces a cross-polarized component and this can be used to distinguish irregular particles and ice crystals from spherical liquid droplets. When multiple scattering occurs, the backscattering reaching the lidar results from a number of sequential scattering events. Even with perfect spheres a cross-polarized component is introduced from the scattering at angles other than 180°. As a result, in water-droplet clouds the cross-polarized backscattered signal is a measure of the multiple scattering contribution.

In lidar studies of multiple scattering in clouds it is also useful to employ spatial filtering techniques along with the polarization measurements. By using focal-plane field stops in the lidar receiver, it is possible to totally block the singly-scattered return. In this way pure multiple scattering can be observed and its polarization characteristics measured. By using such spatial filters along with the range resolution of the lidar it is possible to observe the three-dimensional evolution of the lidar beam as it penetrates the cloud.

Equipment

For our experimental program of cloud measurements we employ both a field lidar system and a laboratory scattering facility. The field lidar has been designed to be a versatile research system and has been developed over a number of years to include many useful features for studying lidar cloud returns. It is a two wavelength (694 and 347nm) ruby system with linearly polarized outputs at both wavelengths. It has four receiver telescopes co-aligned with the transmitted beam and each receiver channel is equipped with a narrow-band interference filter, interchangeable neutral density filters, and polarization optics ($\lambda/4$ plate and linear polarizer) in rotatable mounts so that a polarization analysis of the backscattered signal can be made. Each receiver has a six-position slide in the focal plane so that field of view stops and other spatial filters can be rapidly interchanged during a measurement sequence. With this four-channel receiver system it is possible to make simultaneous measurements of several quantities in the same cloud scattering volume. (For example, it is possible to measure all four components of the Stokes vector for every transmitted pulse.) This is an important feature since sequential measurements are often of little value because of the rapid spatial and temporal changes in the clouds.

In the laboratory we make use of several cloud chambers which permit measurements of the scattering of a laser beam in both the forward and backward directions. The chambers are equipped with diagnostic instrumentation to provide information on the physical properties of the clouds (water content, droplet size distribution and number density). Using cw laser sources (argon, helium-neon and Nd:YAG) measurements have been made at a number of wavelengths between about 440 and 1060 nm. The optical receivers are equipped for polarization measurements and have variable entrance apertures and focal-plane field stops. Ultrasonic nebulizers are used for generating the water droplet clouds which have a size distribution with a modal radius of 2.5μm and a range of radii between about 1 and 10μm. Cloud extinction coefficients of up to 10 m^{-1} and water contents of up to about 10 gm/m^3 are obtainable and laser beam path lengths in the cloud can be varied between about 0.5 and 4 m. The systems permit study of clouds from a very low density up to densities where multiple scattering effects are very large.

Measurements

It is possible using lidar backscattering information to measure a number of cloud parameters of interest. Some of these applications have been described in recent review articles.[2-4] The position, size and extent of cloud coverage can be measured with lidar most readily and this capability has been shown by many lidar systems. To date, however, no operational use is made of lidar for this purpose. For cloud height measurements, the capability of lidar ceilometers has been well demonstrated but as yet they are employed only to a limited extent.

The opacity of clouds as measured by their optical extinction is a very important parameter. Typical clouds can exhibit extinction coefficients over a wide range, (from about 0.1 to 100 km^{-1} in the visible and near IR). This extinction can be measured using lidar methods [3,4]. For thin clouds the measurement accuracies are limited mainly by the spatial variations in the clouds. For dense clouds the interpretation of the lidar data is more difficult because of the limited beam penetration and the multiple scattering. By using polarization information with our lidar we have developed methods whereby the multiple scattering contribution to the extinction coefficient can be measured.[5,6] With multiple scattering, replenishment of the propagating beam occurs and the effective extinction coefficient in the cloud is reduced. Measurements of this variation have been made and show as expected a strong dependence on the field of view of the system.[7,8] Klett[9] has recently described a simple analytical method for inverting the lidar equation to deduce the extinction coefficient in an inhomogenous atmosphere. Although the method assumes the applicability of the single scattering lidar equation, it appears to work quite well even at optical depths much above unity as shown by the recent work of Evans et al.[10,11] Most of the cloud extinction measurements to date have been made in the visible but recent work[12] indicates an increasing interest in measurements out to 10.6μm in the infrared.

Cloud reflectance is another parameter of prime importance which can be measured by lidar systems. This parameter can be expressed in a variety of ways depending on the nature of the incident illumination and on the geometry of the receiver.[13] For lidar systems the volume backscattering coefficient, β, (which is the proportion of the incident energy backscattered per steradian per unit path length) is the most directly measured quantity [3,12]. It is also often useful to express cloud reflectance in terms of the total hemispherical reflectance ρ. For a diffuse Lambertian reflecting cloud, ρ is given by the equation[14,15]

$$\rho = (\pi P_R R^2)/(P_{in}A) \quad ,$$

where P_{in} and P_R are the incident and reflected powers at a cloud surface located a distance R from a lidar with receiver collecting area A.

In our laboratory system ρ can be easily measured by comparing the backscattered signal from the cloud with the reflection from a standard Lambertian reflector. Results of such measurements have been reported earlier [14,15] for visible radiation and in Fig. 1 we show recent measurements made at 1.06μm. The curve of Fig. 1 shows the reflectance of a water droplet cloud (size distribution shown in Ref. 7) up to very high droplet number densities. Such reflectance data are available for a several receiver fields of view and for various cloud depths.[16] For dilute clouds (low attenuation) the reflectance increases rapidly with increasing number density of droplets. At high densities however, the reflectance becomes essentially independent of the cloud optical density apart from a slowly increasing multiple scattering contribution. In this region the cloud has reflecting properties similar to a solid surface with deeper levels contributing very little to the backscattering. Range-resolved measurements of cloud reflectance with lidar and in the laboratory clearly show this behaviour.[5,6,8]

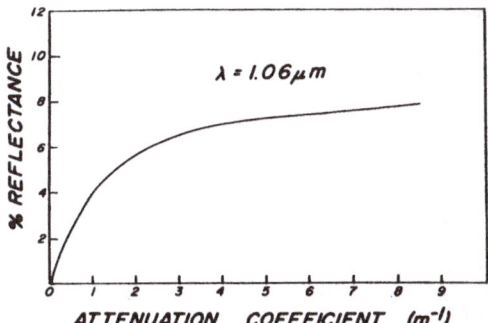

Figure 1.   Cloud reflectance as a function of the cloud attenuation coefficient measured at a wavelength of 1.06μm. with a 10 millirad. field of view

Our measurements have shown that the backscattering from water droplet clouds has some very interesting polarization characteristics. In our studies the linear polarization of the incident beam is strongly preserved in the backscattered signal.[5,6,15,17] The reflection is not depolarized as one might expect from a Lambertian type reflector. Moreover, we have found that the polarization of the cloud back-scattering is spatially anisotropic.[8,18] The multiply-back-scattered component has preferred polarization directions which depend on the position of the scattering volume within the field of view of the receiver. A sample of the backscattered intensity distribution from a laboratory cloud is shown in Fig. 2. In this figure the incident beam polarization is vertical and the photograph is taken through a crossed (horizontal) polarizer. As the polarizer is rotated the cross-shaped pattern also rotates. The spatial intensity distribution of this backscattered polarization field is strongly dependent upon the size parameter (particle radius/scattered wavelength). Since this spatial distribution is imaged in the focal plane of the lidar receiver, this means that focal plane spatial filters can be used to sample the polarized intensity distribution in different regions of the backscattered field. In this way it should be possible to obtain quantitative information on the cloud droplet size from back-scattering measurements at a single wavelength.

We have undertaken such measurements with our lidar system recently and in Fig. 3 is shown the type of spatial filters used. The

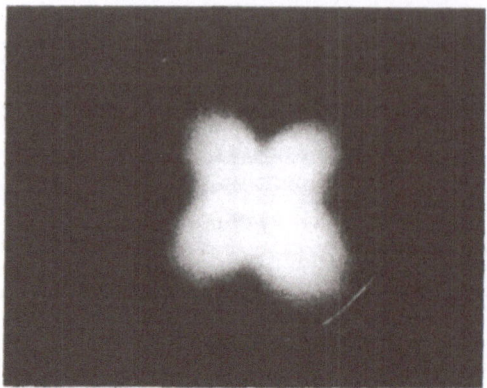

Figure 2.   Photograph of backscattering from cloud illuminated
with linearly polarized laser (polarization direction
vertical) Camera polarizer was horizontal

Figure 3.   Sector spatial filters used in the lidar to measure
the anisotropy of the multiple scattering.   Sectors
are set from 0 to 90° with respect to incident
polarization direction and the central 5 millrad. field
of view is always blocked

sector geometry is chosen such that all single scattering (the central
region) is blocked.   The distribution of the multiple backscattering
around the lidar beam when viewed through a polarizer is measured by
sequentially introducing the sector masks.   In Fig. 4 are shown sample
results from low level cumulus clouds.   In this figure the normalized
intensity ratios (power detected with sector at angle $\emptyset$/power detection
with sector at zero degrees) are shown as a function of the angle at
two pulse penetration depths in the clouds.   The zero direction of the
sector is chosen to be along the direction of the incident linear
polarization.   A number of data samples are shown to indicate the
scatter in the data caused by the changing cloud properties.   Even
with the scatter, however, the data show enhanced scattering in the
90° direction at the cloud base, indicating a predominance of small
particles (size less than about 1 or 2µm).   A more detailed description
of this work will be published elsewhere.

The liquid water content of a cloud is a parameter of considerable
importance which may be obtainable by lidar measurements.   Recent work
by Chylek, Pinnick and co-workers [19],[20] have shown that a linear
relation exists between IR extinction, and liquid water content of
fogs under some circumstances.   They have also shown that a linear
relationship can exist between backscattering and mass content.[21]
Their results have been derived from computations based on measured
size distribution of droplets in the atmosphere.

In our work recently we have been investigating the extent to
which their analysis can be applied to lidar measurements.   In the
laboratory we are measuring simultaneously the extinction, backscattering
and water content of clouds at several wavelengths.   A constant size

322

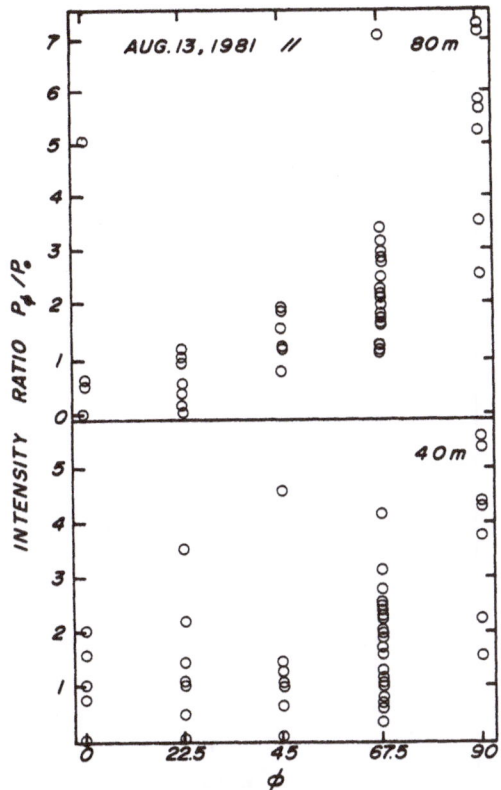

Figure 4. Spatial anisotropy measured in cumulus cloud at two
different penetration depths (40 and 80 m) for
parallel orientation of receiver polarizer. Increase
of intensity towards 90° direction indicates predominance
of small particles

distribution is being maintained while the droplet concentration is
varied over a wide range. In Fig. 5 is shown the measured variation
of the extinction coefficient with the water content at two wavelengths.
It is seen that a linear relationship does exist even up to very large
cloud densities.

Theoretically the shape of this curve is expected to be dependent
upon the droplet size distribution and this aspect will be investigated
in our future measurements. However, even with this variation it may
be possible to determine useful empirical relations between the lidar
observables (extinction, backscattering) and the water content. This
possibility is demonstrated in Fig. 6 in which our curves are plotted
on those of Ref. 20. Although the match to the simplified theoretical
model is not good, all of the data based on measured values is relatively
well grouped together. Making use of the backscattering for water
content measurement may be more difficult because of the very large and
rapid variation of backscattering cross-section with droplet size.

Another area of lidar application to cloud measurements is in the
determination of ice/water mixing in clouds. In our work we have shown

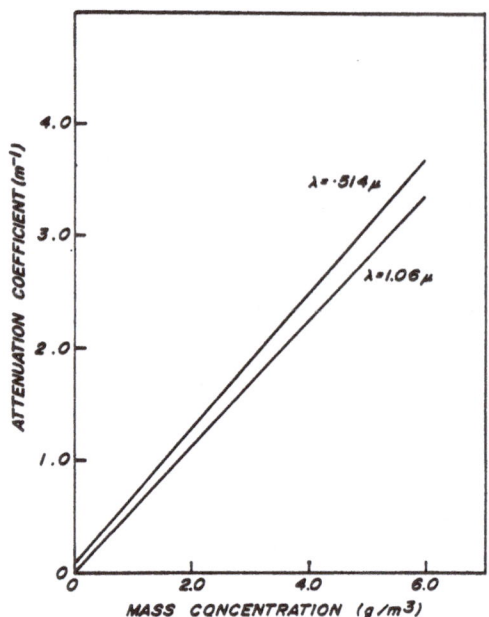

Figure 5.   Measured variation of cloud attenuation with water content
             at two wavelengths (0.514 and 1.06µm)

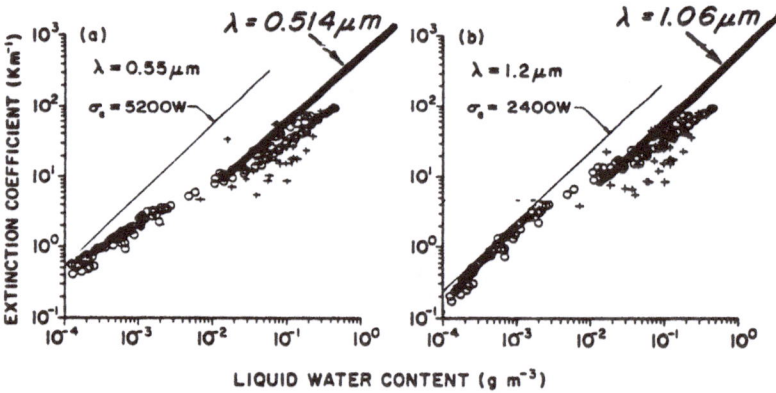

Figure 6.   Comparison of our measured variation (heavy lines) with
             calculated data of Ref. 20

how polarization information can aid in discrimination between the ice
crystals and water droplets [2,22,23] and others have applied similar
techniques.[24] It is also possible to use lidar polarization information
to observe ice crystal orientation in clouds [26]. Although these
measurements have demonstrated the potential of lidar for ice/water
discrimination in clouds much remains yet to be done to reduce the
measurements to operational practice.

CONCLUSION

Our studies of cloud diagnostics with lidar to date have
indicated the value of lidar measurements for providing new data on
atmospheric clouds. A number of very useful techniques have been
demonstrated by several groups working in this field but there is
a need for much more observational data. Information is particularly
scarce in the infrared but increased emphasis recently on CO_2 lidar
applications should improve this situation.

REFERENCES

1.   D. Clarke and J. F. Grainger "Polarized Light and Optical
   Measurement" Pergamon (1971)

2.   A. I. Carswell, pg 363 in "Clouds: Their Formation, Optical
   Properties and Effects" edited by P.V. Hobbs and A. Deepak,
   Academic Press, 1981.

3.   E. D. Hinkley, ed. "Laser Monitoring of the Atmosphere"
   Springer-Verlag, (1976)

4.   A.I. Carswell, Can. J. Phy. 61 No.2 (1983)

5.   S. R. Pal and A. I. Carswell. Appl. Opt. 12, 1530, (1973)

6.   S. R. Pal and A. I. Carswell, App. Opt. 15, 1990, (1976)

7.   J. S. Ryan and A. I. Carswell, J. Opt. Soc. Amer. 68, 900
   (1978).

8.   A. I. Carswell, AGARD Conf. Proc. CP300 Monterey Apr.1981

9.   J. D. Klett, Appl. Opt. 20, 211, (1981)

10.   B. T. N. Evans, Proc 11th International Laser Radar Conf.,
    NASA Conf. Publ. CP 2228 Madison (1982) p,152.

11.   R. E. Kluchert, B. T. N. Evans, J.D. Houston, ibid p.120

12.   Proc. 11th International Laser Radar Conf. NASA Conf. Publ.
    CP2228 (1982)

13.   F. E. Nicodemus, J.C. Richmond and J. J. Hria, NBS Monograph
    160 (1977)

14.   S. R. Pal, J.S. Ryan and A. I. Carswell, Appl. Opt. 17
    2257 (1978)

15.   J. S. Ryan, S. R. Pal and A. I. Carswell, J. Opt. Soc. Amer.
    69, 60 (1979)

16.   A. I. Carswell, Proc. 11th International Laser Radar Conf.
    NASA Conf. Publ. CP 2228, p 236 (1982)

17.   W. R. McNeil and A. I. Carswell, Appl. Opt. 14, 2158 (1975)

18.   A. I. Carswell and S. R. Pal, Appl. Opt. 19, 4123 (1980)

19.   P. Chylek, J. Atmos. Sci. 35, 296, (1978)

20. R. G. Pinnick, S. G. Jennings, P. Chylek and H. J. Auvermann
    J. Atmos, Sci. $\underline{36}$, 1577, (1979)

21. R. G. Pinnick, S. G. Jennings and P. Chylek, J. Geophys.Res.
    $\underline{85}$, 4059 (1980)

22. J. D. Houston and A. I. Carswell, Appl. Opt. $\underline{17}$, 614, (1978)

23. S. R. Pal and A. I. Carswell, J. Appl. Meteor. $\underline{16}$, 70,(1977)

24. K. Sassen, J. Rech. Atmos. $\underline{11}$, 179 (1977)

25. V. E. Derr, N.L. Abshire, R.E. Cupp and R. T. McNice
    J. Appl. Meteor. $\underline{15}$, 1200, (1976)

26. C. M. R. Platt, N.L. Abshire and R. T. McNice, J. Appl. Meteor.
    $\underline{17}$, 1220, (1978)

8.2 Coherent IR Radar Technology[*]

A.B. Gschwendtner, R.C. Harney, and R.J. Hull
Massachusetts Institute of Technology, Lincoln Laboratory
Lexington, MA 02173, USA

INTRODUCTION

For over 15 years Lincoln Laboratory has been engaged in developing coherent infrared radar systems and relevant technology. The first major program was the Firepond laser radar. This program was initiated in the late 1960's and drew heavily on earlier CO_2 laser and infrared detector research at Lincoln Laboratory. Primary emphasis of the program was on satellite tracking and identification. The most recent version of this system was completed several years ago and is still operational.[1,2] In a subsequent section a description of the Firepond radar will be presented and its more important accomplishments will be summarized.

In the mid-1970's another coherent infrared radar program was begun in the area of tactical military applications.[3] Several infrared radar systems have been constructed, the most sophisticated of these being the transportable measurements radar. Initially, the work was aimed at verifying many of the concepts used in early systems studies. To this end a testbed laser radar was built. This system employed a 1W CO_2 laser transmitter, a single-element HgCdTe detector, and a 10 cm aperture transmit/receive telescope. Details of this system and representative results from its operation have been thoroughly desribed in previous publications.[4-7] Concurrent with operation of the testbed radar a significant program of technology development was undertaken. Successful development in the areas of lasers, detectors, telescopes, and laser beam shapers allowed us to proceed to the next phase of the program: construction of a transportable radar system sized according to the results of our systems studies. The most important elements of this system have been assembled and numerous field measurements have been conducted. In later sections of this paper the technology developments are related, the design of the transportable system is described, and representative experimental results are presented. Subsequent to construction of the transportable radar, a cw Doppler radar was built. This system will be described and recent Doppler imagery will be shown.

[*]This work was sponsored by the Department of the Air Force.

The work to date has dealt with conventional "hard" targets of military interest. However, the technology developed in these programs is applicable to remote sensing systems. A baseline coherent DIAL system is described, its performance analyzed, and the technology required is compared with that of the transportable system.

FIREPOND LASER RADAR

The Firepond Laser Radar is an automatic angle, range, and Doppler frequency tracking coherent IR radar operating at a wavelength of 10.6 μm (Fig. 1). It is the primary instrument at the

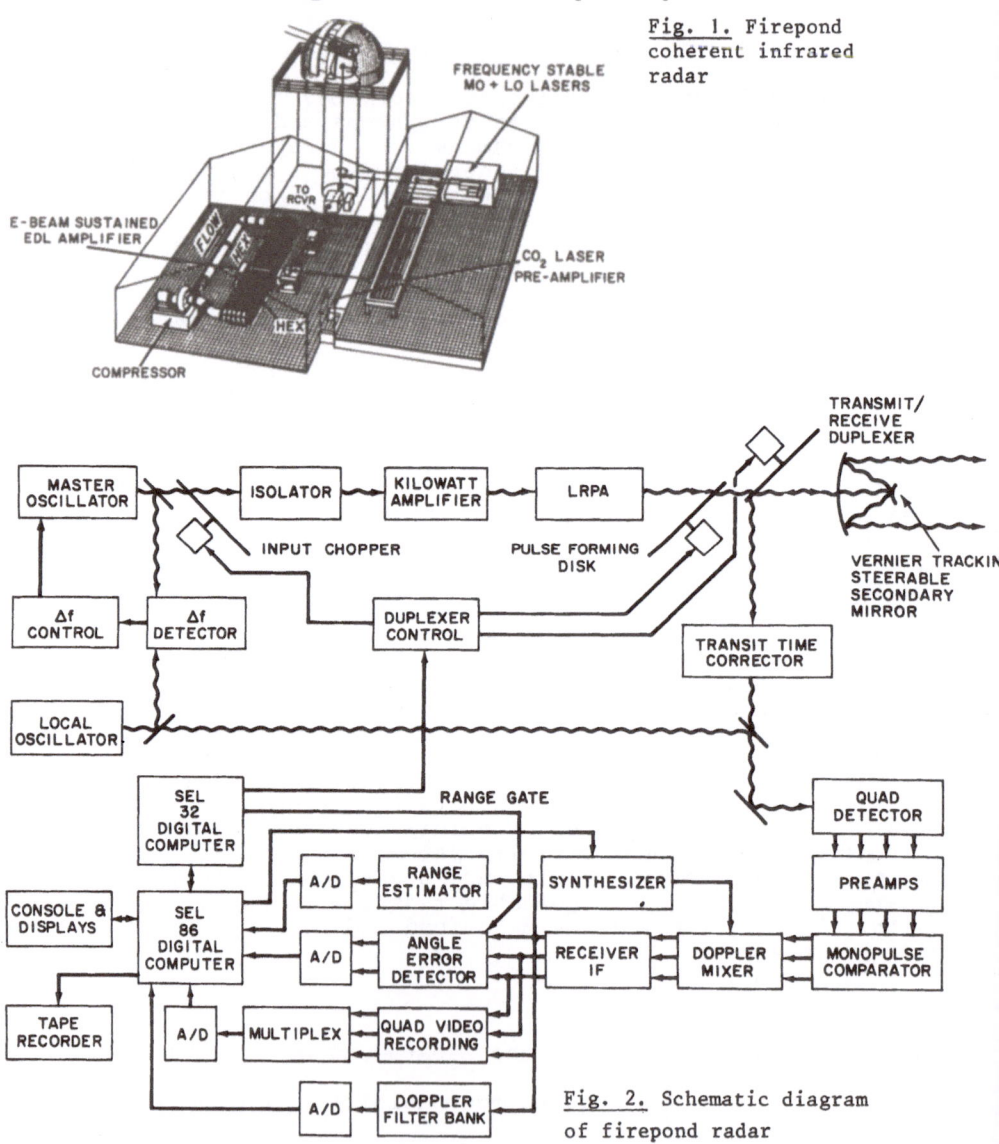

Fig. 1. Firepond coherent infrared radar

Fig. 2. Schematic diagram of firepond radar

TABLE 1

Characteristics of the Firepond Coherent IR Radar

Wavelength	10.6 µm
Aperture	1.2 meters
Beamwidth (λ/D)	~10 µrad
Rayleigh Range (D^2/λ)	100 km
Beam Diameter (at 1000 km)	10 meters
Frequency	3×10^{13} Hz
Doppler Coefficient	2 KHz/cm/sec
Maximum Doppler Frequency (for low earth orbit satellite)	1.4 GHz
Power (peak)	1 to 10 kW
PRF Continuously Variable	20 to 200/sec
Pulse Width Continuously Variable	10 to 1 millisec

Firepond IR facility, located at the Millstone Hill/Haystack microwave radar complex in Westford, Massachusetts. The characteristics of the radar are summarized in Table 1. A simplified block diagram of the infrared radar system in its long-range satellite tracking configuration is shown in Fig. 2.

The Firepond radar has demonstrated a number of significant capabilities. Among the accomplishments of the program are:

- development of a high-frequency-stability, long-range, coherent infrared radar

- demonstration of monopulse angle tracking of enhanced and unenhanced satellite targets

- utilization of high-PRF, short-range mode to perform horizontal wind measurements

- determination of target center-of-gravity motion and rotational dynamics using Doppler measurements capability

- generation of one-dimensional pseudoimages from narrowband time-Doppler-intensity data.

A typical time-Doppler-intenstiy plot of an orbiting Agena-D rocket body obtained at 1350 km range is shown in Fig. 3. Such data clearly indicate the power and sophistication of the Firepond radar.

TECHNOLOGY DEVELOPMENT FOR COMPACT INFRARED RADARS

The Firepond radar is quite large (occupying several rooms). In the Infrared Airborne Radar program compact radars capable of being carried on attack aircraft, helicopters, or ground vehicles

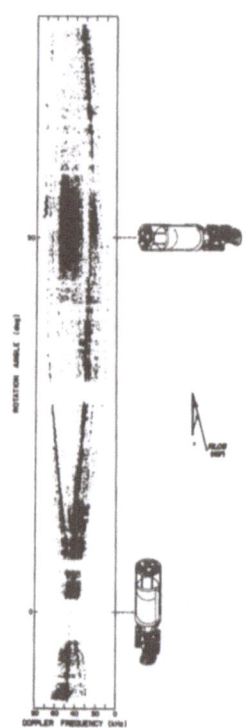

◄Fig. 3. Time-doppler-intensity plot of AGENA-D body
(1350 km)

▼Fig. 4. Compact transmitter and local oscillator
lasers

are being developed. The transportable measurements radar is one
of the major developments of this program.

Several key technology developments had to occur before
construction of the transportable radar was possible. The first
of these concerned the transmitter laser. This laser had to be
small, lightweight, rugged, and efficient as well as being
capable of both pulsed and cw operation with 5-10 W average
power, better than 100 kHz frequency stability in the cw mode,
and narrow pulsewidths (~300 nsec) and high pulse repetition frequency (20-50 kHz). A photograph of the laser finally
constructed is shown in Fig. 4. This system incorporates both
the transmitter and local oscillator lasers into a single
package. Typical performance is shown in Table 2.

Obtaining the high-frame-rate imagery desired of the
transportable system required the use of a large array of heterodyne detectors. To this end we fabricated a 12-element, linear
HgCdTe detector array (Fig. 5). The performance and uniformity
of response of this array are shown in Fig. 6.

Because image plane scanning is employed in conjunction with
heterodyne detection, the transmit/receive telescope had to have
a wide, diffraction-limited field-of-view (>10 mrad).
Furthermore, to facilitate detection of Doppler shifts due to
low-velocity targets, the telescope had to have minimal
backscatter. An eccentric-pupil Ritchey-Chretien design was
chosen and fabricated using diamond turning techniques. The
finished telescope is shown in Fig. 7.

330

TABLE 2

Laser Performance

Transmitter Power	6 W TEM_{oo}
L.O. Power	3.5 W TEM_{oo}
PRF	10–100 kHz
Short Term Stability	<50 kHz
Weight	20 kg
Volume	0.06 m^3
Power Consumption (including ballast)	350 W

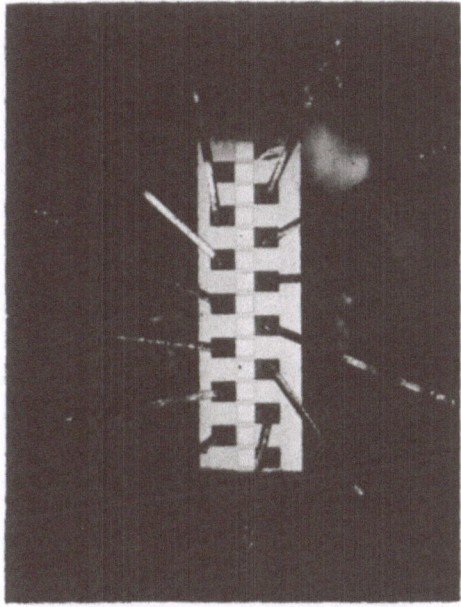

Fig. 5. Microphotograph of the 12-element HgCdTe detector array

CO_2 LASER SCAN OF 12-ELEMENT LINEAR ARRAY

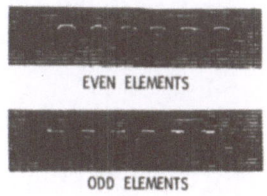

EVEN ELEMENTS

ODD ELEMENTS

12-ELEMENT LINEAR ARRAY

QUANTUM EFFICIENCY (LOW FREQUENCY)	η = 75-80% at 10.6 μm
RESPONSIVITY	R = 6.8 AMP/WATT
MINIMUM DETECTABLE POWER (10 MHz)	8×10^{-20} W/Hz
REQUIRED LOCAL OSCILLATOR POWER	250 μW

Fig. 6. Performance of the 12-element array

Finally, utilization of a linear detector array requires the target plane to be illuminated by a fan beam. A combination (see Fig. 8) of near-field phase modification (achieved by a transparent phase plate with π phase reversals at the zeroes of sinx/x in one dimension) and anamorphic expansion (achieved by a two-prism beam compressor in the near-field) allows the Gaussian input beam to be transformed into a relatively uniform intensity fan beam with the appropriate 12:1 aspect ratio and with little laser energy falling outside the area viewed by the detectors. A liquid crystal photograph of the shaped beam is shown in Fig. 9.

Fig. 7. Eccentric-pupil Ritchey-Chretien telescope

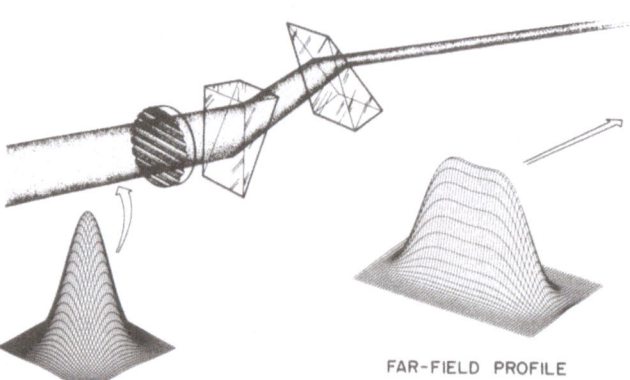

LASER PROFILE

FAR-FIELD PROFILE

Fig. 8. Phase plate-pluse-compressor beam shaper

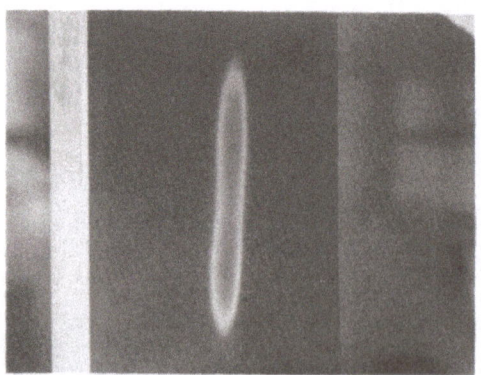

Fig. 9. Liquid crystal photo-graph of the shaped beam

THE TRANSPORTABLE RADAR

A schematic diagram of the transportable radar incorporating these technology developments is shown in Fig. 10. With this system, whose parameters are given in Table 3, we hope to demonstrate automatic moving target detection, real-time target tracking, high frame-rate, high resolution imaging for target identification, and real-time quasi-three-dimensional imaging.

The electronics package for the transportable radar is summarized in Table. 4 Most of the items are self-explanatory. The MTI processor, however, requires further elaboration. For the output of one of the detectors in the array we have decided to implement the processor of Fig. 11. The output from the IF electronics is first mixed with the output of a tunable electronic local oscillator whose frequency is controlled by a device which measures the platform velocity. In an airborne system this might be accomplished by an inertial navigation system or by the infrared radar itself. The platform-motion-compensated signal is

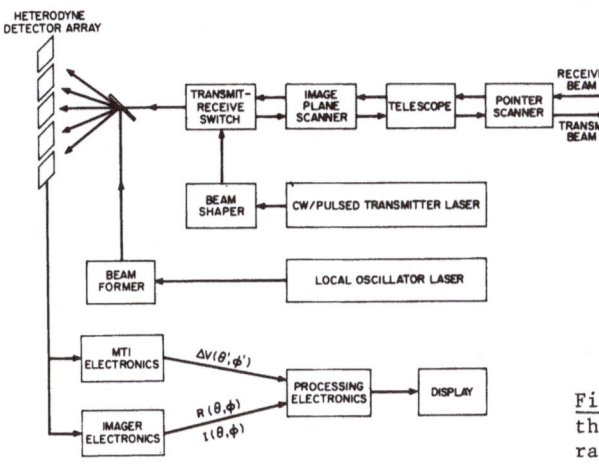

Fig. 10. Schematic diagram of the transportable infrared radar

TABLE 3

Transportable Radar Parameters

Laser Power	8 W (cw) 3 W (average)
Laser PRF	19 kHz
Number of Detectors	12
Telescope Aperture	13 cm
Instantaneous FOV	0.2 mr
Imaging FOV	12 mr (az) x 12 mr (el)
MTI FOV	± 0.25 rad
Estimated Maximum Range (3 dB/km attenuation)	3 km

TABLE 4

Transportable Radar Electronics

A/D Converter	8 Bits Range/8 Bits Intensity
Range Quantization	30 m or 6 m / Digital Count
Image Size	128 (horiz) x 60 (vert)
Real-Time Displays	
Boresighted LLLTV	Monochrome
Radar Image	Monochrome (Intensity Only)
Digital Recording	
Radar Imagery	15 Frames/Sec
TV Imagery	1 Frame/Sec
Time, Temp, and Rel. Humidity	Once per Frame
Instant Tape Replay	
MTI Processor	SAW Transform System on Four Detectors

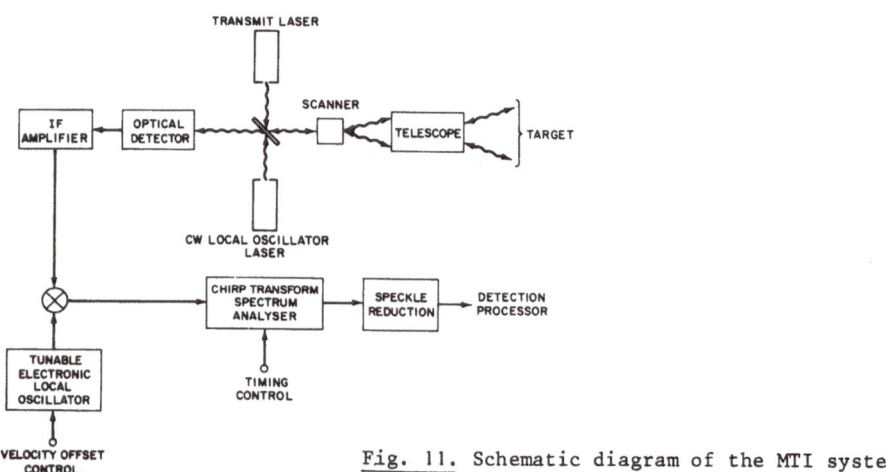

Fig. 11. Schematic diagram of the MTI system

sent into a chirp transform Fourier analyzer. This device, implemented with surface acoustic wave, reflective array compressor, dispersive delay lines, produces a Fourier transform of the signal received from each resolution element in the radar scan. Because fading due to speckle will occur in the MTI signals, adequate P_D/P_F performance cannot be achieved at reasonable signal-to-noise ratios on a single-resolution-element basis. However, the highly resolved nature of most targets allows the returns from a number of resolution elements to be spatially integrated. This speckle reduction by integration allows acceptable P_D/P_F performance to be achieved. Finally, signals in the nonzero velocity channels of the integrator output are threshold detected to determine the presence or absence of a moving target.

Fig. 12. Transportable radar in its measurements van

At the present time, the optical system and the imaging and data recording electronics are complete. Fig. 12 shows the system installed in its measurements van. The MTI processor (minus the speckle reduction integrator) has been assembled and incorporated into the cw Doppler radar mentioned earlier. At some later date it will be transferred to the transportable system. A variety of real-time target tracking and image processing electronics have been designed. However, as will be seen in the next section, much off-line image processing has been done using the recorded digital radar data.[7,8]

COMPACT RADAR MEASUREMENTS

The transportable radar has been operational for some time and has performed field measurements at a number of sites including Ft. Sill, Redstone Aresenal, and Camp Edwards. Figures 13-18 show representative data taken with this and other compact radars. Figure 13 shows testbed radar images of a tank and a truck taken at 2.7 km range indicating the target recognition potential of coherent infrared radars. Each of the images shown is an average of eight frames of 16,000 individual picture elements (pixels) each. Data rate in the testbed radar was one frame per second. Figure 14 shows an eight-frame-averaged testbed radar image of the Lincoln Laboratory Flight Facility at Hanscom AFB. Note the detail on the hangar roof (2.7 km distant) and the utility wires (1.8 km distant) running across the middle of the image. Many infrared radars yield range as well as intensity and az-el data. Using range/intensity color coding quasi-three-dimensional radar images can be displayed. Such a color-coded image of Fig. 14 may be found in Ref. 8. Figure 15 shows four consecutive frames of Doppler imagery of an automobile overtaking a bicyclist. These images were obtained with the cw

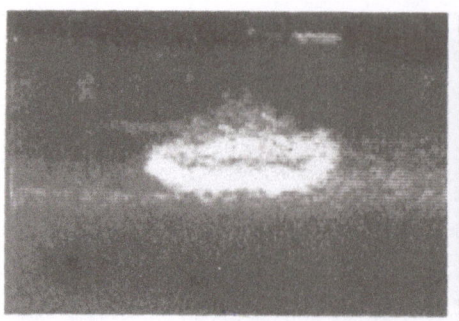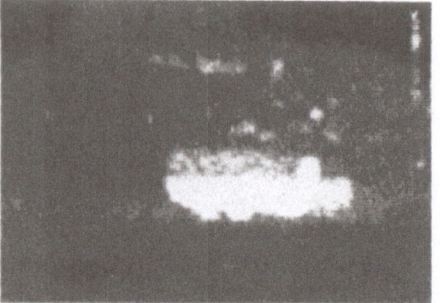

Fig. 13.a,b

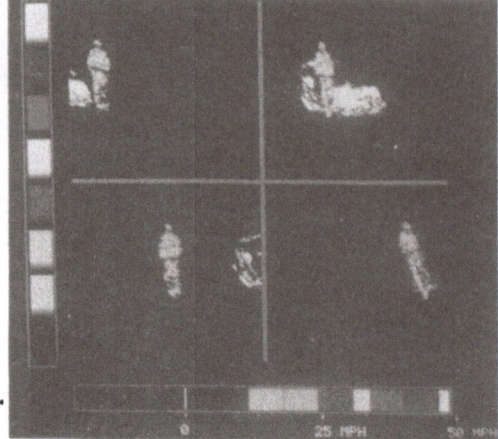

Fig. 14   Fig. 15.

Fig. 13. Testbed radar images of (a) tank and (b) truck at 2.7 km

Fig. 14. Testbed radar image of Lincoln Lab Flight Facility

Fig. 15. Doppler radar images of car and bicyclist

Doppler radar. A frame consists of 16,000 pixels, each made up of a velocity and intensity measurement with 0.5 kmph resolution over a ± 100 kmph acceptance bandwidth. For display purposes a small velocity subrange of 12-50 mph was isolated. Frame rate was 1 Hz. Figures 16-18 were taken with the transportable radar at a field site. Figure 16a shows a scene containing calibration plates (the image is video from the LLLTV) while Fig. 16b shows the same scene with a smoke curtain (of Fog Oil and IR2) between the plates and the radar. Figure 17a shows the radar intensity image (dark = weak return; bright = strong return) of the scene without smoke with 17b shows the radar range image (dark = close range; bright = far range). Note: the field-of-view of the radar images is only half the field-of-view of the video images. Figures 18a and b show the intensity and range images with the smoke curtain present. Note the return at a close and uniform range except for the calibration plate regions. This indicates the radar is detecting the radiation back-scattered from the smoke. This is a true signal and not noise due to excessive

336

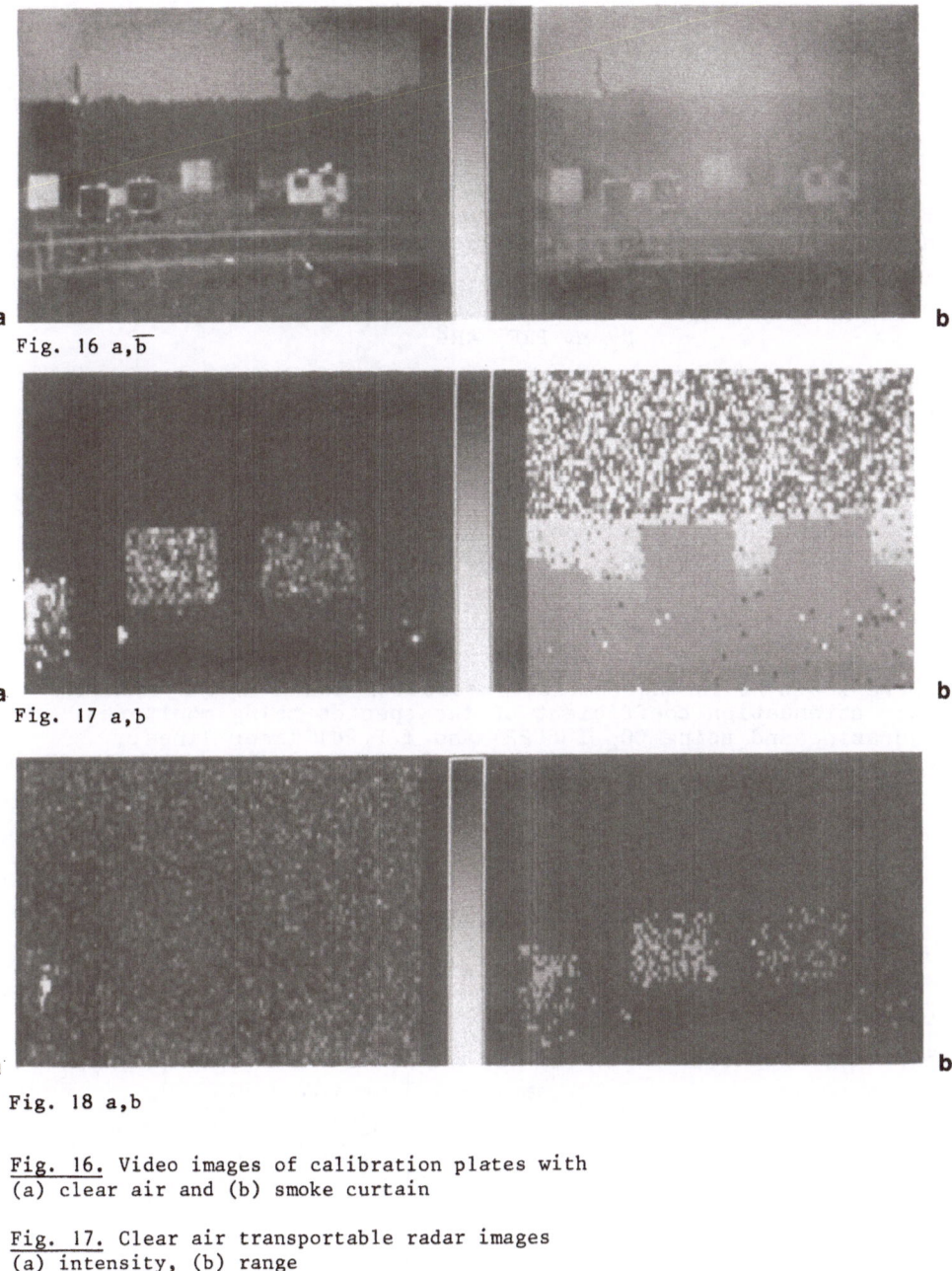

a

Fig. 16 a,b

a

Fig. 17 a,b

a

Fig. 18 a,b

b

b

b

Fig. 16. Video images of calibration plates with
(a) clear air and (b) smoke curtain

Fig. 17. Clear air transportable radar images
(a) intensity, (b) range

Fig. 18. Smoke curtain images (a) intensity, (b) range

attenuation in the smoke as can be ascertained by observing the top portion of Fig. 17b which is due to random noise.

APPLICATION TO RANGE-RESOLVED COHERENT DIAL

Let us consider a baseline differential absorption lidar (DIAL) system employing coherent detection. Reasonable parameters for this system might be those in Table 5. The carrier-to-noise ratio of a coherent dial system is given by

$$CNR = \frac{\eta \; P_{AV}}{N_D \; h\nu \; PRF} \; \frac{\pi D^2}{4R^2} \; \epsilon\beta(c\tau/2)e^{-2KR} \; ,$$

where β is the aerosol backscatter coefficient and K is the atmospheric extinction. For the parameters of Table 5, with $\beta = 10^{-7} \; m^{-1} \; sr^{-1}$ and $K = 3$ dB/km, we obtain

$$CNR = 1.9 \qquad .$$

The concentration measurement accuracy of a DIAL system which integrates the returns from N pulses on each laser is

$$\sigma_c = \left(\frac{N \; CNR/2}{1 + CNR/2 + 1/2 \; CNR} \right)^{-1/2} \frac{1}{\rho x}$$

where $x = c\tau/2$ is the spatial resolution and ρ is the relative mass attenuation coefficient of the species being monitored. For hydrazine and using CO_2 I P(22) and I P(28) laser lines, $\rho = 2.7 \times 10^{-4} \; m^2/mg$. In one second of integration with the baseline system we have N = 25,000 with a resulting sensitivity of

$$\sigma_c = 0.72 \; mg/m^3 = 600 \; ppb.$$

TABLE 5

Baseline Coherent Dial System Parameters

	Baseline	Transportable
Laser Power (P_{AV})	6 W	3 W
Pulse Repetition Frequency (PRF)	50 kHz	19 kHz
Pulse Duration (τ)	350 nsec	350 nsec
Number of Detectors (N_D)	1	12
Transceiver Aperture (D)	13 cm	13 cm
Efficiencies ($\epsilon\eta$)	0.05	0.05
Wavelength	I P(22) I P(28)	I P(20)
Range (R)	1 km	<3 km
Scan Mode	Staring	Raster-Scanned
Processing	Digitization & Recording	Peak Detection & Recording

Performance of this level (50 m range resolution and $\sigma_c < 1$ ppm) indicates a capability for range-resolved measurements of several atmospheric species.

Table 5 also lists the parameters of the transportable radar for direct comparison. Many of the parameters are identical. Of those that are not identical, in most cases the differences turn out to be trivial. For example, if the clock rate on the transportable laser's Q-switch pulser is increased from 19 kHz to 50 kHz, the average laser power increases to 6 W. The transportable radar has a one-detector mode of operation when the beam former and beam shaper are removed (a sample procedure). The local oscillator and master oscillator are grating tunable allowing selection of laser lines. Finally, although normally operated in a raster-scanned mode, the transportable system does have a switch-selectable staring mode. The only significant difference between the baseline DIAL and the transportable system is the electronics package.

CONCLUSIONS

The technology of compact coherent infrared radars is well developed. All significant components have been experimentally demonstrated although many are not yet commercially available. Using the available technology, a number of sophisticated coherent infrared radars have been built and a variety of measurements have been made. Finally, much of the technology and possibly even some of the existing systems have direct applicability in remote sensing.

REFERENCES

1. L. J. Sullivan, "Infrared Coherent Radar" in CO_2 Laser Devices and Applications, Proc. SPIE, Vol. 227, pp. 148-161 (1980).

2. L. J. Sullivan, "Firepond Laser Radar" Electro/81 Conference Record, Session 34 (1981).

3. R. C. Harney, "Military Applications of Coherent Infrared Radar", in Physics and Technology of Coherent Infrared Radar, Proc. SPIE, Vol. 300 (to be published).

4. R. J. Hull and S. Marcus, "A Tactical 10.6 μm Imaging Radar", Proceedings of the IEEE 1980 National Aerospace and Electronics Conference (NAECON), Vol. 2, pp. 662-668.

5. R. C. Harney and R. J. Hull, "Compact Infrared Radar Technology", in CO_2 Laser Devices and Applications, Proc. SPIE, Vol. 227, pp. 162-170 (1980).

6. R. C. Harney, "Infrared Airborne Radar", Proceedings of the IEEE 1980 Electronic and Aerospace Systems Conference (EASCON), pp. 462-471.

7. D. R. Sullivan, "Active 10.6 μm Image Processing", in <u>Imaging Processing and Missile Guidance</u>, Proc. SPIE, Vol. 238, pp. 103-117 (1980).

8. D. R. Sullivan, R. C. Harney, and J. S. Martin, "Real-Time Quasi-Three-Dimensional Display of Infrared Radar Images", in <u>Real-Time Signal Processing II</u>, Proc. SPIE, Vol. 180, pp. 56-65 (1979).

8.3 Tactical and Atmospheric Coherent Laser Radar Technology

Albert V. Jelalian, Wayne H. Keene, and Edwin F. Pearson

Raytheon Company, Equipment Division, Sudbury, MA 01776, USA

1. INTRODUCTION

Infrared laser radars constitute a direct extension of conventional radar techniques to very short wavelengths. Whether they are called Lidar (Light detection and ranging) or Ladar (Laser detection and ranging), they operate on the same basic principles as microwave radars. Because they operate at much shorter wavelengths, Lidars are capable of higher accuracy and more precise resolution than microwave radars. On the other hand, Lidars are subject to the vagaries of the atmosphere and are thus generally restricted to shorter ranges than microwave radar. Rather than supplanting microwave radar, Lidar opens up new applications that exploit the great shift in wavelength. Lidar applications include:

- Tactical imaging systems
- Missile guidance
- Aircraft guidance
- Clear air turbulence and severe storms sensors
- Fire control and line-of-sight command systems
- Remote sensing of atmospheric constituents
- Satellite tracking

Carbon dioxide lasers operate at wavelengths between 9 and 11.5 microns spanning much of the 8-12 micron atmospheric "window." They are very efficient coherent light sources with typical "wall plug" efficiencies of the order of 5-10 percent. These characteristics coupled with the recent emergence of rugged, compact CO_2 laser packages has greatly stimulated the development of versatile and diverse Lidar systems. Figure 1 illustrates the advance in CO_2 laser development over the past decade. Early CO_2 lasers were long glass tubes with heavy mechanical supports to maintain optical alignment. They required both large power supplies and a continuous supply of flowing gas. The latest devices are capable of sealed-off operation and are much smaller. In 1972, for example, a 15-watt CW CO_2 laser was about 1 meter long and was accompanied by a 1.8 meter high, 0.5-meter-wide equipment console that combined the power supplies, cooling unit and laser gas refilling source. Today a 5W, CW CO_2 waveguide laser is only 33 cm long and with its power supply weighs 2.3 kilograms.

In the following sections we will restrict our discussion to CO_2 laser radar systems. It should be noted that historically lasers with proper wavelengths for atmospheric propagation have been adapted for radar use almost as soon as they appeared. The

341

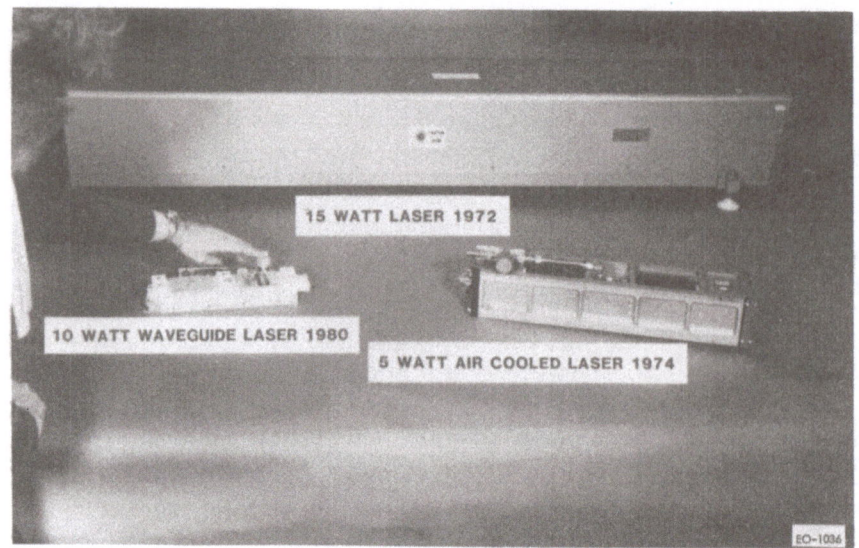

Figure 1. Progress in CO_2 Laser Development

first practical Lidar, a rangefinder demonstrated in 1962, employed
a ruby laser with output in the red portion of the visible spectrum
at 0.69 microns. At this wavelength, the air on a hazy day attenu-
ates the beam at a rate of 5 dB/km, seriously limiting the useful
range. The neodymium-doped YAG laser operating in the near infrared
was applied to radar use a year later. Although its wavelength
of 1.06 microns fits into a less absorptive atmospheric window,
its beam still experiences a 3 dB/km attenuation in haze. Nevertheless, several thousand YAG radars have been built, most of them
rangefinders or target designators. CO_2 lasers experience a 1 dB/km
attenuation in haze although the attenuation due to water vapor
can be higher in tropical environments. CO_2 laser-based systems
are not yet in mass production but their power, efficiency, wavelength and simplicity suggest a very bright future.

2. DETECTION METHODS

LIDAR echoes may be detected either coherently or incoherently
- that is, with or without phase reference to another optical
signal. Incoherent (direct) detection is simpler but coherent
detection yields substantially greater sensitivity as well as
increased information, principally Doppler velocimetry. Figure 2
compares the two types of receivers. The incoherent optical
receiver is analogous to the microwave video or envelope detector.
The significant difference is that in addition to the signal
power, P_S, the optical receiver sees an extra input, P_B, called
the optical background which could be emitted from sunlight,
atmospheric backscatter and flares. These external noise sources
compete with the desired signal, P_S. The total received power is
applied to a square law optical detector which produces a video
bandwidth electrical output.

342

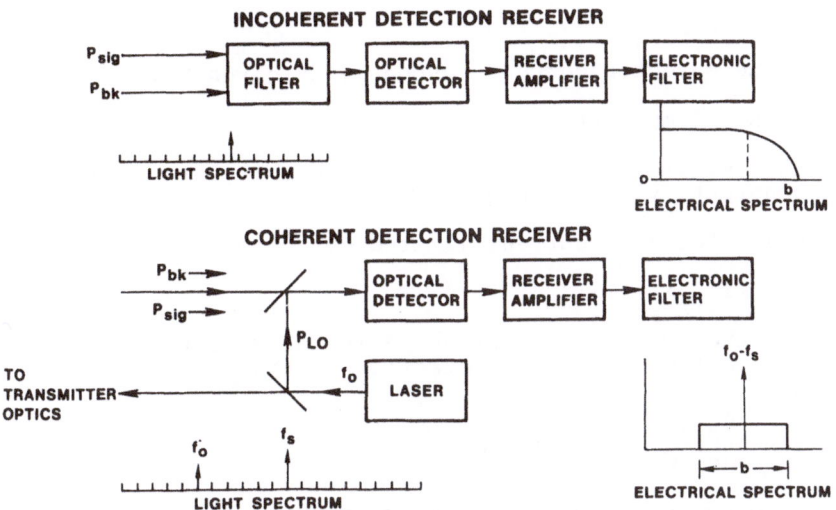

Figure 2. Receiver Systems

In the coherent receiver (homodyne or heterodyne), a local oscillator signal, P_{LO}, is coupled to the optical detector via beamsplitters. The spatial coherence property of the laser permits photo-mixing which is analogous to the signal-mixing in microwave and radio receivers. The matching of the phase fronts of the signal and the L.O. on the photo-mixer gives the coherent receiver antenna properties which determine the field of view. Because the L.O. power is set to be the predominant noise source, the system is typically background immune.

The power signal-to-noise ratio of a Lidar receiver can be calculated by dividing the mean square signal current, $\langle i^2_S \rangle$, by the sum of the mean square noise current terms. As given in Equation 1, the noise terms for most cases of interest include shot noise, i_{SN}, thermal noise, i_{TH}, background noise, i_{BK}, dark current, i_{DK}, and local oscillator-induced shot noise, i_{LO}

1) $$ S/N = \frac{\langle i^2_S \rangle}{\langle i^2_{SN} \rangle \langle i^2_{TH} \rangle + \langle i^2_{BK} \rangle + \langle i^2_{DK} \rangle + \langle i^2_{LO} \rangle} .$$

In the following discussion we will consider only two of the noise terms, i_{DK} for the case of incoherent detection and i_{LO} for coherent detection. The dark current is a leakage current which exists in the detector when it is blocked from having radiation applied to it. It therefore provides a criterion for estimating the optimum performance of an incoherent detector. The local oscillator shot noise, i_{LO}, is a quantum effect due to the fluctuation in the rate of arrival of photons at the detector.

The signal current for direct (incoherent) detection is simply:

2) $\langle i^2{}_S \rangle_I = \rho^2 P^2{}_S$,

where ρ is the detector current responsivity given by:

3) $\rho = \eta e / h\nu$ (amp/watt)

and η is the detector quantum efficiency, e = electron charge, h = Planck's constant, ν = frequency.

The noise for dark current limited detection is:

4) $\langle i^2{}_N \rangle = \langle i^2{}_{DK} \rangle = 2eI_{DK}B$,

where i_{DK} is the DC dark current and B is the detection bandwidth. The signal-to-noise for direct detection is then:

5) $S/N|_I = \dfrac{\rho^2 P^2 S}{2eI_{DK}B}$.

It is common practice to describe an optical detector in terms of a noise equivalent power (NEP). This can be calculated by setting Equation 5 equal to unity and solving for P_S=NEP.

6)   NEP = $(2eI_{DK}B)^{1/2}/\rho$

A normalized figure of merit that increases as sensitivity increases is called the specific detectivity and is defined to be:

7) $D^\star = \dfrac{(AB)^{1/2}}{NEP}$,

where A is the area of the detector element. Combining Equations 5, 6, and 7, the signal-to-noise expression can be written in the useful form:

8) $S/N|_I = \dfrac{P_S{}^2 D^{\star 2}}{AB}$.

In the heterodyne (coherent) detection case, the signal current is given by:

9.) $\langle i^2{}_S \rangle_C = 2\rho^2 P_S P_{LO}$.

The LO shot noise is:

10) $\langle i^2{}_{LO} \rangle = 2eP_{LO}\, \rho B$

Substituting Equations 9 and 10 into Equation 1, it is seen that when the L.O. power is raised to a level such that Equation 10 dominates the other noise terms in the denominator, the L.O. power explicitly disappears from the signal to noise. Recalling the definition of the responsivity given by Equation 3, Equation 1 can be written as:

11) $\quad S/N|_C = \langle i2_S\rangle_C / \langle i2_{LO}\rangle = \dfrac{\eta P_S}{h\nu B}$.

To obtain a measure of the increased sensitivity offered by heterodyne detection, we can solve Equation 8 and Equation 11 for the received signal power. The heterodyne advantage can be defined as:

12) $\quad HA = \dfrac{P_{SI}}{P_{SC}} = \left[\dfrac{A}{(S/N)B}\right]^{1/2} \dfrac{\eta}{D^* h\nu}$

where we are comparing two detection cases, each with the same signal to noise. For typical operation at 10 microns:

$h\nu = 2.0 \times 10^{-20}$ joules

$\eta = 0.5$

$A = 6.0 \times 10^{-4}$ cm^2

$D^* = 2 \times 10^{10}$ Hz$^{1/2}$ cm W^{-1} .

With these values, Equation 12 is plotted in Figure 3. As an example, if 20 dB is a minimum acceptable S/N for either case,

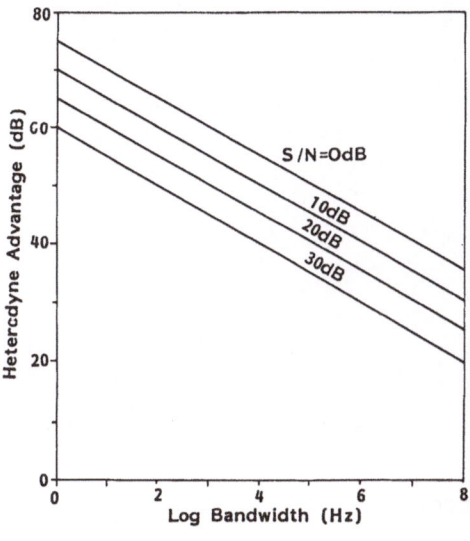

Figure 3. Comparison of the sensitivity advantage of a heterodyne receiver to a direct detection receiver with the same signal to noise

345

then a heterodyne system with 10 MHz bandwidth will have 35 dB greater sensitivity than a direct detection system. As bandwidth decreases, the advantage can become many orders of magnitude.

To estimate the achievable coherent system performance, it is necessary to consider the effects of speckle and heterodyne losses related to phase matching. The Rayleigh statistical fading of the target due to speckle generally increases the coherent system's S/N requirement by 10 dB. Optimum heterodyne performance is achieved when the phase fronts of the received signal and the local oscillator signal are perfectly matched at the detector. Experimental measurements show losses of up to 5 dB which are attributed to optical matching limitations. These two considerations are compensated, however, by the high sampling rates attainable in moderate power heterodyne systems. Whether CW or pulsed, coherent systems with average powers of the order of 5 watts have data rates in the range of 10 to 100 kHz. Rates for comparable incoherent systems are in the range of 1 to 100 Hz. Integration of 1000 samples would then recover the 15 dB of signal-to-noise increase required for the two effects discussed. It should also be noted that in some cases it may be possible to raise the D* of the incoherent detector to values approaching 10^{11} Hz$^{1/2}$ cm W^{-1}, thus increasing the sensitivity of the incoherent detector by up to 7 dB. For the case of 20 dB S/N and 10 MHz bandwidth cited above, the sensitivity of a heterodyne system would still be 28 dB greater than the comparable direct detection system and retain the advantage of the larger number of target samples.

3. LASER RANGEFINDER

A simple but useful type of tactical laser radar is the CO_2 laser rangefinder. This portable instrument displays to the user the distance to a target, such as a tank or armored vehicle.

Figure 4. CO_2 TEA Laser Rangefinder

346

Although heavier than YAG versions, the CO_2 rangefinder is less affected by weather and is therefore usable more often in passive infrared fire-control systems. Figure 4 illustrates the first U.S. 10.6 micron laser rangefinder developed for the Army by Raytheon. It is a direct detection system employing a miniature TEA laser source. It consists of two modules: an optics package (14.4 kg) and an electronics package (4 kg). It has a range of 5 km and is accurate to within ± 10 meters. It produces 40-ns pulses of 10.6 μm light at a rate adjustable from 1 to 5 pps and is eye safe.

4. TACTICAL IMAGING SYSTEMS

Of all laser radar characteristics, perhaps the most remarkable and one of the most useful is its capability to form sharply defined images of objects. Lidars can produce images that discriminate between stationary and moving targets. A small, tripod-mounted CO_2 imaging Lidar developed by Raytheon is shown in Figure 5. The 5 W continuous-wave Lidar scans a target in a raster pattern. The homodyne receiver can be tuned to process only those returns with a certain frequency shift. The result is an image of a moving target which contrasts sharply with the background. The shape of an approaching tank made with the unit shown in Figure 5 is seen in Figure 6. Alongside is a photograph of the vehicle. The strong return from the narrow gun and the very thin antenna are particularly noticeable. It is also

Figure 5.
Coherent Laser Radar

Figure 6.
Coherent Lidar Image

interesting to note that only the foremost section of the track tread, moving at exactly the same speed as the body of the vehicle, is seen.

The small apertures of laser radars make them attractive to the military for weapons guidance. Cruise missiles, designed for low-altitude missions, are special beneficiaries. Flying at low altitude, such missiles and their CO_2 laser radars will be little affected by haze, fog, or rain. They potentially will be able to strike preselected targets, fixed or moving, and will be able to classify targets on the basis of three-dimensional shape and size.

5. ATMOSPHERIC MEASUREMENTS

The ability of a laser to detect the backscattered energy from aerosols naturally suspended in air stands the radar in good stead in the area of atmospheric research. In a coherent detection laser radar the backscatter energy is Doppler-shifted according to the velocity of the particles, which travel at the air speed.

Raytheon's pulsed-Doppler laser radar, developed for the National Aeronautics and Space Administration, exploits this phenomenon. The airborne CO_2 laser radar detects clear air turbulence 2 to 16 km in front of the plane.

The experimental laser radar produces light pulses, adjustable in length from 2 to 10 μs, at a rate adjustable up to 200 pulses per second. Pulse power is 10 kW. Pulses are transmitted forward through the atmosphere and are scattered by aerosols moving with the air. The receiver coherently detects the backscattered energy and measures its frequency shift to determine the velocity of the aerosol. If a sudden shift in velocity or broad increase in velocity with range is measured, there is clear air turbulence ahead.

Recently this equipment was utilized by NASA as part of its severe storms and weather research program to help scientists to understand storms and air pollution better and to harness wind energy. The equipment was flown on board a NASA-owned Convair 990 research aircraft this summer to gather data at various altitudes near the edge of storms. The data were computerized and displayed as plots of wind speed and direction in horizontal cross sections at selected altitudes over a region from 1000 feet to 12 miles from the plane. The equipment was also utilized to map a wind field near the San Gorgonio Pass of California to collect site selection data for wind-turbine electrical generators.

The complementarity of optical and microwave radars is particularly evident in the storms program. Microwave systems give a detailed picture of the cloud portion of a storm which the laser cannot penetrate. The laser, on the other hand, maps the clear air field around the clouds which is invisible to microwave radar.

6. SMOKESTACK MONITOR

The Environmental Protection Agency (EPA) monitors emissions from industrial smokestacks to obtain such information as to how

much pollutant material is being discharged into the atmosphere,
how rapidly it is being discharged, and how the smoke plume
disperses after leaving the stack. Until recently, this was a
costly and time-consuming job. The EPA had to make arrangements
with plant officials to avoid disrupting operations, install
sensors on the smokestacks, take samples for later analysis, then
remove the sensors.

A mobile laser Doppler velocimeter system built by Raytheon
can now be used to monitor plumes from as far away as 3000 feet.
This instrument is an offshoot of the clear air turbulence
monitoring system described above.

Figure 7. Mobile LDV Smokestack Monitor

The photo in Figure 7, taken at a power station during
acceptance tests, illustrates the concept of the system. The
Lidar beam is projected from the roof-mounted scanning pod near
the rear of the vehicle. Particles in the emission from the
stack provide a signal in strong contrast to the normal atmospheric backscatter.

The system offers multiple advantages; it is economical,
provides instant data, is self-contained and easily movable, and
can be used at any time without interfering with plant operations.

8.4 Atmospheric Remote Sensing Using the NOAA Coherent Lidar System

R.M. Hardesty

NOAA/ERL/Wave Propagation Laboratory, Boulder, CO 80303, USA

During the previous 10 months the Infrared Doppler Lidar program area of NOAA's Wave Propagation Laboratory has used a pulsed coherent CO_2 lidar system as a remote sensing tool to study various atmospheric phenomena. The operational feasibility of the system was initially demonstrated by United Technologies Research Center under contract from NOAA. At the conclusion of the contract, WPL purchased the TEA laser, telescope, and locking loop hardware used in the UTRC effort and incorporated it into the current semi-trailer-mounted lidar system.

The NOAA system was designed to address two basic objectives. The first consists of demonstrating the basic feasibility of a satellite-mounted coherent Doppler lidar for measuring atmospheric wind fields around the globe. This concept, known as WINDSAT, was found in an earlier analysis to have the potential to significantly increase available wind-field data, especially in regions over the oceans.[1] A second application is to use the system as a tool for atmospheric research to remotely measure such phenomena as turbulence, winds and backscatter (β). In this paper we summarize the results of some of the atmospheric measurements made in 1981 using the pulsed system.

2.0 SYSTEM DESCRIPTION

Since a mobile capability is essential for a practical remote sensor, the system was designed to fit in a transportable semi-trailer. Figure 1 shows the optical layout. All principle optical components, with the exception of the TEA laser are mounted on the 4' x 9' optical table. The TEA section is acoustically isolated from the remainder of the system to diminish vibrationally-induced instabilities.

Table 1 lists the primary system characteristics. The hybrid TEA transmitter is grating-tunable across most of the P and R lines of the 00°1, 10°0 CO_2 transition. To produce a single longitudinal mode output, the low pressure continuous-wave mode control laser operates within the cavity to seed the UV-preionized

PULSED DOPPLER LIDAR OPTICAL CONFIGURATION

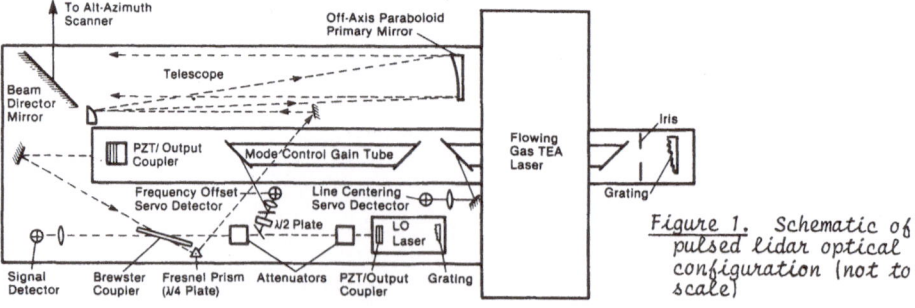

Figure 1. Schematic of pulsed lidar optical configuration (not to scale)

Table 1. System Parameters

Wavelength	Selectable P and R lines, nominally $\lambda = 10.6$ µm
Pulse energy	0.065–0.150 joules
Pulse duration	2–8 µs, nominal 3 µs
Pulse repetition frequency	1–25 Hz
Telescope	Off-axis paraboloid 28 cm diameter optics f/6
Detector	HgCdTe, 24 hour dewar
Scanning capability	Elevation −5° to 90° Azimuth 0° to 360° 1° accuracy
Digital processing	10 MHz digitizing rate Complex covariance frequency estimation

pulsed TEA laser. A hill-climbing servo-loop adjusts the piezoelectric transducer to maintain operation on line-center of the low-pressure laser.

After exiting the laser cavity, the S-polarized transmit pulse passes through the wedge-shaped germanium Brewster plate which acts as a transmit/receive switch, and into the Fresnel's rhomb, which changes the transmit polarization of the radiation from linear to circular. After passing through the prism the pulse is directed into the 28 cm diameter, f/6, off-axis telescope, which provides approximately 75 dB of optical isolation between transmitter and receiver, and out through the 2-axis scanner.

The backscattered signal is gathered by the telescope and starts back through the system on the inverse path of the transmitted pulse. Upon exiting the Fresnel's rhomb the radiation is P polarized; thus it reflects from the germanium plate and is directed through the focusing lens onto the HgCdTe detector where it mixes with the signal from the CW local oscillator (LO) laser. A second system servo loop maintains the LO frequency at an offset of 20 MHz relative to the transmit pulse frequency. Incorporation of a frequency-offset LO permits discrimination between approaching and receding scatterers.

The system signal processing is controlled by a Nova 3 minicomputer. A 20 MHz reference signal mixes with the signal from the detector in a complex demodulator, producing in-phase and quadrature baseband components of the backscattered signal. These signals are digitized at a 10 MHz rate and read into the computer. The computer also samples, on a pulse-to-pulse basis, voltages relating to LO frequency offset and scanner position. During the interval between pulses the computer calculates and displays velocity or received power at a single range. All raw data is archived to magnetic tape.

3.0 CALIBRATION AND PULSE CHARACTERISTICS

Periodically during periods of operation the system calibration is examined for changes in operational capabilities. For this task we employ a large diameter, sandpaper-surfaced disk whose reflecting characteristics are well established.[2] Frequency and amplitude properties of reflected returns from the disk are measured and compared with those from previous measurements.

System calibration is performed by comparing the measured signal-to-noise ratio (SNR) of the returns with a theoretical value calculated from the lidar equation. The theoretical computation must include values for system parameters such as telescope size, transmit energy, receiver bandwidth, and target reflections to accurately predict the SNR. With coherent systems, average SNR is usually difficult to measure from a single return because of speckle.[3] This effect, which is the result of relative phase changes of the signals scattered by the individual particles, produces random fluctuations in irradiance at the receiver. Thus a number of samples must be averaged to accurately estimate the mean signal level as illustrated in Fig. 2. While the individual returns show large fluctuation, the 50 pulse-average accurately represents the pulse shape.

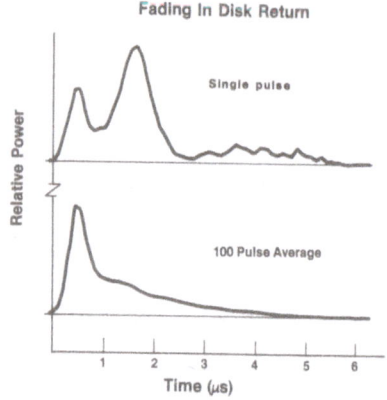

Figure 2.--Reflected pulse from disc target showing instantaneous fluctuation due to speckle

Our calibration shows that the SNR as measured from the averaged disk return is 7 dB less than that predicted by the lidar equation. This difference, which represents unaccounted-for system losses, may result from such factors as mismatching of signal and LO wavefronts, beam aberrations, or beam spot size differences. As a verification we measured SNR on a spectrum analyzer using the CW mode control laser to irradiate the sandpaper disk. As in the pulsed case the measured value was 7 dB less than predicted.

Frequency-domain properties of the transmitted pulse are important in both wind and irradiance measurements. While increased bandwidth of the signal degrades the accuracy of some frequency estimates,[4] it also produces more rapid speckle-induced fluctuations in the backscattered irradiance. As a result, averag-

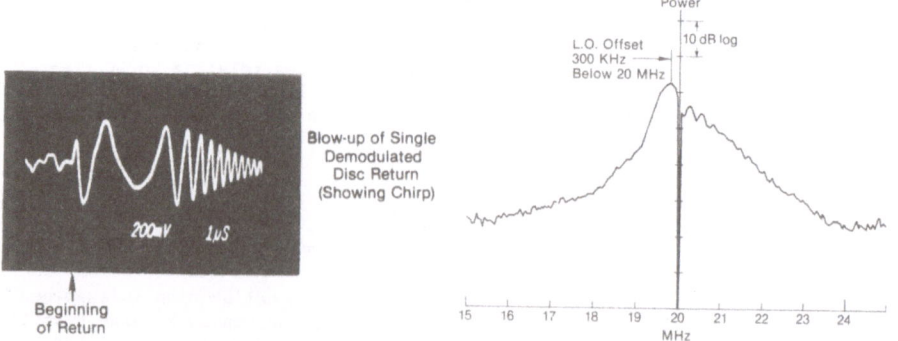

Figure 3. Baseband disc return showing chirp (left); average power spectrum of disc returns (right)

ing of intensity over a single range gate is more effective with wider bandwidth signals. The effectiveness comes about at a cost of decreased SNR because of the necessary corresponding enlargement of the receiver bandwidth.

We used the return from the sandpaper disk to measure the frequency chirp in the transmit pulse. Figure 3 shows the time series of the detected, demodulated return. The transmit frequency initially is pulled sharply lower (down chirp) then gradually increases to beyond the line-center frequency. In the tail of the pulse the frequency increases approximately quadratically with time. Figure 3 also shows a digital Fourier transform of the pulse return. Although chirp in the pulse tail exceeds 1 MHz, the 3-dB bandwidth of the return signal is seen to be approximately 375 kHz. The first moment of the spectrum in Fig. 3 is at about 20.2 MHz, which produces a slight (1 m s^{-1}) bias in wind measurements. This can easily be eliminated during processing.

4.0 ATMOSPHERIC MEASUREMENTS

4.1 Characteristics of Returns

We examined the statistical properties of the aerosol backscattered signal in order to determine the type of signal processing required for accurate Doppler and irradiance measurements. One expects the distributed, nonhomogeneous nature of the atmospheric target to modify the return properties relative to those from hard target returns.

Figure 4 shows a measured probability density function (pdf) and autocovariance function of irradiance measurements from the aerosol-backscattered signal at a height of 2 km. The nearly exponential distribution of irradiance indicates that the Rayleigh-phasor model of aerosol returns is a reasonable approximation for pulsed systems. Since this distribution is characterized by a standard deviation equal to the mean value, averaging is required to reduce the uncertainty in the individual irradiance measurement. We see from the signal autocovariance function that the time scale of the speckle fluctuations is approximately .6 µs, or about one-half that measured from disk returns. This bandwidth increase, which probably results from atmospheric effects such as shear and turbulence, improves the speckle averaging such that ten independent samples exist over 6 µs.

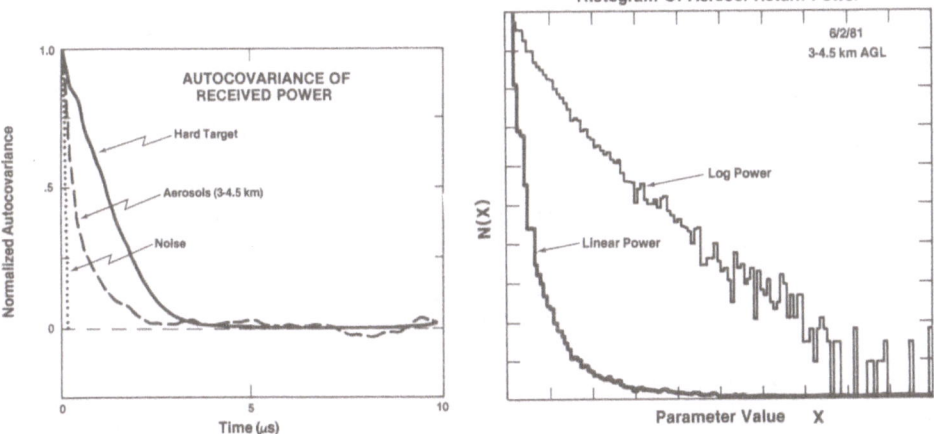

Figure 4. Measured autocovariance of returns from disc and aerosol targets indicating time scale of fluctuation (left); histogram of aerosol returns showing exponential distribution (right)

4.2 Aerosol Profiling

Since a scarcity of data exists on the variability of the lidar backscattering coefficient (β) over time and space, we measured β profiles on a daily basis throughout the summer. Backscatter coefficient is calculated by measuring the SNR in the return and inverting the lidar equation to solve for β. We typically averaged returns from 500 pulses to reduce speckle-caused irradiance fluctuations.

Throughout this investigation we consistently observed returns to the tropopause, and often into the stratosphere. Figure 5 is an example of these data, where the need to correct for absorption is illustrated. During late autumn we transported the lidar to Edwards Air Force Base to support the STS-2 orbiter landing. Preliminary analysis of β measurements taken there seems to indicate much clearer air in the upper troposphere than measured during the summer in Boulder.

When measurements of 10.6 μm aerosol backscatter taken over a three month period are plotted on a Gaussian probability graph, we obtain the results shown in Fig. 6. The nearly linear nature of the log β plot seems to indicate that the long-term aerosol distribution is lognormally distributed, as might be expected.

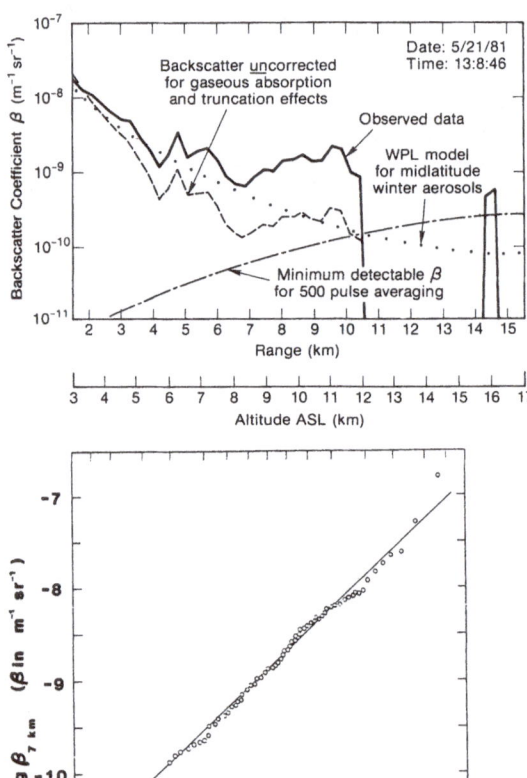

Figure 5. Typical measured summer β profile. Note stratospheric dust layer

Figure 6. Cumulative probability of summer log β measurements plotted on Gaussian scale

4.3 Velocity Measurements

Estimates of radial wind velocity are produced from measurements of the Doppler shift in the backscattered signal. In order to measure the mean 3-dimensional wind profile over a volume, we employ a VAD (velocity-azimuth-display) concept, where the beam is conically scanned in azimuth around a vertical axis. If the flow in the volume is relatively nonturbulent, the measured velocity should vary sinusoidally as a function of azimuth. We process VAD scans by finding the sinusoid which has the best least-squares fit to the observed velocity data. Figure 7 shows the measured and best-fit waveforms for a typical high SNR measurement. Horizontal wind speed is proportional to the amplitude of the sinusoid, while direction is proportional to its phase.

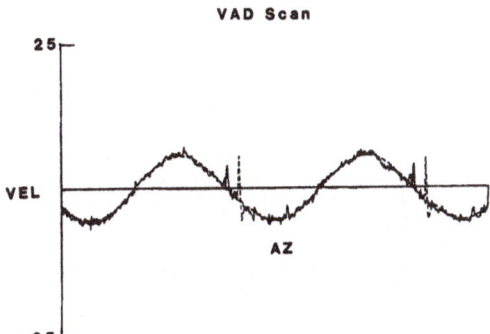

Figure 7. VAD scan showing measured velocity vs. azimuth (solid line) and least-squares fit sinusoid (dotted line)

The system capability for wind speed measurement was demonstrated during support of the STS-2 landing. Comparisons of lidar-measured winds with rawinsonde and jimsphere measurements showed good correlation.

4.4 Acknowledgment

The work described in this paper resulted from the combined and individual efforts of personnel in the Infrared Doppler Lidar Program Area. The author wishes to particularly thank F.F. Hall, Jr., R.M. Huffaker, T.R. Lawrence, M.J. Post, and R.A. Richter for their contributions to this summary of program area research.

REFERENCES

1. Huffaker, R.M. (Editor) 1978: Feasibility study of satellite-borne lidar global wind monitoring system, NOAA Technical Memorandum ERL WPL-37.

2. Post, M.J., R.A. Richter, R.J. Keeler, R.M. Hardesty, T.R. Lawrence, and F.F. Hall, Jr., 1980: Calibration of coherent lidar targets, Appl. Opt. 19: 2828-2832.

3. Hardesty, R.M., R.J. Keeler, M.J. Post, and R.A. Richter, 1981: Characteristics of coherent lidar returns from calibration targets and aerosols, Appl. Opt. 20: 3763-3768.

4. Zrnic, D., 1977: Spectral moment estimation from correlated pulse pairs, IEEE Trans. of Aerosp. and Electron. Syst., 7:344.

8.5 Coherent CO₂ Lidar Systems for Remote Atmospheric Measurements

James W. Bilbro

National Aeronautics and Space Administration
Marshall Space Flight Center, Huntsville, AL 35812, USA

Coherent CO_2 Doppler lidar systems have been applied to a wide variety of phenomena since the first measurement of atmospheric winds (Jelalian and Huffaker, 1967). A few examples of the applications sponsored over the last decade by the Marshall Space Flight Center (MSFC) have been selected to illustrate the tremendous potential of these systems. Detailed information on these examples is readily available, therefore I intend to provide only a brief summary of each program addressing the initial concept, the resulting hardware, and a representative sample of the data. Additional information on these and other efforts can be obtained from an overview of atmospheric laser Doppler velocimeters in the publication Optical Engineering (Bilbro, 1980).

Following the summarization of previous efforts, a more detailed description of current programs at MSFC will be given. These programs involve investigations into two-dimensional wind field measurement, atmospheric backscatter measurement, transverse velocity measurement, and the feasibility of space operations.

Let us begin the applications summary with the conceptual drawing shown in Fig. 1a. This was a continuous wave (CW) system designed to detect and track aircraft wake vortices and is shown at the JFK Airport in New York in Fig. 1b. A sample of the real-time generated track of a vortex pair is shown in Fig. 7. The same CW system was used with a Bragg cell to provide the bidirectional velocity profiles of dust devils seen in Fig. 8. The next data (Fig. 9) shows the results of this system performing transverse velocity measurements. Work in this area is still in progress. Fig. 2a shows the conceptual drawing of a system specified for the EPA in a technology transfer program. The resulting system is shown in Fig. 2b during acceptance testing. The purpose behind this effort was the measurement of mass flow rates of high stack emissions.

Turning now to pulsed coherent lidars, Fig. 3a depicts the artist's concept of an airport wind monitoring system. The trailer shown in Fig. 3b was deployed at the Kennedy Space Center to obtain thunderstorm gust front data for use in this program. The bull's eye type plot also shown in Fig. 3b is a plan view color display of the velocity distribution obtained by scanning through the storm gust front. Operation of the ground-based pulsed system in a velocity azimuth display (VAB) mode produced the wind profile which is shown in comparison with Rawinsonde data in Fig. 10. Fig. 4a shows the airborne operation of this system for clear air turbulence detection. The aircraft fairing used to direct the beam forward is shown in Fig. 4b. An example of the lidar derived true air speed data and its spectral distribution are shown in Fig. 11.

Figs. 1a/b

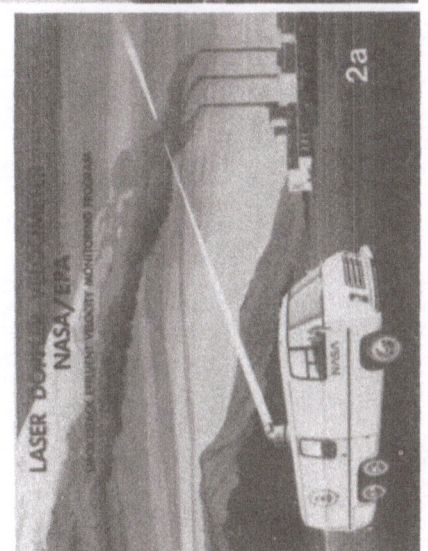

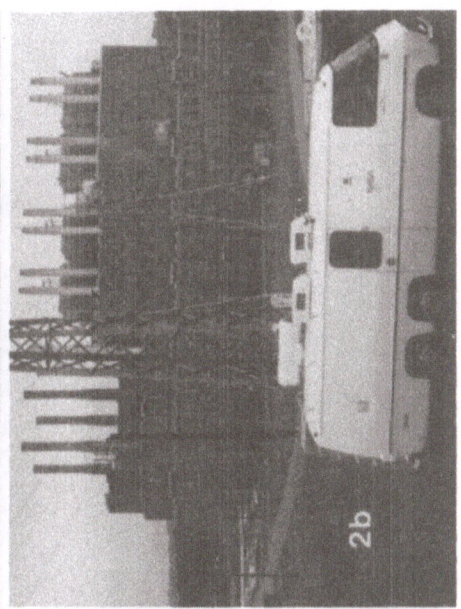

Fig. 2a/b

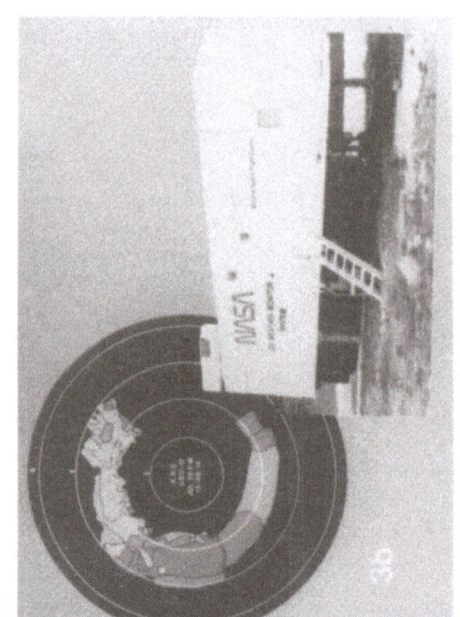

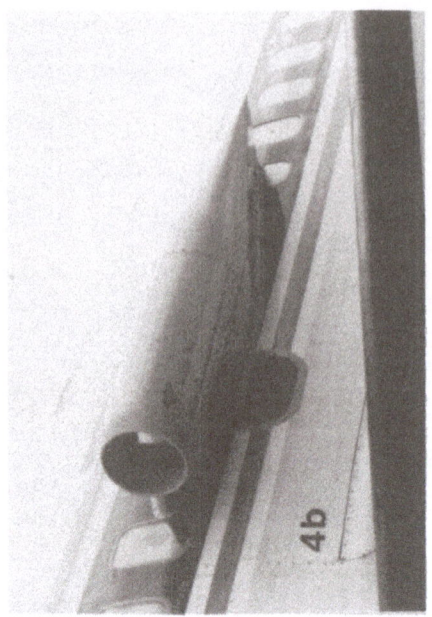

Fig. 3a/b

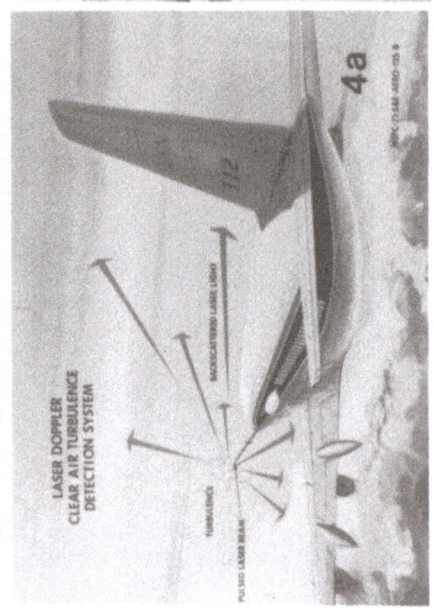

Fig. 4a/b

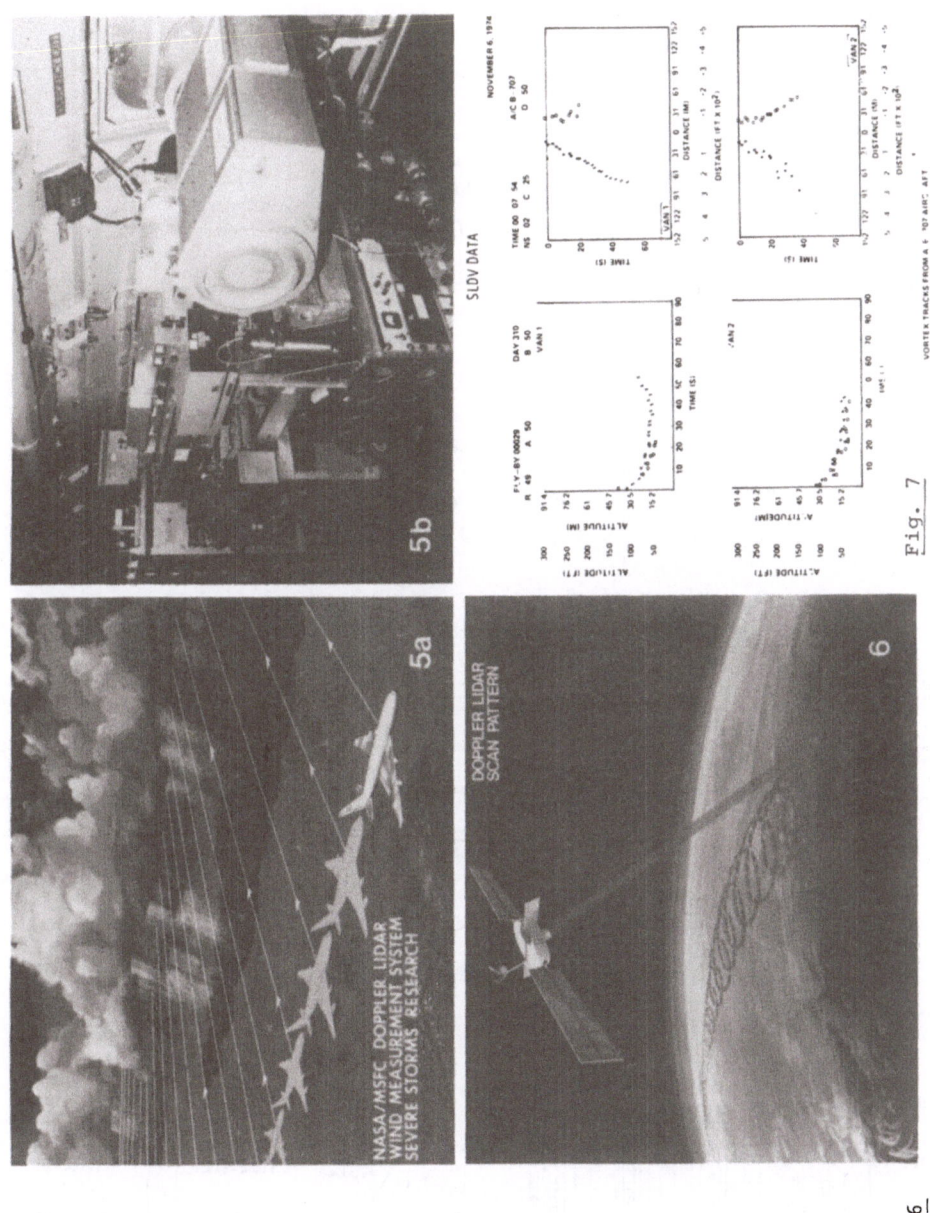

Fig. 5a/b

Fig. 6

Fig. 7

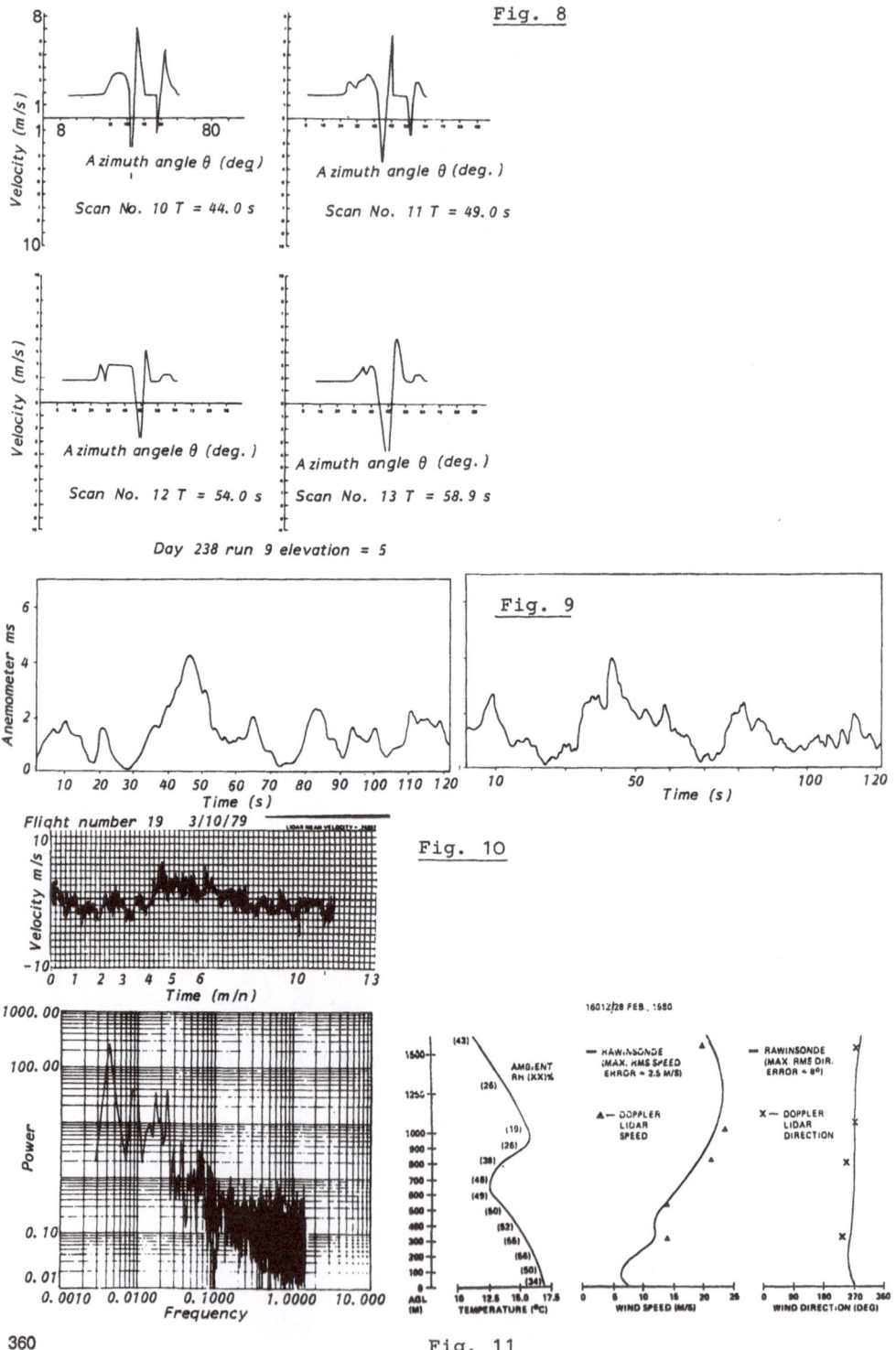

Fig. 8

Scan No. 10 T = 44.0 s

Scan No. 11 T = 49.0 s

Scan No. 12 T = 54.0 s

Scan No. 13 T = 58.9 s

Day 238 run 9 elevation = 5

Fig. 9

Fig. 10

Flight number 19   3/10/79

Fig. 11

360

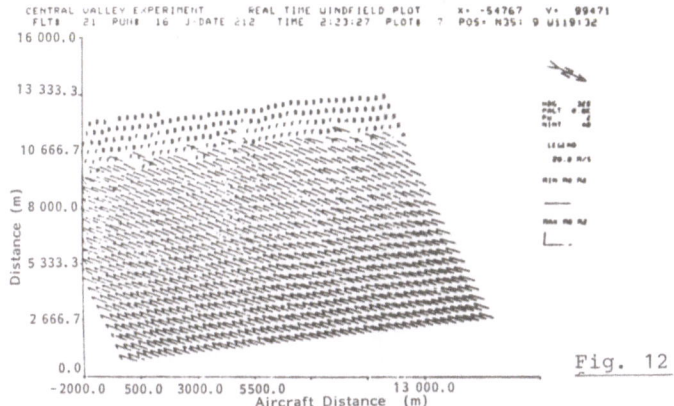

Fig. 12

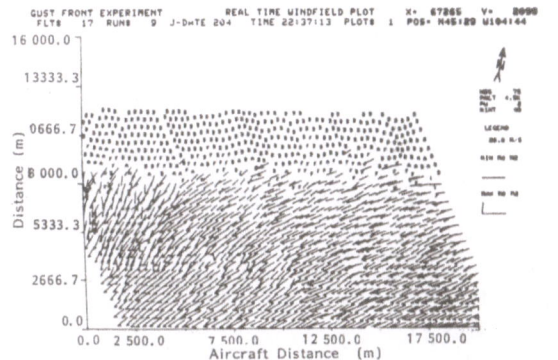

Fig. 13

One of our main programs currently underway at MSFC is depicted in Fig. 5a. This artist's concept demonstrates the manner in which two-dimensional wind fields are constructed. As may be seen from this figure, the beam is first directed forward and then aft. As the aircraft advances, aft scans intersect with previous forward scans providing velocity data from two different view angles at the point of intersection. Two simultaneous equations are then solved to produce the horizontal vector velocity at that point. This vector is then drawn on a computer controlled display producing the real-time generated plots shown in Figures 12 and 13. Fig. 12 shows a highly uniform wind field measured in the Central Valley region of California while Fig. 13 shows a typical thunderstorm gust front obtained in Montana. The lidar shown in Fig. 5a produced numerous plots of wind fields in an 80-hour flight program conducted in regions of California, Oklahoma, and Montana during the past summer. A summary of these tests has been submitted for publication in the Bulletin of the American Meteorological Society (Bilbro et al., 1982).

Another of our major areas of interest is that of atmospheric backscatter variation. Data was also collected in this area during last summer's measurements. Two different types of displays are shown for the data in Figures 14 and 15. In Fig. 14, the pulsed lidar intensity is displayed as a function of range for successive shots. This figure demonstrates the variability encountered in many of these measurements. Fig. 15 depicts the preliminary comparison of back-scatter altitude profile data obtained with four different lidars.

Fig. 14

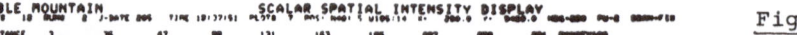

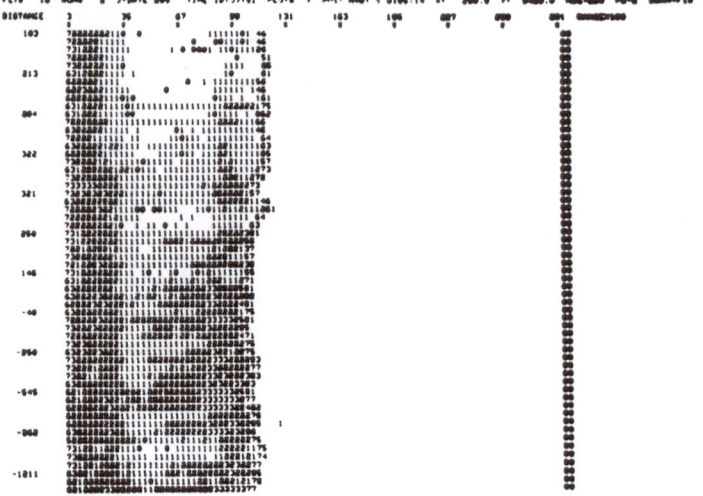

TABLE MOUNTAIN   SCALAR SPATIAL INTENSITY DISPLAY

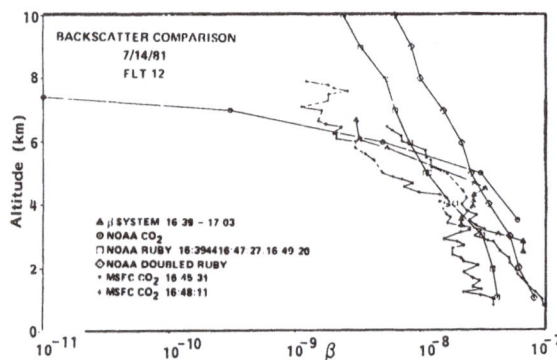

Fig. 15

Backscattering Coefficient as a Function of Altitude

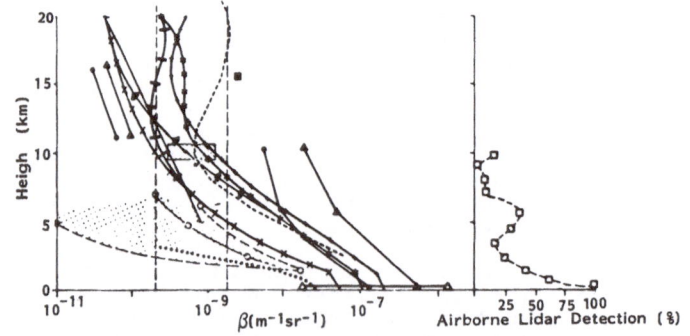

Fig. 16

Two of the lidars were ground-based systems operated by NOAA - a pulsed coherent CO_2 system and a ruby system. The other two lidars were MSFC systems - the pulsed lidar previously mentioned and a special CW lidar built specifically for backscatter measurements. Analysis of this data is still in process (Jones et al., 1981).

The interest in backscatter information is largely the result of investigations into the feasibility of a space-based Doppler lidar such as that depicted in Fig. 6. The system proposed by NOAA as WINDSAT (Huffaker, 1978) has been extensively investigated by NOAA and is under assessment by NASA at their request (Steincamp, 1981). The ultimate determination of the feasibility of such a system will in part rest on the viability of the backscatter models depicted in Fig. 16. It is toward the verification of these models that our backscatter measurement efforts are directed.

In summary, MSFC continues to maintain an active research and development program in the area of coherent CO_2 Doppler lidars. Many applications have been demonstrated with these systems and it is anticipated that even broader applications will result from increased sophistication and the extension of capabilities through enhanced measurement techniques.

REFERENCES

Bilbro, James W., "Atmospheric Laser Doppler Velocimetry: An Overview," Optical Eng. 19(4), 533-542, July/Aug. 1980.

Bilbro, J., G. Fichtl, D. Fitzjarrald, and M. Krause, "Airborne Doppler Lidar Wind Field Measurements," submitted for publication in the Bull. of the American Meteorological Society, 1982.

Jelalian, A. V., R. M. Huffaker, "Laser Doppler Techniques for Remote Wind Velocity Measurements," Specialist Conference on Molecular Radiation, 345, Huntsville, AL, Oct. 1967.

Jones, W. D., J. W. Bilbro, S. C. Johnson, H. B. Jeffreys, L. Z. Kennedy, R. W. Lee, C. A. DiMarzio, "Design and Calibration of a Coherent Lidar for Measurement of Atmospheric Backscatter," Paper presented at the SPIE's 25th Annual International Technical Symposium and Exhibit, San Diego, CA, Aug. 24 - 28, 1981.

8.6 Wide-Area Air Pollution Measurement by the NIES Large Lidar

N. Takeuchi, H. Shimizu, Y. Sasano, N. Sugimoto, I. Matsui, and H. Nakane

The National Institute for Environmental Studies, Yatabe, Tsukuba, Ibaraki 305, Japan

1. Introduction

As the study of air pollution progressed, the aim of research moved from pollutant concentration measurements in highly contaminated areas (e.g., emission sources and their vicinity) to long-range transportation and its wide-area influence. Presently, air pollution in Japan is measured by monitoring stations distributed around the country. This type of monitoring is based on point sampling, and it is difficult to grasp the general features of the region in question from it. In order to obtain these features, temporally and spatially continuous measurement for a wide area is required. Remote monitoring methods meet this requirement, and provide new information on the atmosphere, as well as the two-dimensional distribution of pollutants near the ground. In Japan, populated areas are separated by mountain ranges, and the long-range transportation of urban plumes from Tokyo to its suburbs [1], for example, in the scale of 50 to 100 km, indicates the desired detection range of the remote monitoring.

Lidar is suitable for the spatially continuous measurement of air pollution, and it is desirable to develop a lidar which can cover a wide area. For such measurement, the effort has been mainly directed towards the development of airborne lidar [2]. However, airborne lidar is not useful for the instantaneous measurement over an area and its running cost is great compared to ground-based lidar. For these reasons, the National Institute for Environmental Studies(NIES) decided to develop a ground-based lidar to cover a 50 km-in-radius area, i.e., to construct a large-scale station-type lidar [3]. The purpose of this lidar is (1) the development of measurement and analysis methods to be applied when using a lidar for wide-area air-pollution studies, and (2) spatial data collection of meteorological parameters, as well as aerosol concentration distribution, based on Mie and Raman-scattering measurements. The system at NIES has been named the "LAMP lidar(Large Atmospheric Multi-Purpose lidar)". Presently, the LAMP lidar has finished the test stage of its measurement ability and the preparation of computer programs for control and analysis, and is now being used intensively to study air pollution phenomena through the behavior of the aerosol spatial distribution. The main subjects of the research are

(1) measurement of wide-area pollution distribution using aerosols as tracers,
(2) study of the long-range (wide-area) transportation of pollutants, and measurement of the wind vector field,
(3) study of the atmospheric boundary layer structure and its behavior, and
(4) measurement of meteorological parameter profiles, such as temperature and humidity based on Raman-scattering.

2. Design of the LAMP lidar

2.1 Features

Based on the considerations in the previous section, the LAMP lidar was planned to have the following features:

1) wide-area observation capability to monitor aerosol distribution over a range of about 50 km under typical meteorological conditions in approximately ten minutes,
2) multi-functional measurements using Raman scattering techniques for meteorological parameters, such as temperature, humidity and visibility (the typical detection range is several km), (wind derivation is possible from time-sequential aerosol distribution patterns),
3) mini-computer program control of system operation, data acquisition, data recording and real-time display,
4) high response to rapidly varying phenomena, such as stack plume dispersion, by the use of a highly repetitive pulsed-laser and a rapid scanning pedestal,
5) positioning with high angular accuracy (0.3 mrad), which enables the pinpointing of an emission source on a map,
6) reliability which enables long-period continuous operation.

2.2 Evaluation of detection range

The system was designed to include
(1) a high averaged-power laser,
(2) a large aperture telescope, and
(3) high-performance optics.

In order to determine the specifications [4], the NIES mobile lidar system [5] was actively used. Assuming a Nd:YAG laser, the parameters for theoretical estimation are listed in Table 1. An example of this estimation is shown in Fig.1 for the case when the detected aerosol concentration (S/N = 10 for a range resolution of 100 m at a wavelength of 532 nm) is assumed to be equal to the background aerosol concentration (which determines visibility). For the estimation, the lidar equation

$$P_r(R) = P_o K \ell A_r \beta(R) \, \exp[-2\int_0^R \alpha(r)dr] Y(R)/R^2 \tag{1}$$

Symbol	Definition	Value	
		1.06 μm	532 nm
n_0	Photon number of laser (photons/pulse)	4×10^{18}	1×10^{18}
K	Efficiency of optics (%)	27	16
A_r	Effective aperture(m^2)	1.7	1.7
Y(R)	Geometric form factor	1	1
η	Quantum efficiency(%)	0.1	8
S_B	Background intensity (W/nm·sr·m^2)	3×10^{-3}	2×10^{-2}
κ	Mie scattering parameter = α/β	30	30
Q	Receiving angle (sr)	7×10^{-8}	7×10^{-8}
f	Band width of filter (nm)	0.5	0.1

Table 1 Parameters used for the estimation of lidar detection range

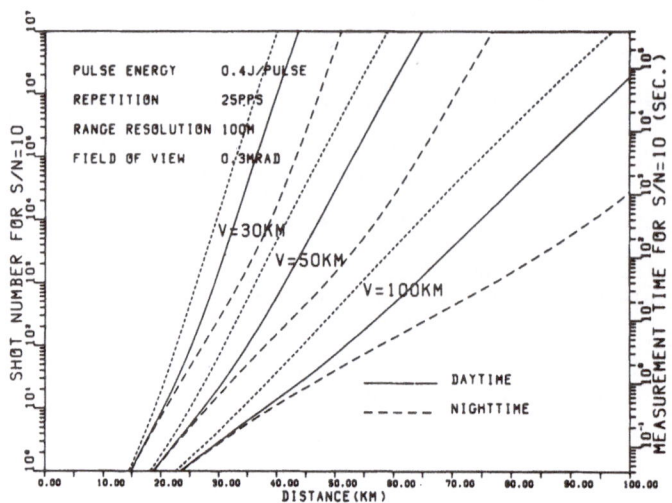

Figure 1. Estimation of necessary shot number (measurement time) for S/N=10 at the wavelength of 532 nm as a function of detection range under different visibilities. Solid line for daytime measurement; dashed line for nighttime measurement; dotted line for a case of random noise ten times greater than daytime shot noise

and S/N equation

$$S/N = \sqrt{M} \, n_s / (\sqrt{\mu} \, \sqrt{n_s + 2n_B}) \tag{2}$$

are used. In Eq. (1), $P_r(R)$ is the signal power received from distance R, P_o is the transmitted peak power with pulse width τ_p, K is the efficiency of the receiving optics, ℓ is one-half the pulse spatial-width ($\ell = c\tau_p/2$; c is the light velocity), A_r is the aperture area of the receiving optics, $\beta_r(R)$ is the Mie backward scattering coefficient at distance R, $\alpha(R)$ is the extinction coefficient, and the geometrical form factor $Y(R)$ is the fraction of the laser beam cross section involved in the receiver optics field of view. In Eq. (2), M is the repetition time, μ is the noise factor of a photomultiplier(PM), and n_s and n_B are photoelectron numbers for signal and background radiation, respectively. The quantities n_s, n_B are given by

$$n_i = P_i \tau / h\nu \quad (i = s,B) \quad , \tag{3}$$

where τ is the sampling time, h is the Planck's constant, is the light frequency, and P_B is given by

$$P_B = K S_B \Omega f A_r. \tag{4}$$

Here, S_B is the background radiation intensity, Ω is the receiver-optics field of view. and f is the bandwidth of the narrow band interference filter. For covering a quarter-circle area up to 50 km in about ten minutes in moderate visibility (V = 30　50 km), a larger range resolution (e.g., L = 500 m) and measurement at the fundamental wavelength (1.06 μm) are required.

3. Construction of the LAMP lidar

The block diagram of the LAMP lidar is shown in Fig.2. It occupies the top two floors of an eight-story tower. The main part of the system

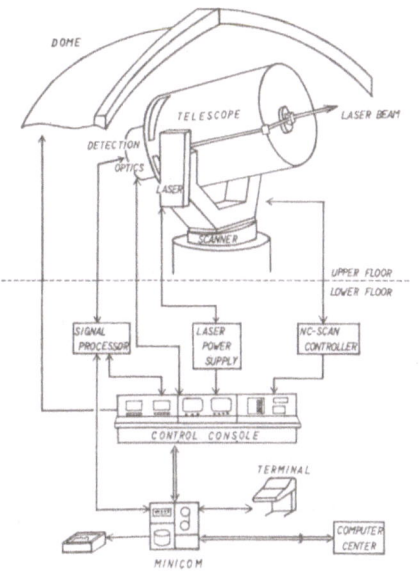

Figure 2. Block diagram of the "LAMP lidar"

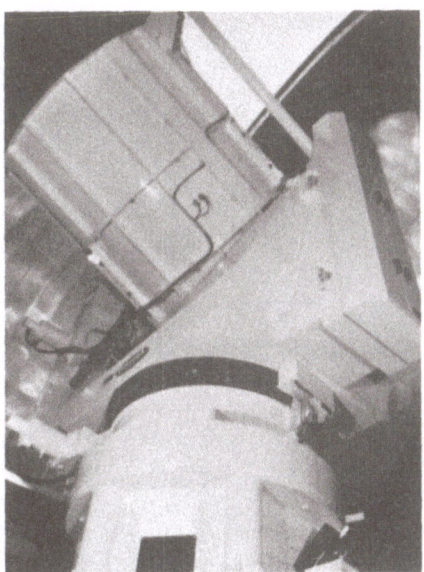

Figure 3. "LAMP Lidar"

Table 2 Specification of the "LAMP Lidar" system

Method:	Mie/Raman Scattering
Object:	Wide-area distribution of aerosol, pollutant gases, and meteorological parameters
Manufacturer:	Tokyo Shibaura Electric Co., Ltd.

Laser: Nd:YAG laser (1 osc. + 3 amp.)
   Wavelength: 1.06 μm   0.532 μm
   Average power.   30 W   10 W
   Repetition: 25 pps
   Pulse duration: 14 ns
   Beam divergence: 0.3 mrad (after 10 time expansion)

Receiving telescope (Cassegrainian)
   Effective aperture: 1500 mm
   Effective focal length: 8000 mm
   Field of view: 0.15 - 4 mrad
   Range of detection: up to >50 km

Pedestal (Azimuth - Elevation type)
   Scanning speed: 1 turn/day - 10 deg/sec

Data Processor
   Transient degitizer: Iwatsu DM-902
   Minicomputer: TOSBAC 7/40

(laser and telescope, Fig.3) is located on the top floor covered by a dome-type roof. Control and data processing apparatus are located one floor below. The system consists of a laser transmitter, signal receiving optics, a scanning pedestal, signal processing electronics, a control console and a mini-computer. Brief specifications of the entire system are shown in Table 2. The detail of the construction was discussed in [4]. Here the summary of features of each subsystem are described below.

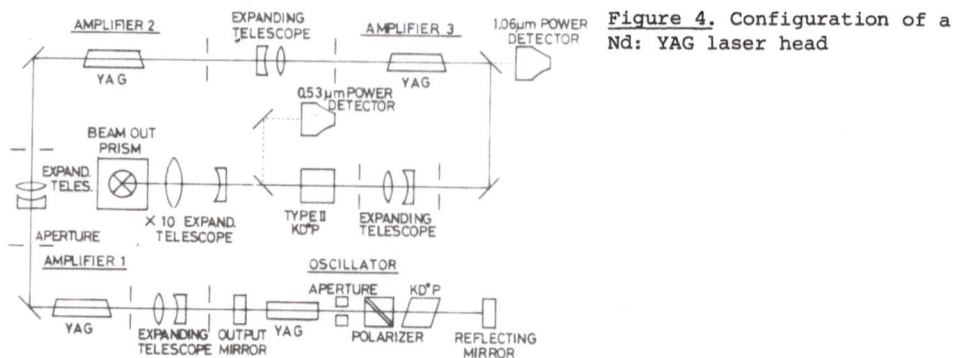

Figure 4. Configuration of a Nd: YAG laser head

3.1 Laser transmitter

For reliable operation and high averaged-power pulse operation, and also for the purpose of obtaining the information on particle size distribution, two frequencies (1.06 μm in the infrared(IR) range and its harmonics 532 nm in the visible range) of a Nd:YAG laser, which are emitted simultaneously, were selected. From the standpoint of simple optical alignment and reliable laser operation, a one-axis Coude-configuration is used for transmitting. The laser system, which consists of one oscillator and three stages of amplifiers, is shown in Fig.4. The repetition rate is fixed at the optimum of 25 pps. The laser energy per pulse is 1.2J and 0.4J for 1.06 μm and its second-harmonic, 532 nm, respectively.

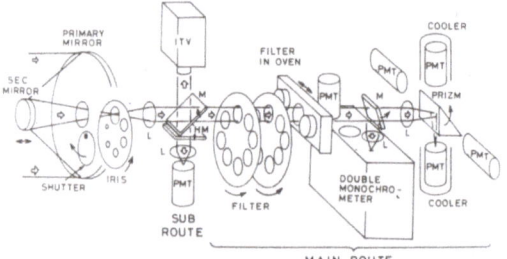

Figure 5. Schematic diagram of signal-receiving optics

3.2 Signal receiving optics

The receiving telescope was designed to have a large aperture area for gathering a large amount of light and a short cylinder for versatility in scanning. The Cassegranian-type reflecting telescope was adopted, and super-duralumin was selected as the mirror material. The optics system is shown in Fig.5. About 4 % of the received signal light is introduced into the auxiliary channel, and is used for monitoring the Mie-scattering at 532 nm. The remaining 96 % of the light is processed by being passed through various selecting mechanisms: iris-size selection for a suitable field-of-view setting, narrow-band interference filter selection, double-monochromator selection, switching of a PM for the IR or visible range. Raman-scattering measurement can be made by selecting a suitable interference filter. Also, an ITV monitors the focus plane of the telescope (iris position), which makes the alignment of laser light to the telescope optical axis easy and also is useful to position the optical axis to the true north direction, using the polar star as the reference.

The scanning pedestal has an elevation-azimuth-type configuration. The scanning speed is controlled by a micro-computer in the range from one rotation per day to 10°/sec, which is enough to rapidly follow varing pollution phenomena such as plume dispersion. The absolute angular accuracy is 1 minute of arc, which corresponds to an ambiguity of 15 m at 50 km.

3.3 Signal processing electronics
The signal detected by a PM is amplified, band-pass filtered, and is digitized by a two-channel transient digitizer (IWATSU DM902), which has 2048 segments per channel with 8—bit accuracy and the minimum gate width of 10 ns (corresponding to 1.5 m range resolution). The digitized signal is transferred to a mini-computer (TOSBAC 7/40), which, after accumulating a given number of shots, analyzes and stores the data on a magnetic tape recorder, and displays the result in real time, if necessary.

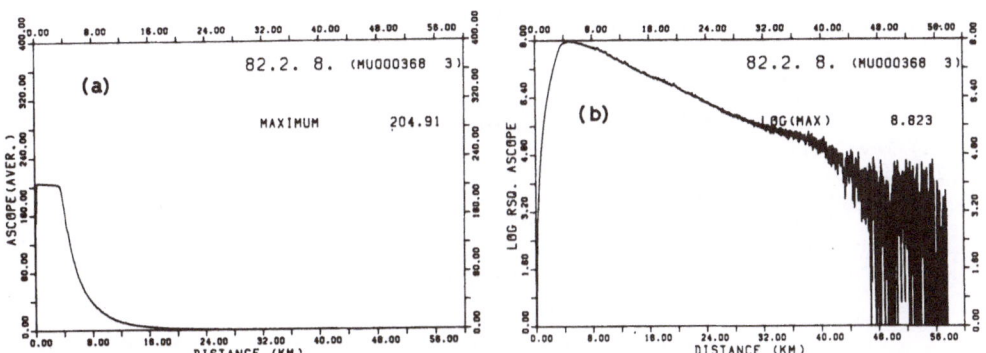

Figure 6. Examples of a) A-scope and b) logarithmic range-squared A-scope for the average of scan range over 90°. The initial part is affected by the overflow to the transient-dizitizer input level

3.4 Control console and minicomputer
The whole system is controlled from the central console. Control is either manual or computer-programmed. The latter is usual in the measurement.
The mini-computer system, which is used for the operation control, has four functions:
(1) programmed operation of the lidar system, (2) data recording on magnetic tape, (3) real-time data processing and display, and (4) transfer of the data to and from the large computer at the institute computer center.

3.5 Safety
To protect airplanes flying in the range of the lidar, the lidar is equipped with a microwave radar mounted on the side of the telescope with the optical axis parallel to the laser beam. The wide divergence of the microwave radiation (4° FWHM) detects aircraft before they reach the laser beam (1´ FWHM), and the laser is automatically switched off if an aircraft is detected.

4. Results and examples of display
4.1 Processing of lidar data

In order to display the raw
data of a lidar measurement as
image data, the following procedures are necessary:
1) subtraction of the DC component,
2) normalization by monitored
   laser power (A-scope),
3) elimination of anomalous signals created by obstacles,
4) range-squared correction
   (R-squared A-scope),
5) correction for the geometrical
   form factor $Y(R)$ [6],
6) correction for atmospheric attenuation (R-corrected A-scope),
7) smoothing,
8) display area setting,
9) interpolation of eliminated signals,
10) coordinate transformation from polar to Cartessian,
11) level slicing,
12) print out (image data).

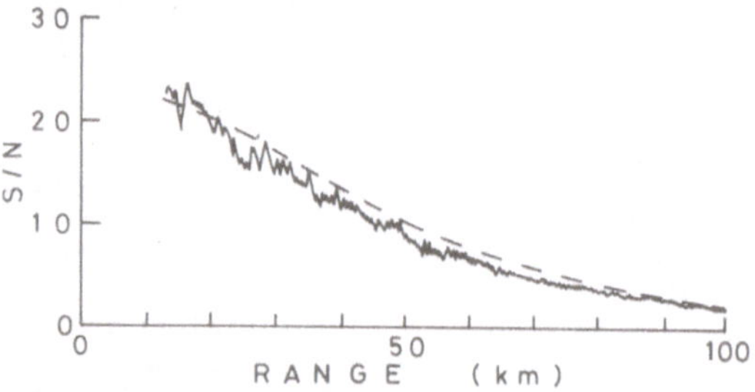

Processing is carried out with the use of a large computer (HITAC
M-180). According to the purpose of a given measurement, the processed
image-data are displayed in RHI (Range Height Indication) or PPI (Plan
Position Indication) mode on a line-printer or a color graphic display.

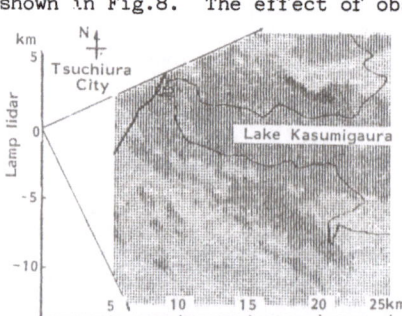

Figure 7. Measured S/N value as a function of
detection range. Measurement conditions: wavelength
at 532 nm, nighttime, clear sky, range resolution
of 60 m, and an average of 100 shots. The broken
line is a theoretical fit [4]

4.2 A-scope
Examples of A-scope and range-squared A-scope data are shown in
Fig.6. This data was taken in the daytime(14:30 JST), and was averaged
over a scan range of 90°(34,000 shots). The initial part is limited
by the saturation of the input signal to a transient digitizer. In this
case, the data is significant up to over 40 km In Fig.7, an A-scope
measurement (100 shot average) taken in a clear-sky conditions at night
is shown. In this case, an S/N value of over 10 was obtained up to 50
km.

4.3 Aerosol distribution map
As an example of wide-area aerosol distribution, horizontally scanned
data over 20 km x 20 km are shown in Fig.8. The effect of obstacles
was eliminated and the data
was smoothed by a spatial
filter. In this region an
urban area and an industrial
area are located at the
north-west corner along a
railroad. The pattern in
Fig.8 shows that a high-
density aerosol spreads from
the urban and industrial
areas to the suburbs. The
direction of the aerosol
spread agrees with the wind
direction (north-west).
It is planned to study the
relation of lidar pattern

Figure 8. An example of aerosol distribution map measured over
a region of 20 km x 20 km. Measurement conditions——time:10:46-
(JST), wind:north west, 5.3 m/s, extinction coefficient derived
from logarithmic range-corrected A-scope:4.92 x $10^{-5}m^{-1}$(visibil-
ity:79 km)

370

data with the behavior of other air pollutant gas concentrations, meteorological factors and emission sources, and it is also planned to use lidar image data for the simulation and prediction of air pollution.

4.4 Boundary layer structure

The behavior of the atmospheric boundary layer [7] is an important factor in air pollution simulation. The structure of the boundary layer has been intensively measured by lidar [8]. In Fig.9, the structure of a convective mixed layer horizontally scanned by lidar at the elevation angle of 1.9° is displayed, after being processed by an edge enhancement technique [9]. This is very similar to those previously obtained by sodar [10], radar [11] and FW-CW radar [12].

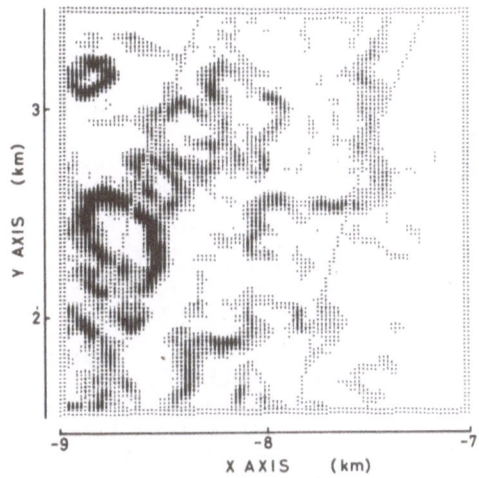

Figure 9. Convective cell structure displayed by edge-enhancement technique. Arcs in the figure correspond to the horizontal distances of 8 km and 9 km from the lidar [9]

4.5 Wind velocity [13]

The wind vector field is necessary for air pollution modeling. The horizontal wind vector was obtained from successively scanned lidar data. Figure 10 shows two successive scans with an interval of about 50 sec. The wind vector was obtained

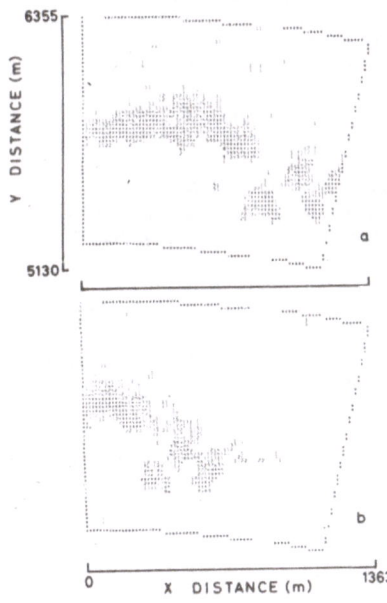

Figure 10. An example of two successive measurements of horizontal aerosol distribution. Time interval is about 50 sec [13]

371

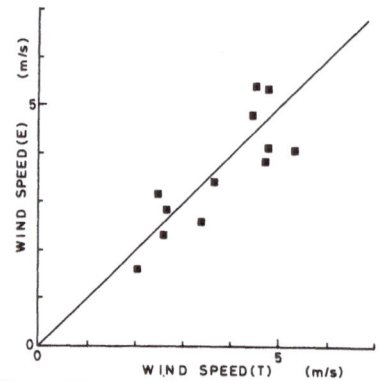

Figure 11. Comparison between the estimated wind speed by lidar observation(E) and aerovane data on the meteorological observation tower (T). The both data are averaged over ten minutes [13]

by spatial correlation. A comparison with meteorological tower data is shown in Fig.11; they show good agreement. If this method is extended to divided sub-areas, the wind vector field over a wide area can be obtained.

5. Concluding remarks

In this paper, the significance of wide-area air pollution measurement was noted, and the capabilities and a variety of measurement results of a large, station-type lidar system were described.

This large lidar can measure the wide-area aerosol distribution continuously at desired time. This is suitable to understand the actual three-dimensional aspects of wide area pollution and to validate air-pollution simulation and prediction models. A large lidar is also effective to routinely monitor the synoptic features of wide area air pollution in short time, and such a system at the center of each populated area would be very useful in the understanding and control of air pollution.

Acknowledgement

The authors thank Mr. F. Sakurai for his assistance in the lidar operation.

References

1. S.WAKAMATSU, Y.OGAWA. K.MURANO, M.OKUDA, H.TSURUTA, K.GOI and Y.ABURAMOTO: Aircraft survey of photochemical smog in Tokyo metropolitan area, J. Japan Soc. Air Pollution 16 (1981) 199-214(in Japanese).
2. E.E.UTHE, N.B.NIELSEN and W.L.JIMISON: Airborne lidar plume and haze analyzer (ALPHA-1), Bull. Am. Meteor. Soc. 61 (1980) 1035-43.
3. N.TAKEUCHI, H.SHIMIZU, Y.SASANO, N.SUGIMOTO, I.MATSUI and M.OKUDA, LAMP lidar for wide-area air pollution monitoring, Proc. of 10th ILRC C-2, Silver Spring, October 1980.
4. H.SHIMIZU, N.TAKEUCHI, Y.SASANO, N.SUGIMOTO, I.MATSUI and M.OKUDA: The design and construction of a large-scale laser radar for monitoring air pollution over a wide range, Oyo-butsuri 50 (1981) 1154-64(in Japanese).
5. H.SHIMIZU, Y.SASANO, N.TAKEUCHI, O.MATSUDO and M.OKUDA: A mobile computerized laser radar system for observing rapidly varying meteorological phenomena, Opt. Quantum Electr. 12 (1980) 159-167.
6. Y.SASANO, H.SHIMIZU, N.TAKEUCHI and M.OKUDA: Geometrical form factor in the laser radar equation: an experimental determination, Appl. Optics 18 (1979) 3908-10.
7. Y.SASANO, H.SHIMIZU, N.SUGIMOTO, I.MATSUI, N.TAKEUCHI and M.OKUDA: Diurnal variation of the atmospheric planetary boundary layer observed by a computer-controlled laser radar, J. Meteor. Soc. Japan 58 (1980) 143-8.
8. R.T.H.COLLIS and P.B.RUSSELL, Laser monitoring of the atmosphsere, Chapt.3, E.D.HINKLEY ed. Topics in Appl. Phys. vol.14, Springer-Verlag, Berlin Heidelberg New York, (1976).

9. Y.SASANO, H.SHIMIZU and N.TAKEUCHI: Convective cell structures revealed by Mie laser radar observations and image-data processing, to be published in Appl. Optics (1982).

10. B.A.CREASE, S.J.CAUGHEY and D.T.TRIBBLE: Information on the thermal structure of the atmospheric boundary layer from acoustic sounding Meteorol. Magazine 106 (1977) 42-52.

11. T.G.KONRAD: The dynamics of the convective process in clear air as seen by radar, J. Atm. Sci. 27 (1970) 1138-47.

12. V.R.NOONKESTER: The evolution of the clear air convective layer revealed by surfacebased remote sensors J. Appl. Meteor. 15 (1976) 594-606.

13. Y.SASANO, H.HIROHARA, T.YAMASAKI, H.SHIMIZU, N.TAKEUCHI and T.KAWAMURA: Horizontall wind vector determination from the displacement of aerosol distribution patterns observed by a scanning lidar, submitted to J. Appl. Met.

8.7 ALPHA-1/Alarm Airborne Lidar Systems and Measurements

E.E. Uthe

SRI International, Menlo Park, CA 94025

I INTRODUCTION

Laser radar (lidar) systems provide the means to observe extended atmospheric regions remotely with high spatial and temporal detail (Collis and Russell, 1976; Uthe, 1981). Operation of lidar systems from aircraft platforms greatly extends the sensor's capabilities by providing mobility over large regional areas. Moreover, because of returns from the earth's surface, atmospheric transport and diffusion over complex terrain can be evaluated, and the surface returns may supplement observed atmospheric backscatter with additional information on the optical and physical properties of atmospheric constituents.

Under funding from the Electric Power Research Institute (EPRI), SRI International (SRI) has constructed and operated a two-wavelength airborne lidar (Uthe et al., 1980). The ALPHA-1 (Airborne Lidar Plume and Haze Analyzer) is now being routinely used for studies of aerosol plume transport and diffusion and boundary layer behavior. Multiple-wavelength techniques are being investigated by use of ALPHA-1 and a new four-color van-mounted lidar system. Under consideration is an extension of ALPHA-1 using line-tunable CO_2 lasers. The ALARM (Airborne Lidar for Agent Remote Measurement) system may provide the means by which to map low concentrations of atmospheric gas species over large regional areas with high spatial and temporal detail.

This paper presents some data examples collected with existing systems and discusses possible system extensions for a greater variety of measurement capabilities.

II ALPHA-1 SYSTEM DESCRIPTION

The ALPHA-1 system was designed to observe smoke plume and boundary layer conditions over large regional areas. Two wavelengths — 1.06 and 0.53 μm — are used; the near-infrared 1.06-μm wavelength is most suitable for boundary layer particle observations and plume tracking operations; the 0.53-μm wavelength supplements the 1.06-μm data with information on particle size characteristics and closely relates to eye-response (visibility) quantities. For optimum eye safety, the system uses large receiver optics and solid-state detector circuits with greater sensitivity than used on previous surface-based lidar systems.

Because lidar systems have extremely high data rates, their performance is directly related to the speed of the data processing system and its peripherals. Two high-speed digital converters (10-ns sample intervals) simultaneously sample the two-wavelength backscatter signatures, a 16-bit microprocessor (LSI-11/2) formats and writes data on a nine-track magnetic tape, and a second microprocessor formats and writes data on a hard-copy gray-scale facsimile recorder. The dual microprocessors also provide for range-squared correction and real-time analysis for aerosol and terrain parameters.

Figure 1 shows a block diagram of the ALPHA-1 system; system characteristics are presented in Table 1. The ALPHA-1 is flown on the SRI Queen Air aircraft shown in Figure 2. Dual area

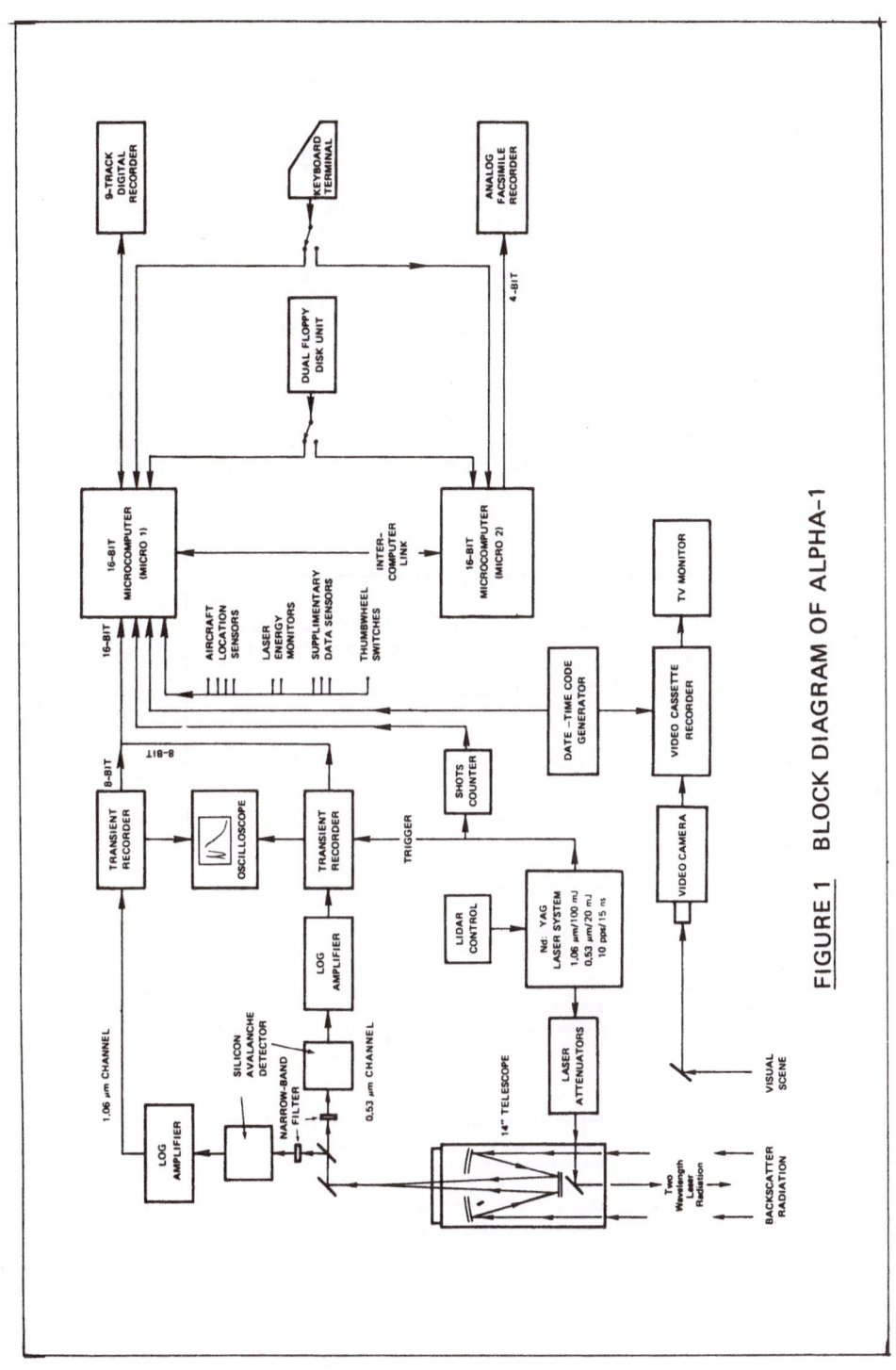

FIGURE 1 BLOCK DIAGRAM OF ALPHA-1

375

navigation (RNAV) systems provide for both accurate aircraft navigation relative to selected waypoints and recorded aircraft position for subsequent data analysis programs. The aircraft also is instrumented with a downlooking video monitor/recorder and supplementary radiometric and meteorological sensors.

III ALPHA-1 DATA EXAMPLES

Los Angeles Boundary Layer — The ALPHA-1 can operate well above the level of restricted flight operations and still observe detailed particulate concentration distributions near the surface level with excellent resolution — about 2 m in the vertical and 6 m in the horizontal. Figure 3 presents aerosol cross sections collected during initial flights of the system that were made over Los Angeles on 16 December 1979. These data, collected at a wavelength of 1.06 μm, show development of the urban pollution layer and suggest an elevated wind flow reversal near the San Gabriel Mountains.

Downwind Plume Transport in Complex Terrain — The ALPHA-1 was used to observe downwind transport of the particulate plume from the Navajo Power Plant (near Page, Arizona). Although the plume is subvisible at distances beyond 1 km from the source, it could be observed by ALPHA-1 at downwind distances as far as 100 km. Figure 4 presents plume cross sections observed at several

Table 1. ALPHA-1 SYSTEM SPECIFICATIONS

COMPONENT	DESCRIPTION
Transmitter	
Wavelength	Simultaneous 1.06 μm (infrared) and 0.53 μm (green)
Pulse energy	100 mJ at 1.06 μm; 20 mJ at 0.53 μm (adjustable from 0 to 20 mJ)
Pulse length	15–17 ns at 50 percent peak amplitude
Pulse repetition frequency	10 pps maximum
Beam divergence	2 mrad
Laser attenuation	0, 5 dB at 1.06 μm; 0, 5, 10, 20 dB at 0.53 μm
Receiver	
Telescope	14-inch Schmitt-Cassegrainian
Field of view	Adjustable 1 to 5 mrad

Optical filtering	Dichroic filter/Long wave pass narrowband filters:
	    1.06 μm: 45 percent max. transmission, 4.6–nm wide
	    0.53 μm: 45 percent max. transmission, 0.86–nm wide

Detectors	Silicon avalanche
Logarithmic amplification	80–dB dynamic range, 40 MHz
Data System	
Backscatter digitization	Simultaneous two–wavelength sampling Sample interval: 0.01-μs minimum Resolution: 8 bits
Processing	Dual microcomputers (LSI–11/2)
Program storage	Dual floppy disk unit
Recording	9–track magnetic tape, 4–bit analog facsimile hard copy
Display	Real–time two–channel A-scope, real–time 16 gray–scale analog facsimile
Visual scene	Recording: video cassette Display: 9–inch video monitor
Data	Two–wavelength lidar backscatter signatures, aircraft location, data and time, laser energy; real–time program control switch readings; 16–channel 10–bit low–speed A/D input
Supplementary data sensors	Pyranometers (upward and downward looking), nephelometer, turbulence

downwind distances. This example shows that the plume was being channeled by the complex terrain to the west of the plant. Other data collected show mountain restrictions and barriers to plume transport (Uthe et al., 1980).

Convective Processes — On 14 May 1980, ALPHA-1 was repetitively flown across the plume of the Kincaid Power Plant (Illinois) at 10-km south of the plant. Early morning observations showed a well-defined plume above the developing mixing layer, while subsequent passes across the plume showed the structure of clear-air convective cells and entrainment of the plume particulates. Lidar cross sections collected on the west-to-east passes of the plume between about 0900 and 1000 Central Standard Time (CST) are shown in Figure 5. At 0856 CST, the top of the mixing layer has just reached the base of the plume, causing distortion from its nighttime geometry. At 0907 the mixed layer has risen higher than the plume base, and plume particulates are starting to mix within the layer. The 0917 cross section shows plume particulates almost reaching the surface, while at 0928 the mixed lower part of the plume appears to have impacted on the surface. At 0938 convective cells are seen to rise above the plume top, and concentrated plume may be reaching the ground level within downward-compensating air motions. The convective cells have risen to much greater heights at 0948, with the highest cells being in the vicinity of the plume position — possibly indicating the buoyancy effects of thermal energy released from the plant. The plume pattern seen at 0959 CST also indicates active convection, with a clear air parcel penetrating downward to within 500 m of the surface.

Two-Wavelength Observations — During a flight test of the ALPHA-1, observations were made downwind of a forest fire located near the California coast (Uthe et al., 1982). Visually, the fire appeared to be contained within a small area; the resulting smoke plume towered over the source but was transported downwind at lower altitudes.

Figure 6 shows plume cross sections derived from infrared and visible backscatter signatures recorded during the second pass of ALPHA-1 across the plume. For these data, receiver gain was reduced so that surface returns did not saturate receiver electronics — as a result, clear air haze layers were not as well observed on this

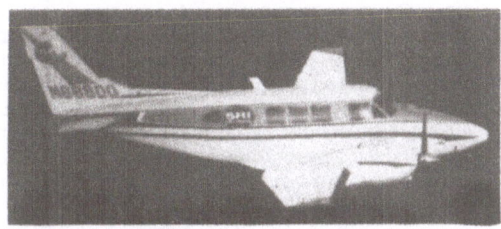

FIGURE 2   SRI QUEEN AIR USED TO SUPPORT THE ALPHA-1 SYSTEM

(a) 1120–1150 PST, 1.06 μm, 190° TRUE COURSE

(b) 1602–1634 PST, 1.06 μm, 190° TRUE COURSE

San Gabriel Mountains   Coast Line

FIGURE 3   LOS ANGELES BOUNDARY LAYER STRUCTURE DERIVED FROM ALPHA-1 MAGNETIC TAPE RECORDS, 16 DECEMBER 1979

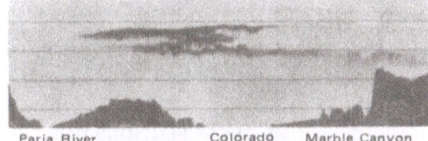

300 m

5 km

Colorado River

Paria River   Colorado River   Marble Canyon

Vermillion Cliffs   Marble Canyon   Colorado River

FIGURE 4   AIRBORNE LIDAR OBSERVATIONS OF THE CROSS-PLUME STRUCTURE OF A SUBVISIBLE POWER PLANT PLUME AT DIFFERENT DOWNWIND DISTANCES FROM THE PLANT

377

data. Greater plume attenuation of the visible energy than of the infrared energy is evident by the absence of plume visible backscatter following penetration by the laser pulse of the denser plume elements. Plume transmissions in the vertical were evaluated by normalizing to 100 percent observed surface returns in the absence of plume backscatter and by assuming constant surface reflectivity and laser–transmitted peak power for each laser firing made along the cross–plume path. Lidar response information required for quantitative analysis of backscatter signatures was derived using standard calibration techniques with neutral density filters of known attenuation. Vertical transmissions for one–way passage of the laser energy through the smoke plume are plotted in Figure 6. The results show minimum plume transmissions of about 50 percent at 1.06 μm and about 6 percent at 0.53 μm. The strong wavelength dependence of attenuation suggests submicron particle sizes, although strong absorption in the visible and weak absorption in the infrared by plume constituents could also explain the observations. Assuming the wavelength difference results entirely from particle size effects, the laboratory data presented in Figure 7 indicate particle diameters of about 0.1 μm, which is consistent with in–situ measurements reported by various investigators in the literature. Inference of particle size is important to evaluate lidar backscatter quantitatively in terms of aerosol extinction and concentration.

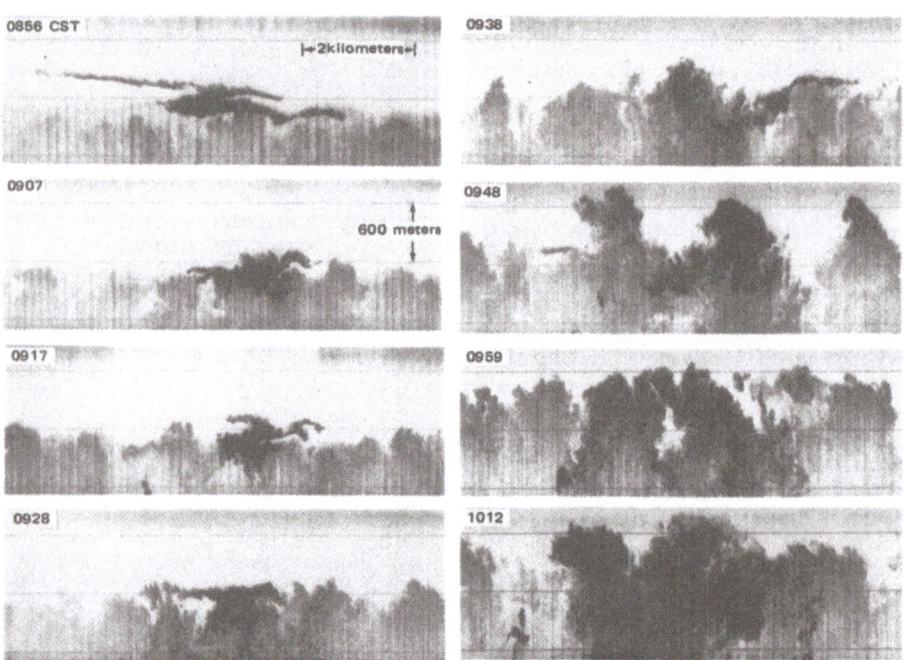

FIGURE 5 PLUME CROSS SECTIONS DURING CONVECTION, 10-km DOWNWIND OF KINCAID POWER PLANT, 4 MAY 1980

IV   MARK XI MULTIPLE WAVELENGTH LIDAR

To further investigate capabilities of multiple–wavelength lidar systems for remote measurement of aerosol optical and physical densities and of particle characteristics, a four–color lidar system is being constructed under a project funded by the U.S. Army Research Office (ARO). As shown in Figure 8, the system transmits energy at 0.53 μm and 1.06 μm using a neodymium–yag laser, 3.8 μm using a DF laser, and 10.6 μm using a CO_2 laser. Pulses from the three lasers are transmitted coaxially with a 12–inch Newtonian telescope. Therefore, backscatter and transmission

378

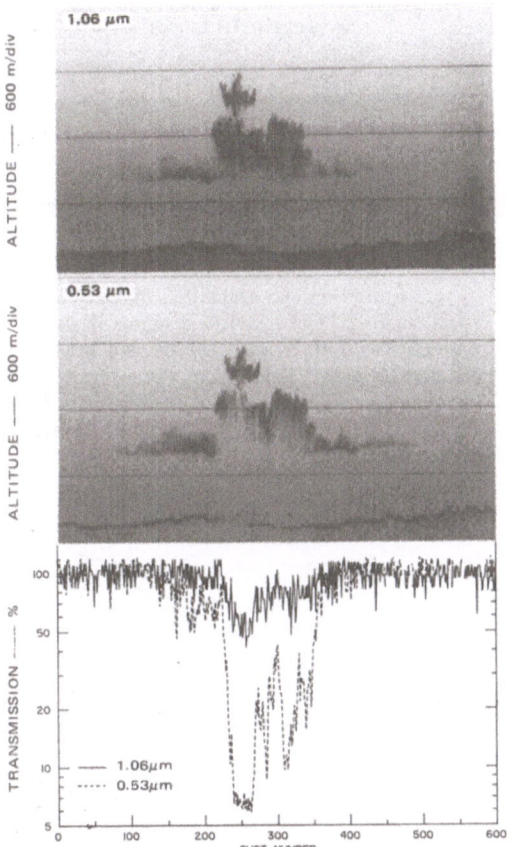

FIGURE 6 CROSS-PLUME STRUCTURE DERIVED
FROM ALPHA-1 PLUME BACKSCATTER
AND VERTICAL TRANSMISSIONS
DERIVED FROM SURFACE RETURNS
AT 1.06- AND 0.53-μm WAVELENGTHS
(Aircraft pass 2)

measurement of gas concentrations (for example,
see the apper presented at this workshop by
Hawley et al., 1982). We have proposed to
extend the capabilities of the ALPHA-1 airborne
lidar for remote gas measurement by adding two
line-tunable CO_2 lasers. The proposed system
is named ALARM, for "Airborne Lidar for
Agent Remote Measurement." Although the
ALARM system has not yet been constructed,
the system design has been completed so that
expected system capabilities can be evaluated
using simulation programs.

As shown by Figure 9, ALARM will
operate similarly to ALPHA-1. Surface returns
at two closely spaced wavelengths from laser
transmissions only a few microseconds apart can
be evaluated in terms of column content of a
gas that absorbs greater energy at one of the two
transmitted wavelengths. High sensitivity to low
gas concentrations is obtained because of the
double passage of the laser pulses through the
gas cloud being observed and the large-amplitude
signal return from the earth's surface. The
ALPHA-1 data presented in Figure 6 illustrate

measurements are made along the same viewing
path to facilitate analysis of the information content of multiple-wavelength lidar signatures. The
lidar is now being installed within a van similar
to that of the SRI Mark IX lidar, and a field program to evaluate system capabilities is scheduled
for this summer. Findings from this program may
be used to extend ALPHA-1 airborne lidar capabilities for remote aerosol measurements.

V. ALARM SYSTEM DESCRIPTION

Differential absorption lidar (DIAL) systems
have been constructed and applied for remote

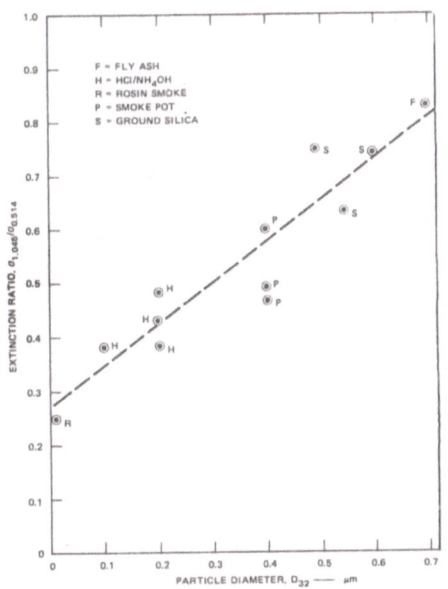

FIGURE 7
RELATIONSHIP OF EXTINCTION RATIO
AT WAVELENGTHS OF 1.045- AND
0.514-μm AND MEAN PARTICLE DIAM-
ETER DERIVED FROM LABORATORY
RESULTS

379

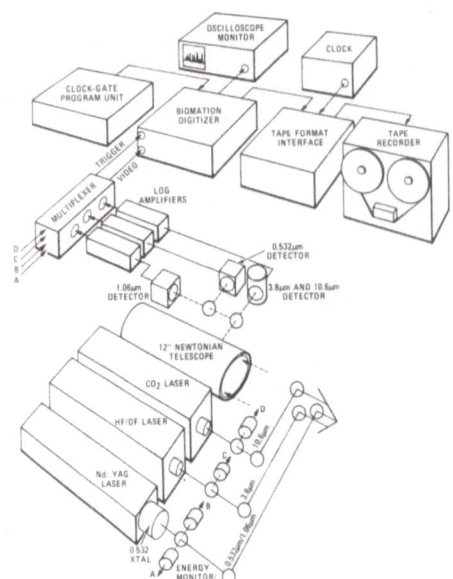

FIGURE 8 SRI MARK XI MULTIPLE-
WAVELENGTH LIDAR

possible applications of ALARM. In this figure,
the two wavelength transmissions derived from
ALPHA-1 surface returns could have been interpreted as if the wavelength differences in transmission were associated with gas absorption at
the green wavelength.

Construction and evaluation of ALARM
await funding. The ALARM system will use
1-J lasers with 40-ns pulse widths and a 14-inch-
diameter receiver. Simulations using these parameters indicate that range-resolved information on
gas concentrations for several gas species can be
expected under a wide range of atmospheric
conditions.

Airborne Lidar for Agent Remote Measurement – ALARM

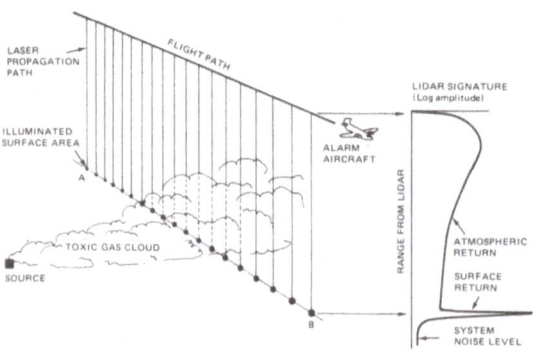

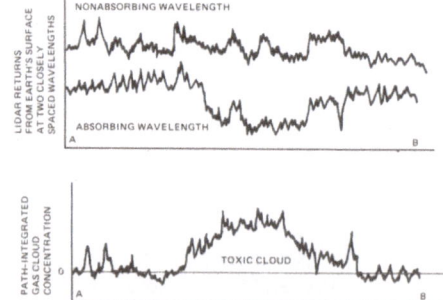

FIGURE 9 AIRBORNE LIDAR OBSERVATION OF A TOXIC GAS CLOUD

VI CONCLUSIONS

The unique capabilities of the ALPHA-1 airborne lidar for evaluating transport and diffusion of aerosol plumes and boundary layer behavior have been demonstrated. The system is currently being applied routinely to air quality research programs.

Initial two-wavelength analysis of ALPHA-1 data indicates evaluation of mean particle size and thus improved inferences of aerosol optical and physical densities are possible. Multiple-wavelength lidar techniques are being more fully investigated with surface-based systems for possible incorporation of developed technique into ALPHA-1. An extension of ALPHA-1 to remote gas measurement using multiple-wavelength DIAL methods is under consideration. Therefore, we expect that the capabilities already demonstrated by ALPHA-1 will greatly be extended in the near future.

The research programs discussed in this paper were supported by the Electric Power Research Institute and the U.S. Army Research Office, Geosciences Division.

REFERENCES

Collis, R.T.H., and P.B. Russell, 1976: "Lidar Measurement of Particles and Gases by Elastic Back-scattering and Differential Absorption," *Laser Monitoring of the Atmosphere*, E.D. Hinkley, ed., Springer, New York.

Hawley, J.G., L.D. Fletcher, and G.F. Wallace, 1982: "Ground-Based UV-DIAL System and Measurements," Paper presented at the *Workshop on Optical and Laser Remote Sensing*, Monterey, California, 9-11 February.

Uthe, E.E., 1981: "Lidar Evaluation of Smoke and Dust Clouds," *Applied Optics*, Vol. 20, pp. 1503-1510.

Uthe, E.E., 1982: "Particle Size Evaluations Using Multiwavelength Extinction Measurements," to appear in *Applied Optics*, February 1982.

Uthe, E.E., N.B. Nielsen, and W.L. Jimison, 1980: "Airborne Lidar Plume and Haze Analyzer (ALPHA-1)," *Bull. Amer. Met. Soc.*, Vol. 61, pp. 1035-1043.

Uthe, E.E., B.M. Morley, and N.B. Nielsen, 1982: "Airborne Lidar Measurements of Smoke Plume Distribution, Vertical Transmission and Particle Size," to appear in *Applied Optics*, February 1982.

Index of Contributors

Aerosol Microphysics I

Particle Interaction
Editor: **W. H. Marlow**
1980. 35 figures, 1 table. XI, 160 pages.
(Topics in Current Physics, Volume 16). ISBN 3-540-09866-6

Contents: *W. H. Marlow:* Introduction: The Domains of Aerosol Physics. – *J. R. Brock:* The Kinetics of Ultrafine Particles. - *J. D. Doll:* Classical and Statistical Theories of Gas-Surface Energy Transfer. - *P. J. McNulty, H. W. Chew, M. Kerker:* Inelastic Light Scattering. – *W. H. Marlow:* Survey of Aerosol Interaction Forces.

Aerosol Microphysics II

Chemical Physics of Microparticles
Editor: **W. H. Marlow**
1982. 50 figures. XI. 189 pages
(Topics in Current Physics, Volume 29). ISBN 3-540-11400-9

Contents: *W. H. Marlow:* Aerosol Chemical Physics. – *H. P. Baltes, E. Šimànek:* Physics of Microparticles. – *I. P. Batra:* Electronic Structure Studies of Overlayers Using Cluster and Slab Models. – *B. J. Berne, R. V. Mikkilineni:* Computer Experiments on Heterogeneous Systems. – *P. E. Wagner:* Aerosol Growth by Condensation.

Laser Monitoring of the Atmosphere

Editor: **E. D. Hinkley**
With contributions by numerous experts
1976. 84 figures. XV, 380 pages
(Topics in Applied Physics, Volume 14).
ISBN 3-540-07743-X

Contents: Introduction. - Remote Sensing for Air Quality Management. - Laser-Light Transmission Through the Atmosphere. - Lidar Measurement of Particles and Gases by Elastic Backscattering and Differential Absorption. - Detection of Atoms and Molecules by Raman Scattering and Resonance Fluorescence. - Techniques for Detection of Molecular Pollutants by Absorption of Laser Radiation. - Laser Heterodyne Detection Techniques.

The Stratospheric Aerosol Layer

Editor: **R. C. Whitten**
1982. 62 figures. XI, 152 pages
(Topics in Current Physics, Volume 28). ISBN 3-540-11229-4

Contents: *R. C. Whitten, P. Hamill:* Introduction. - *E. C. Y. Inn, N. H. Farlow, P. B. Russell, M. P. Mc Cormick, W. P. Chu:* Observations. - *R. G. Keesee, A. W. Castleman, Jr.:* The Chemical Kinetics of Aerosol Formation. - *R. P. Turco:* Models of Stratospheric Aerosols and Dust. - *O. B. Toon, J. B. Pollack:* Stratospheric Aerosols and Climate.

Springer-Verlag
Berlin
Heidelberg
New York

W. Demtröder

Laser Spectroscopy

Basic Concepts and Instrumentation
2nd corrected printing. 1982. 431 figures. XIII, 696 pages
(Springer Series in Chemical Physics, Volume 5)
ISBN 3-540-10343-0

This book helps close the gap between classical works on optics
and spectroscopy and more specialized publications on modern
research in this field. It is addressed to graduate students in physics
and chemistry as well as scientists just starting out in this field.

From the reviews: "The scope of this book is most impressive. It is
authoritative, illuminating and up-to-date. The 650 pages of text are
supplemented by 34 pages of references, and many of the chapters
are furnished with a selection of problems. It is strongly recom-
mended for all spectroscopists of the laser era and will be valuable
for research students entering spectroscopic laboratories. ..."

Contemporary Physics

Inverse Source Problems

in Optics
Editor: **H. P. Baltes**
With a foreword by J.-F. Moser
1978. 32 figures. XI, 204 pages
(Topics in Current Physics, Volume 9). ISBN 3-540-09021-5

Contents: *H. P. Baltes:* Introduction. – *H. A. Ferwerda:* The Phase
Reconstruction Problem for Wave Amplitudes and Coherence
Functions. – *B. J. Hoenders:* The Uniqueness of Inverse Problems. –
H. G. Schmidt-Weinmar: Spatial Resolution of Subwavelength
Sources from Optical Far-Zone Data. – *H. P. Baltes, J. Geist,
A. Walther:* Radiometry and Coherence. – *A. Zardecki:* Statistical
Features of Phase Screens from Scattering Data.

Inverse Scattering Problems

in Optics
Editor: **H. P. Baltes**
With a foreword by R. Jost
1980. 49 figures, 2 tables. XIV, 313 pages
(Topics in Current Physics, Volume 20). ISBN 3-540-10104-7

Contents: *H. P. Baltes:* Progress in Inverse Optical Problems. –
G. Ross, M. A. Fiddy, M. Nieto-Vesperinas: The Inverse Scattering
Problem in Structural Determinations. – *E. Jakeman, P. N. Pusey:*
Photon-Counting Statitics of Optical Scintillation. – *A. Selloni:*
Microscopic Models of Photodetection.– *M. Bertero, C. De Mol,
G. A. Viano:* The Stability of Inverse Problems. – *R. Goulard,
P. J. Emmerman:* Combustion Diagnostics by Muliangular Absorp-
tion. – *W.-M. Boerner:* Polarization Utilization in Electromagnetic
Inverse Scattering.

V. E. Zuev, I. E. Naats

Inverse Problems of Lidar Sensing
of the Atmosphere

1983. 71 figures. XI, 260 pages. (Springer Series in Optical Sciences,
Volume 29). ISBN 3-540-10913-7

Contents: Introduction. – Theory of Optical Sensing in Aerosol
Polydispersed Systems. – Determination of the Microphysical Para-
meters of Aerosol Ensembles by the Method of Multi-Frequency
Laser Sensing. – Methods for Inverting Polarization Data. The
Theory of Bistatic Lidars. – Light Scattering by Aerosols and Lidar
Sensing of the Atmosphere. – References. – Subject Index.

Springer-Verlag
Berlin
Heidelberg
New York